Lecture Notes in Mathematics

Edited by A. Dold and B. Eckmann

501

Spline Functions

Proceedings of an International Symposium
Held at Karlsruhe, Germany, May 20–23, 1975.

Edited by K. Böhmer, G. Meinardus, and W. Schempp

Springer-Verlag
Berlin · Heidelberg · New York 1976

Editors
Prof. Dr. Klaus Böhmer
Institut für Praktische
Mathematik der Universität Karlsruhe
Englerstraße
D-7500 Karlsruhe

Prof. Dr. Günter Meinardus
Prof. Dr. Walter Schempp
Lehrstühle für Mathematik
Universität Siegen
Hölderlinstraße 3
D-5900 Siegen 21

AMS Subject Classifications (1970): 41A05, 41A10, 41A15, 41A50, 41A63, 41A65, 65D30, 65L05, 65R05

ISBN 3-540-07543-7 Springer-Verlag Berlin · Heidelberg · New York
ISBN 0-387-07543-7 Springer-Verlag New York · Heidelberg · Berlin

Printed in Germany
Offsetdruck: Julius Beltz, Hemsbach/Bergstr.

Vorwort

Im Frühjahr 1973 fand im Mathematischen Forschungsinstitut in Oberwolfach die erste Tagung über Spline-Funktionen im deutschsprachigen Raum statt. Anläßlich der 150-Jahr-Feier der Universität Karlsruhe bot sich die Möglichkeit, in der Zeit vom 20.5.-23.5.75 ein internationales Symposium über Approximationstheorie unter besonderer Berücksichtigung der Spline-Funktionen zu veranstalten.

Die ungebrochene Bedeutung der Spline-Funktionen innerhalb der Approximationstheorie sowie für numerische und außermathematische Anwendungen wurde durch zahlreiche wichtige Vorträge und Diskussionsbeiträge unter Beweis gestellt. Die Herausgeber, denen die Tagungsleitung oblag, sind daher sehr erfreut, daß auch diesmal eine größere Zahl der dort vorgelegten Ergebnisse veröffentlicht werden können.

Den Autoren sei für die Bereitstellung ihrer Beiträge herzlichst gedankt. Ebenso danken wir dem Herausgeber der Lecture Notes und dem Springer-Verlag für die schnelle Erstellung des vorliegenden Bandes.

Die Abhaltung der Tagung wurde ermöglicht durch Unterstützung der Stiftung Volkswagenwerk sowie der Universität Karlsruhe und des dortigen Instituts für Praktische Mathematik. Die Herstellung der Druckvorlage koordinierte Herr Dr. F.J. Delvos, Siegen. Ihnen allen sei an dieser Stelle gedankt.

K. Böhmer, G. Meinardus, W. Schempp

INHALTSVERZEICHNIS

C. de BOOR and I.J. SCHOENBERG
Cardinal interpolation and spline functions VIII. The Budan-Fourier theorem for splines and applications 1

F.J. DELVOS and W. SCHEMPP
An extension of Sard's method 80

G. HÄMMERLIN
Zur numerischen Behandlung von homogenen Fredholmschen Integralgleichungen 2. Art mit Splines.. 92

G. JENTZSCH, G. LANGE, O. ROSENBACH
Anwendung der Spline-Funktionen zur Bearbeitung geophysikalischer Meßreihen 99

H. JOHNEN and K. SCHERER
Direct and Inverse Theorems for Best Approximation by Λ-Splines 116

H.-W. KÖSTERS
Computation of periodic M-Splines with equispaced nodes 132

T. LYCHE
Discrete polynomial spline approximation methods.. 144

G. MEINARDUS
Periodische Splinefunktionen 177

G. MICULA
Bemerkungen zur numerischen Lösung von Anfangswertproblemen mit Hilfe nichtlinearer Spline-Funktionen 200

H. ter MORSCHE
On the relations between finite differences and derivatives of cardinal spline functions 210

A. SARD
On optimal approximation 220

W. SCHÄFER and W. SCHEMPP
Splineapproximation in intermediären Räumen 226

K.-H. SCHLOSSER
Mehrdimensionale Spline-Interpolation mit Hilfe der Methode von Sard 247

L.L. SCHUMAKER
Toward a constructive theory of generalized spline functions 265

F. SCHURER and F.W. STEUTEL
On an inequality of Lorentz in the theory of Bernstein polynomials 332

M. SOMMER
Alternanten bei gleichmäßiger Approximation mit zweidimensionalen Splinefunktionen 339

H. STRAUSS
Approximation mit Splinefunktionen und Quadraturformeln 371

Anschriften der Autoren 420

CARDINAL INTERPOLATION AND SPLINE FUNCTIONS VIII.
THE BUDAN-FOURIER THEOREM FOR SPLINES AND APPLICATIONS

Carl de Boor and I. J. Schoenberg

Dedicated to M. G. Krein

Introduction. The present paper is the reference [8] in the monograph [15], which was planned but not yet written when [15] appeared. The paper is divided into four parts called A, B, C, and D. We aim here at three or four different results. The unifying link between them is that they all involve the sign structure of what one might call a "Green's spline", i.e., a function which consists of two null-splines pieced together at a certain point to satisfy at that point several homogeneous conditions and just one inhomogeneous condition, much as (any section of) a (univariate) Green's function consists of two solutions to a homogeneous ordinary differential equation which are pieced together at a point in just that way. The different results are further linked by the fact that we use an extension of the Budan-Fourier theorem to splines in an essential way. In each of our applications of this theorem, the circumstances are such that the inequality furnished by the theorem becomes taut, i.e., must be an equality, and this provides an unexpected amount of precise information.

In Part A, we state and prove the Budan-Fourier theorem for splines with simple knots in the form in which we need it. We also apply it right away to the "Green's function" for odd-degree spline interpolation at arbitrarily spaced knots in a finite interval, i.e., to the Peano kernel for the error in that interpolation process.

In Part B, we develop the information about the sign structure of cardinal nullsplines required for later applications, using the Gantmacher-Krein Theory of oscillation matrices in an essential way.

Part C: The study of the remainder of cardinal spline interpolation for odd degree $n = 2m - 1$, as given in [17], depended on the behavior of the remainder $K(x, t)$ of the interpolation of the function $(x-t)_+^{2m-1}$ where t is a parameter, $0 < t < 1$. The assertion (Theorem 3 of [17]) was that $\operatorname{sgn} K(x, t) = (-)^m \operatorname{sgn} \sin \pi x$ for all real x, and this was stated in [17] without proof. A proof is given in Part C, where we also discuss the remainder of <u>even degree</u> cardinal spline interpolation as well as the fundamental function of this interpolation process.

Part D: The elementary cases of the Landau-Kolmogorov problem were discussed in [16] by means of appropriate formulae of approximate differentiation with integral remainders. However, [16] was restricted to the orders $n = 2$ and $n = 3$, when only finitely many of the ordinates of the function appear in the differentiation formula. In [16], and also in [15, Lecture 9, §1], the first non-elementary case $n = 4$ was briefly mentioned. In Part D, we study the general case. Cavaretta gave in [4] an elegant proof of Kolmogorov's theorem that uses only Rolle's theorem. Our approach is much more elaborate, but provides information on the extremizing functions

Part A. The Budan-Fourier theorem for splines and spline interpolation on a finite interval.

1. The Budan-Fourier theorem for splines with simple knots.

We begin with the introduction of some standard notation.

For $v = (v_i)_1^n \in \mathbb{R}^n$, $S^- v$ and $S^+ v$ denote the minimal, respectively maximal, number of sign changes in the sequence v achievable by appropriate assignment of signs to the zero entries (if any) in v. Hence, always $S^- v \leq S^+ v$. Further,

$$S^- v \leq \liminf_{u \to v} S^- u \leq \limsup_{u \to v} S^+ u \leq S^+ v .$$

Should $S^- v = S^+ v$, then it is customary to denote their common value by Sv. The identity

(1) $$S^-(v_i)_1^n + S^+((-)^i v_i)_1^n = n-1$$

will be used repeatedly.

If v is, more generally, a real valued function on some domain $G \subseteq \mathbb{R}$, then, with $*$ standing for $-$ or $+$,

$$S^* v := \sup\{S^*(v(t_i)) \mid (t_i)_1^n \text{ in } G, \ n \in \mathbb{N}, \ t_1 < \ldots < t_n\} .$$

If $E \subseteq G$, then we will write $S_E^* v$ for $S^*(v|_E)$.

Induction establishes the following useful lemma which is essentially Lemma 1.2 of Karlin and Micchelli [7].

Lemma 1. If $f \in C^{(n)}[0, \delta]$ and $f^{(n)}(0) \neq 0$, then, for some positive ε, $f^{(j)}$ does not vanish on $(0, \varepsilon]$, $j = 0, \ldots, n$, and

$$S^{-}(f(0), \ldots, f^{(n)}(0)) = \lim_{t \downarrow 0} S^{+}(f(t), \ldots, f^{(n)}(t)) .$$

Therefore, with (1),

$$S^{+}(f(0), \ldots, f^{(n)}(0)) = \lim_{t \uparrow 0} S^{-}(f(t), \ldots, f^{(n)}(t)) .$$

in case $f^{(n)}(0) \neq 0$ and $f \in C^{(n)}[-\delta, 0]$.

Next, we define the multiplicity of a zero of a spline function f of degree n on $[a, b]$ with simple knots, i.e., f is composed of polynomial pieces of degree $\leq n$ in such a way that $f \in C^{(n-1)}[a, b]$. Then $f \in \mathbb{L}_{\infty}^{(n)}[a, b]$. If $n = 0$, i.e., if f is piecewise constant, then we say that the (possibly degenerate) interval $[\sigma, \tau]$ in (a, b) is a zero of f of multiplicity $\{{0 \atop 1}\}$ iff f vanishes on (σ, τ) and $f(\sigma^{-})f(\tau^{+})\{{> \atop <}\}\, 0$. With this definition, the number of zeros counting multiplicity of a piecewise constant function in (a, b) equals the number of its strong sign changes on (a, b). If $n > 0$, then we say that the (possibly degenerate) interval $[\sigma, \tau]$ in (a, b) is a zero of f of multiplicity r iff either $r = 0$ and $\sigma = \tau$ and $f(\sigma) \neq 0$ or else $r > 0$ and f vanishes on $[\sigma, \tau]$ and $[\sigma, \tau]$ is a zero of $f^{(1)}$ of multiplicity $r-1$. We denote the total number of zeros, counting multiplicities, of f in (a, b) by

$Z_f(a, b)$.

To give an example, $Z_f(0, 12) = 6$ for the linear spline f drawn

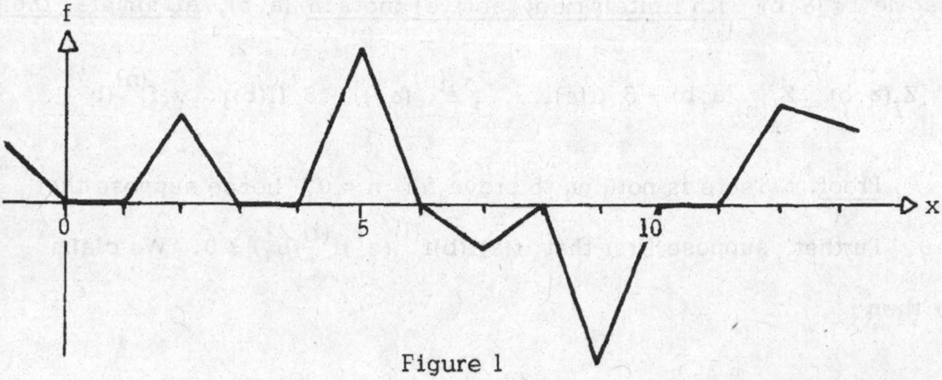

Figure 1

in Figure 1, with a double zero at $[3, 4]$, a simple zero at $[6, 6]$, a double zero at $[8, 8]$ and a simple zero at $[10, 11]$, and no other zeros of positive multiplicity in $(0, 12)$. In particular, the interval $[0, 1]$ is not counted as a zero in $(0, 12)$ for this f. Its first derivative has simple zeros at $[2, 2]$, $[3, 4]$, $[5, 5]$, $[7, 7]$, $[8, 8]$, $[9, 9]$ and no other zeros of positive multiplicity in $(0, 12)$, so $Z_{f^{(1)}}(0, 12) = 6$. In particular, $[10, 11]$ is a zero of 0 multiplicity for $f^{(1)}$, and, again, $[0, 1]$ is not a zero in $(0, 12)$ for $f^{(1)}$. For this particular f, we would have equality in (2) below.

The number $Z_f(a, b)$ is necessarily finite if f has only finitely many knots in (a, b). Also, if $f(a)f(b) \neq 0$, then $f(a)f(b) < 0$ iff $Z_f(a, b)$ is odd. Further, $Z_f(a, b) + Z_f(b, c) \leq Z_f(a, c)$ for $a < b < c$, and, as we said earlier, $Z_{f^{(n)}}(a, b) = S^-_{(a,b)} f^{(n)}$.

Theorem 1 (Budan-Fourier for Splines). If f is a polynomial spline function of exact degree n on (a,b) (i.e., of degree n with $f^{(n)}(t) \neq 0$ for some $t \in (a,b)$) with finitely many (active) knots in (a,b), all simple, then

$$(2)\quad Z_f(a,b) \le Z_{f^{(n)}}(a,b) + S^-(f(a), \ldots, f^{(n)}(a^+)) - S^+(f(b), \ldots, f^{(n)}(b^-)).$$

Proof. There is nothing to prove for $n = 0$, hence suppose that $n > 0$. Further, suppose first that $f(a)f(b)f^{(1)}(a^+)f^{(1)}(b^-) \neq 0$. We claim that then

$$(3)\quad Z_f(a,b) \le Z_{f^{(1)}}(a,b) + S(f(a), f^{(1)}(a^+)) - S(f(b), f^{(1)}(b^-)).$$

Indeed, if $Z_f(a,b) = 0$, there is nothing to prove unless also $Z_{f^{(1)}}(a,b) = 0$, but then $S(f(a), f^{(1)}(a^+)) = S(f(b), f^{(1)}(b^-))$ and (3) holds trivially. On the other hand, if $Z_f(a,b) > 0$, then we can find σ and τ so that $(\sigma,\tau) \supset \{t \in [a,b] \mid f(t) = 0\}$ while $f(\sigma)f(\tau) \neq 0$, and $f(a)/f(\sigma)$ and $f(b)/f(\tau)$ are both greater than 1. But then

$$Z_{f^{(1)}}(a,\sigma) \ge 1 - S(f(a), f^{(1)}(a^+)), \quad Z_{f^{(1)}}(\tau,b) \ge S(f(b), f^{(1)}(b^-))$$

while, by Rolle's theorem and our definition of multiplicity of zeros,

$$Z_f(a,b) - 1 \le Z_{f^{(1)}}(\sigma,\tau) .$$

Hence, (3) holds in this case, too.

If now $f^{(j)}(a^+)f^{(j)}(b^-) \neq 0$ for $j = 0, \ldots, n$, then

$$S(f(x), \ldots, f^{(n)}(x)) = \sum_{j=1}^{n} S(f^{(j-1)}(x), f^{(j)}(x))$$

for $x = a^+, b^-$, while by (3),

$$Z_{f^{(j-1)}}(a,b) \leq Z_{f^{(j)}}(a,b) + S(f^{(j-1)}(a), f^{(j)}(a^+)) - S(f^{(j-1)}(b), f^{(j)}(b^-)),$$

$$j = 1, \ldots, n,$$

which proves (2) for this case. From this, a limit process establishes (2) with the aid of Lemma 1 in case merely $f^{(n)}(a^+)f^{(n)}(b^-) \neq 0$.

If, finally, $f^{(n)}(a^+) = 0$, then, as f is of exact degree n on (a,b), there exists $\sigma \in (a,b)$ so that $f^{(n)}$ vanishes on (a,σ), but $f^{(n)}(\sigma^+) \neq 0$. Note that $[a,\sigma]$ is not counted as a zero of $f^{(n)}$ in (a,b) by our definition. By Lemma 1, we can find $\hat{\sigma} > \sigma$ so that $f^{(j)}$ does not vanish on $(\sigma, \hat{\sigma}]$ for $j = 0, \ldots, n$ and so that

$$S^-(f(\sigma), \ldots, f^{(n)}(\sigma^+)) = S(f(\hat{\sigma}), \ldots, f^{(n)}(\hat{\sigma})) . \tag{4}$$

Then

$$Z_f(a,b) = Z_f(a,\hat{\sigma}) + Z_f(\hat{\sigma},b), \quad Z_{f^{(n)}}(a,b) = Z_{f^{(n)}}(\hat{\sigma},b) . \tag{5}$$

We claim that

$$Z_f(a,\hat{\sigma}) \leq S^-(f(a), \ldots, f^{(n)}(a^+)) - S(f(\hat{\sigma}), \ldots, f^{(n)}(\hat{\sigma})) . \tag{6}$$

For this, let

$$j := \begin{cases} 0 & \text{if } f \text{ vanishes identically on } [a,\sigma], \\ \max\{i \in [0,n-1] \mid f^{(i)}(a) \neq 0\} & \text{otherwise.} \end{cases}$$

Then

$$(7) \qquad S^-(f^{(j)}(a), \ldots, f^{(n)}(a^+)) = 0 = S^-(f^{(j)}(\sigma), \ldots, f^{(n)}(\sigma^+)) .$$

Hence, (4) implies (6) in case f vanishes identically on $[a, \sigma]$ since then $[a, \sigma]$ is not counted as a zero of f in $(a, \hat{\sigma})$, therefore $Z_f(a,\hat{\sigma}) = 0$. Otherwise, $f^{(j)}$ is a nonzero constant on $[a,\sigma]$, therefore

$$Z_f(a,\hat{\sigma}) \leq S^-(f(a),\ldots,f^{(j)}(a)) - S^+(f(\sigma),\ldots,f^{(j)}(\sigma))$$

$$+ \text{ mult. of } \sigma \text{ as a zero of } f$$

$$\leq S^-(f(a),\ldots,f^{(j)}(a)) - S^-(f(\sigma),\ldots,f^{(j)}(\sigma))$$

$$= S^-(f(a),\ldots,f^{(n)}(a^+)) - S^-(f(\sigma),\ldots,f^{(n)}(\sigma^+)),$$

the last equality by (7), and (4) now gives again (6).

These considerations also imply, by going from t to $-t$, that, for some $\hat{\tau} \in (\hat{\sigma},b)$, $f^{(j)}(\hat{\tau}) \neq 0$ for $j = 0, \ldots, n$, and

$$Z_f(\hat{\sigma},b) = Z_f(\hat{\sigma},\hat{\tau}) + Z_f(\hat{\tau},b), \qquad Z_{f^{(n)}}(\hat{\sigma},b) = Z_{f^{(n)}}(\hat{\sigma},\hat{\tau}) ,$$

$$Z_f(\hat{\tau},b) \leq S(f(\hat{\tau}), \ldots, f^{(n)}(\hat{\tau}^-)) - S^+(f(b), \ldots, f^{(n)}(b^-)) ,$$

and combining these facts with (5) and (7) and with (2) as already proved establishes (2) in the general case. |||

It is possible to derive from this theorem appropriate statements concerning splines with multiple knots, e.g., monosplines, by an appropriate limit process. But we will not pursue this further here, as the theorem is sufficient for the purposes of this paper.

If f has its support in (a,b), then we obtain that

$$Z_f(a,b) \leq S^-_{(a,b)} f^{(n)} - n .$$

Since $Z_f(a,b) \geq 0$, we recover in this way well known facts about B-splines, such as their sign structure or the fact that B-splines have minimal support. Further, we obtain $Z_f(a,b) \leq S^-_{(a,b)} f^{(n+1)} - (n+1)$, which is the polynomial case of a more general result for Chebyshev splines due to H. Burchard [3].

We note that the particular choice of an f without any active knots in (a,b) and the replacement of S^+ by (the weaker) S^- leads to the classical Budan-Fourier Theorem [14]. The further specialization $a = 0$ and $b = \infty$ produces Descartes' familiar Rule of Signs.

2. <u>A simple application to complete spline interpolation.</u> If n is odd and greater than 1,

$$n = 2m - 1 > 1,$$

and $(x_i)_0^N$ is a sequence in $[a,b]$ with

$$a = x_0 < \ldots < x_N = b,$$

then there exists for given $f \in C^{(k-1)}[a,b]$ exactly one spline Sf of degree n with simple knots $x_1, \ldots, x_{N-1}$ in (a,b) which agrees with f in the sense that

(1) $$Sf(x_i) = f(x_i), \quad i = 1, \ldots, N-1$$

(2) $$(Sf)^{(j)}(x_i) = f^{(j)}(x_i), \quad j = 0, \ldots, m-1 \text{ and } i = 0, N .$$

This spline has been called the <u>complete spline interpolant</u> (of degree $2m-1$ with knots $x_1, \ldots, x_{N-1}$) <u>to</u> f .

An imitation of the error analysis carried out for cubic spline interpolation in [1] leads directly to the statement that, for $f \in \mathbb{L}_1^{(n+1)}[a,b]$,

(3) $$f(x) - Sf(x) = \int_a^b K(x,t) f^{(n+1)}(t)\, dt/n!$$

with the Peano kernel K also equal to the interpolation error when applying complete spline interpolation to $(x-t)_+^n$ for fixed x as a function of t, i.e.,

(4) $$K(x,t) = (x-t)_+^n - S_{(t)}(x-t)_+^n .$$

Hence, if, for fixed x,

$$g(t) := K(x,t) \quad \text{for} \quad t \in [a,b] ,$$

then g vanishes at least once at $x_1, \ldots, x_{N-1}$ and vanishes m-fold at a and at b. Because of this latter fact,

$$(5) \qquad S^-(g(a), \ldots, g^{(n)}(a^+)) = S^-(g^{(m)}(a), \ldots, g^{(n)}(a^+)) \le n-m = m-1$$

and

$$(6) \qquad S^+(g(b), \ldots, g^{(n)}(b^-)) \ge S^+(g(b), \ldots, g^{(m)}(b)) = m .$$

If $x = x_i$ for some $i \in [1, N-1]$, then $g = 0$ since then $(x-t)_+^n$ is its own spline interpolant. Otherwise, g is a spline of exact degree n (since $g^{(n)}$ has a nonzero jump at x) with simple knots at $x_1, \ldots, x_{N-1}$ and at x and nowhere else in (a,b), and vanishes (at least) at all but one of these. Now let $[\hat{a}, \hat{b}]$ be an interval in $[a, b]$ which is maximal with respect to the property that it contains only isolated zeros of g. Then either $\hat{a} = a$ or else $S^-(g(\hat{a}), \ldots, g^{(n)}(\hat{a}^+)) = 0$, and either $\hat{b} = b$ or else $S^+(g(\hat{b}), \ldots, g^{(n)}(\hat{b}^-)) = n$. Therefore, by the Budan-Fourier theorem for splines and by (5) and (6),

$$\begin{aligned}
Z_g(\hat{a},\hat{b}) &\le Z_{g^{(n)}}(\hat{a},\hat{b}) + S^-(g(\hat{a}), \ldots, g^{(n)}(\hat{a}^+)) - S^+(g(\hat{b}), \ldots, g^{(n)}(\hat{b}^-)) \\
&\le Z_{g^{(n)}}(\hat{a},\hat{b}) + S^-(g(a), \ldots, g^{(n)}(a^+)) - S^+(g(b), \ldots, g^{(n)}(b^-)) \\
&\le Z_{g^{(n)}}(\hat{a}, \hat{b}) + m-1 - m \\
&\le \text{number of knots of } g \text{ in } (\hat{a}, \hat{b}) - 1 \\
&\le Z_g(\hat{a}, \hat{b}),
\end{aligned}$$

the last inequality since g fails to vanish at only one knot and since the preceding inequality already shows that $(\hat{a},\hat{b})$ contains at least one knot. It follows that all inequalities used to establish this string of inequalities must have been equalities.

In particular, $S^-(g(\hat{a}), \ldots, g^{(n)}(\hat{a}^+)) = m-1$, hence, as $m > 1$ by our assumption that $n > 1$, we must have $\hat{a} = a$, and, similarly, $S^+(g(\hat{b}), \ldots, g^{(n)}(\hat{b}^-)) = m$ and so $\hat{b} = b$. Further, these equalities produce a wealth of information about the x-section $g(t) = K(x,t)$ of the Peano kernel in case $x_i = x < x_{i+1}$:

(i) Since $Z_{g^{(n)}}(a,b) = N$, $g^{(n)}$ changes sign strongly across each of its simple knots. Since $\text{jump}_x g^{(n)} = (-)^n n! < 0$, this implies that

$$(-)^{i-j} g^{(n)} > 0 \text{ on } (x_j, x_{j+1}) \text{ for } j = 0, \ldots, i-1$$

$$(-)^{j-i} g^{(n)} < 0 \text{ on } (x_j, x_{j+1}) \text{ for } j = i+1, \ldots, N-1.$$

(ii) $S^-(g(a), \ldots, g^{(n)}(a^+)) = S^-(g^{(m)}(a), \ldots, g^{(n)}(a^+)) = n-m$, therefore $g^{(j)}(a^+) \neq 0$ for $j=m, \ldots, n$, and, with (i),

$$(-)^{i+n-j} g^{(j)}(a^+) > 0 \text{ for } j=m, \ldots, n.$$

Since also $g^{(j)}(a) = 0$ for $j=0, \ldots, m-1$, it follows that

$$(-)^{i+1-m} g(a+\varepsilon) > 0 \text{ for positive } \varepsilon \text{ near } 0.$$

(iii) g has a simple zero at $x_1, \ldots, x_{N-1}$ and vanishes nowhere else in (a,b), hence changes sign across each x_j and nowhere else. Therefore, with (ii),

$$(-)^{i+1-m+j} g(t) > 0 \ \underline{\text{for}} \ x_j < t < x_{j+1}, \quad j = 0, \ldots, N-1 .$$

(iv) $S^+(g(b), \ldots, g^{(n)}(b^-)) = S^+(g(b), \ldots, g^{(m)}(b)) = m,$

therefore, with (i) or (iii),

$$(-)^{i+N} g^{(j)}(b^-) > 0 \ \underline{\text{for}} \ j=m, \ldots, n .$$

It follows, in particular, that, in the usual pointwise estimate

$$|f(x) - Sf(x)| \le \int_a^b |K(x,t)| dt \, \|f^{(n+1)}\|_\infty / n! \tag{8}$$

obtained from (3), we have equality iff $|f^{(n+1)}| = \|f^{(n+1)}\|_\infty$ and $f^{(n+1)}$ changes sign across each of the interpolation points $x_1, \ldots, x_{N-1}$, i.e., f is a perfect spline of degree $n+1$ with simple knots at $x_1, \ldots, x_{N-1}$ and nowhere else. If such a spline has a positive $(n+1)$st derivative in (a, x_1), then

$$(-)^{i+1-m}(f(x) - Sf(x)) > 0 .$$

The sign structure of the fundamental functions for complete spline interpolation can be completely analysed in the same way. More interestingly, such an analysis of the sign structure of Peano kernel and fundamental functions can be carried out just as easily for spline interpolation with a variety of other side conditions, such as matching of value and odd derivatives at the boundary, matching of value and even derivatives at the boundary, matching of value and the m-th through (n-1)st derivatives at the boundary etc. The essential feature shared by these side conditions is that they are of the form

$$\lambda_i Sf = \lambda_i f, \quad i = 1, \ldots, 2m$$

with $(\lambda_i)_1^{2m}$ a sequence of linear functionals linearly independent over $\mathbb{P}_{2m} = \ker D^{2m}$:= polynomials of degree $< 2m$, and $(\lambda_i)_1^m$ a "good" m-sequence at a and $(\lambda_i)_{m+1}^{2m}$ a "good" m-sequence at b.

Here, we call an m-sequence $(\mu_i)_1^m$ of linear functionals a "good" m-sequence at α, provided $(\mu_i)_1^m$ has the following properties:

(i) $\mu_i f = \sum_{j=0}^{2m-1} a_{ij} f^{(j)}(\alpha)$ for appropriate a_{ij}'s, $i = 1, \ldots, m$.

Further, with

$$\ker(\mu_i) := \{g \in C^{(2m-1)} \text{ near } \alpha \mid \mu_i g = 0 \text{ for } i=1, \ldots, m\},$$

(ii) $g \in \ker(\mu_i)$ implies that $S^+(g(\alpha), \ldots, g^{(2m-1)}(\alpha)) \geq m$,

(iii) $g \in \ker(\mu_i)$ implies that $g^* \in \ker(\mu_i)$, with $g^*(\alpha+x) = g(\alpha-x)$, all

(iv) $f, g \in \ker(\mu_i)$ implies that

$$\sum_{j=0}^{2m-1} (-)^j f^{(j)}(\alpha) g^{(2m-1-j)}(\alpha) = 0 .$$

We note that (ii) and (iii) together give

(ii)' $g \in \ker(\mu_i)$ implies that $S^-(g(\alpha), \ldots, g^{(2m-1)}(\alpha)) \leq m-1$

because of (1.1).

A particularly simple way of choosing a "good" m-sequence $(\mu_i)_1^m$ at α is to choose a strictly increasing subsequence $\underline{\underline{r}} = (r_i)_1^m$ of $(0, \ldots, 2m-1)$ so that, for every $j = 0, \ldots, 2m-1$, either j or $2m-1-j$ occurs in $\underline{\underline{r}}$ and then to take $\mu_i f = f^{(r_i)}(\alpha)$, $i=1, \ldots, m$. All of the examples mentioned earlier are of this simple form. For more complicated examples, we note that property (ii) is insured by having all m-minors of the $m \times 2m$ matrix (a_{ij}) of (i) of one sign with at least one of them nonzero.

Theorem 2. Let $m > 1$, $n := 2m-1$, let $a = x_0 < \dots < x_N = b$, and let $(\lambda_i)_1^{2m}$ be a sequence of linear functionals, linearly independent over $\mathbb{P}_{2m} = \ker D^{2m}$ and so that $(\lambda_i)_1^m$ is a "good" m-sequence at a. and $(\lambda_i)_{m+1}^{2m}$ is a "good" m-sequence at b. Then

(i) For every $f \in C^{(n)}[a,b]$ there exists exactly one spline Sf of degree n with simple knots $x_1, \dots, x_{N-1}$ in (a,b) which agrees with f in the sense that

(9) $$(Sf)(x_i) = f(x_i), \quad i = 1, \dots, N-1,$$

(10) $$\lambda_i Sf = \lambda_i f, \quad i = 1, \dots, 2m.$$

(ii) If L_j is the unique spline of degree n with simple knots $x_1, \dots, x_{N-1}$ in (a,b) for which

$$L_j(x_i) = \delta_{ji}, \quad i=1, \dots, N-1,$$

$$\lambda_i L_j = 0, \quad i=1, \dots, 2m,$$

with $j \in [1, N-1]$, (i.e., if L_j is a fundamental function for the interpolation process), then L_j has simple zeros at x_i for $i \neq j$ and vanishes nowhere else in (a, b), and its n-th derivative changes sign strongly across each knot.

(iii) For $f \in \mathbb{L}_1^{(2m)}[a,b]$ and for $x \in (a,b)$,

(11) $$f(x) - Sf(x) = \int_a^b K(x,t) f^{(2m)}(t)\, dt/n!$$

where $g := K(x,\cdot)$ is the error in interpolating $(x-\cdot)_+^n$, i.e.,

(12) $$K(x,t) = (x-t)_+^n - S_{(t)}(x-t)_+^n .$$

This x-section $g = K(x,\cdot)$ of the kernel K is a spline of degree n with simple knots $x_1, \ldots, x_{N-1}$ and x. If $x \notin \{x_1, \ldots, x_{N-1}\}$, then g has simple zeros at $x_1, \ldots, x_{N-1}$ and vanishes nowhere else in (a,b), and its n-th derivative changes sign strongly across each knot in (a,b); otherwise g vanishes identically. Hence, for $f \in \mathbb{L}_\infty^{(2m)}[a,b]$,

(13) $$|f(x) - Sf(x)| \le \|f^{(2m)}\|_\infty \int_a^b |K(x,t)|\,dt/n!$$

with equality if and only if f is a perfect spline of degree $2m$ with simple active knots at $x_1, \ldots, x_{N-1}$ in (a,b) and no other active knots in (a,b), i.e., $f^{(2m)}$ is absolutely constant and changes sign strongly at $x_1, \ldots, x_{N-1}$ and nowhere else in (a,b).

Proof. The proof parallels closely the earlier argument for the special case

$$\lambda_i f = \begin{cases} f^{(i-1)}(a), & i=1, \ldots, m, \\ f^{(i-m-1)}(b), & i=m+1, \ldots, 2m . \end{cases}$$

Property (iv) of a "good" m-sequence insures the selfadjointness of the problem

$$D^{2m} f = y, \quad \lambda_i f = 0, \quad i=1, \ldots, 2m,$$

which then gives (11) and (12), and the sign structure of L_j and of $K(x, \cdot)$ follows from the Budan-Fourier theorem for splines and properties (ii) and (ii)' of a "good" m-sequence. We omit the details.

Remark. For the particular side conditions of matching even derivatives at a and b, (iii) of the theorem was stated by C. Hall and W. W. Meyer in [6], with the proof of the sign structure of $K(x, \cdot)$ deferred to an as yet unpublished paper (see Lemma 3 of [6]).

We add here that we learned only recently of a paper by Avraham A. Melkman entitled "The Budan-Fourier theorem for splines" which will appear eventually in the Israel Journal of Mathematics. In it, the author establishes such a theorem even for splines with multiple knots.

Part B. The sign structure of cardinal null splines

1. Introduction. A cardinal spline function of order k is a piecewise polynomial function of degree $< k$ in $C^{(k-2)}(\mathbb{R})$ with knots $\alpha + m$, for all $m \in \mathbb{Z}$ and some α. We denote their collection by

$$\$_{k,\alpha+\mathbb{Z}}.$$

A cardinal null spline is a cardinal spline which vanishes at all points of the form $\tau + m$ for $m \in \mathbb{Z}$ and some τ. Cardinal null splines have been analysed by Schoenberg [15] who showed them to be linear combinations of finitely many eigensplines. Here, an eigenspline is a nontrivial solution in $\$_{k,\alpha+\mathbb{Z}}$ of the functional equation

$$f(x+1) = zf(x),$$

shown by Schoenberg to exist for certain exceptional values of z called eigenvalues.

We wish to note in passing the work of Nilson [12] and earlier work by Ahlberg, Nilson and Walsh referred to therein where this eigenvalue problem is considered from another point of view.

Schoenberg used methods from the theory of difference equations for his analysis. We will proceed here somewhat differently and without reference to Schoenberg's earlier work. We start from the observation that a cardinal nullspline is completely determined by any one of its polynomial pieces, and study the linear transformation which carries such a polynomial piece into its neighboring polynomial piece. A convenient matrix representation for this linear transformation can be shown to be an oscillation matrix, and Gantmacher and Krein's [5] theory of such matrices then provides the detailed information about the sign structure of mull-splines needed in the later parts of this paper.

We wish to bring to the reader's attention the beautiful recent work by C. Micchelli [11] in which he, too, uses oscillation matrices in the analysis of cardinal nullsplines and eigensplines, but covers much more general splines and much more general interpolation conditions. We became aware of his work after we had completed the following sections and decided then to retain our arguments as that seemed more convenient for the reader than being told how to specialize Micchelli's more general results.

2. Cardinal splines which vanish at all knots. With the usual convention that $\binom{j}{i} = 0$ for $j<i$, we have, for a polynomial p of degree $<k$,

$$p^{(i)}(x+h)/i! = \sum_{j=0}^{k-1} \binom{j}{i} h^{j-i} p^{(j)}(x)/j!, \quad i=0, \ldots, k-1 .$$

Hence, if such a polynomial vanishes at x and at $x+h$, then, for $i = 0$,

$$0 = \sum_{j=1}^{k-1} \binom{j}{0} h^j p^{(j)}(x)/j!$$

and, on subtracting $\binom{k-1}{i}/h^i$ times this equation from the i-th equation, we obtain

$$p^{(i)}(x+h)/i! = \sum_{j=1}^{k-2} \left(\binom{j}{i} - \binom{k-1}{i}\right) h^{j-i} p^{(j)}(x)/j!, \quad i=1,\ldots,k-2.$$

This we write as

$$\text{(1)} \qquad \underline{\underline{p}}(x+h) = -H_h^{-1} A_k H_h \underline{\underline{p}}(x)$$

with $\underline{p}(x)$ the vector

(2) $$\underline{p}(x) := (p^{(1)}(x),\ p^{(2)}(x)/2,\ \ldots,\ p^{(k-2)}(x)/(k-2)!),$$

H_h the diagonal matrix

(3) $$H_h := \operatorname{diag} \ulcorner 1, h, \ldots, h^{k-3} \lrcorner = H^{-1}_{1/h},$$

and A_k the matrix

(4) $$A_k := \left(\binom{k-1}{i} - \binom{j}{i}\right)_{i,j=1}^{k-2}.$$

If p is a polynomial of degree $< k$ vanishing at x and at $x+1$, then it vanishes also at $x+1$ and at $(x+1) + (-1)$, therefore (1) implies

$$1 = -H^{-1}_{-1} A_k H_{-1} (-H^{-1}_{1} A_k H_1)$$

or, with $H_1 = 1$,

(5) $$A_k^{-1} = H^{-1}_{-1} A_k H_{-1},$$

showing A_k to be invertible and similar to its own inverse. Therefore,

(6) $$\operatorname{spectrum}(A_k) = 1/\operatorname{spectrum}(A_k) = \{1/\lambda \mid \lambda \in \operatorname{spectrum}(A_k)\}.$$

Suppose that $\underline{u}$ is a (nonzero) eigenvector of A_k belonging to some nonzero eigenvalue λ. Let p be the polynomial of degree $< k$ which vanishes at 0 and at 1 and for which $\underline{p}(0) = \underline{u}$. Then $\underline{p}(1) = -A_k\underline{u} = -\lambda\underline{p}(0)$, hence,

$$p^{(j)}(1) = -\lambda p^{(j)}(0),\ \ j = 0, \ldots, k-2.$$

The rule

$$S(x+n) := (-\lambda)^n p(x), \quad n \in \mathbb{Z}, \quad x \in [0, 1)$$

therefore defines a spline of degree $< k$ with simple knots at $\mathbb{Z}$, and this spline is evidently an eigenspline belonging to the eigenvalue $-\lambda$. This explains our interest in the eigenvalue-eigenvector structure of the matrix A_k.

In [2], the matrix A_k was claimed to be totally positive. Actually, A_k <u>is an oscillation matrix</u>, i.e., A_k is totally positive and some power of A_k is strictly totally positive. To see this, observe that

$$A_k = -B_k' \begin{pmatrix} 1,\ldots,k-2 \\ 1,\ldots,k-2 \end{pmatrix}$$

with B_k' the matrix obtained from

$$B_k := \left(\binom{j}{i}\right)_{i,j=0}^{k-1}$$

by subtracting the last column of B_k from all other columns. This implies that

$$\det B_k \begin{pmatrix} 0, i_1, \ldots, i_r \\ j_1, \ldots, j_r, k-1 \end{pmatrix} = \det B_k' \begin{pmatrix} 0, i_1, \ldots, i_r \\ j_1, \ldots, j_r, k-1 \end{pmatrix}$$

and that the first row of B_k' vanishes except for a 1 in column $k-1$. Hence, further,

$$\det B_k'\begin{pmatrix}0,i_1,\ldots,i_r\\ j_1,\ldots,j_r,k-1\end{pmatrix} = (-)^r \det B_k'\begin{pmatrix}i_1,\ldots,i_r\\ j_1,\ldots,j_r\end{pmatrix}$$

$$= \det A_k\begin{pmatrix}i_1,\ldots,i_r\\ j_1,\ldots,j_r\end{pmatrix}$$

provided $0 < i_1,\ldots,i_r,j_1,\ldots,j_r < k-1$. This shows A_k to be totally positive since B_k is known to be totally positive (cf., e.g., [14]). Since A_k is also invertible, A_k is proven to be an oscillation matrix once we show that none of its entries on the first subdiagonal and the first superdiagonal vanishes (Gantmacher and Krein [5, Theorem 10 in Chap. 2, par. 7]). The numbers in question are

$$\binom{k-1}{i} - \binom{i+1}{i}, \quad i=1,\ldots,k-3, \quad \text{and} \quad \binom{k-1}{i}, \quad i=2,\ldots,k-2,$$

and clearly all are positive.

The fact that A_k is an oscillation matrix allows the following conclusions (cf. [5, Theorem 6 in Chap. 2, par. 5]):

Theorem 1. (i) The spectrum of A_k consists of $k-2$ (different) positive numbers which we will think of as ordered:

$$0 < \lambda_{k-2} < \ldots < \lambda_1 .$$

Further, by (6),

$$\lambda_i\lambda_j = 1 \ \text{for} \ i+j = k-1 . \tag{7}$$

(ii) <u>If</u> $(\underline{\underline{u}}^{(i)})_1^{k-2}$ <u>is a corresponding (necessarily complete) sequence of (nonzero) eigenvectors for</u> A_k <u>and</u> $c_p, \ldots, c_q$ <u>are numbers not all zero, then</u>

$$p - 1 \leq S^-(\sum_{i=p}^{q} c_i \underline{\underline{u}}^{(i)}) \leq S^+(\sum_{i=p}^{q} c_i \underline{\underline{u}}^{(i)}) \leq q-1 .$$

As a particular consequence of (ii), we have

(8) $$S^-(\underline{\underline{u}}^{(i)}) = S^+(\underline{\underline{u}}^{(i)}) = i-1, \quad i=1,\ldots,k-2,$$

therefore

(9) $$u_1^{(i)} \neq 0 \text{ and } u_{k-2}^{(i)} \neq 0$$

(since otherwise $S^-(\underline{\underline{u}}^{(i)}) < S^+(\underline{\underline{u}}^{(i)})$.) Further, let p be the polynomial of degree $< k$ which vanishes at 0 and at 1 and for which

$$\underline{\underline{p}}(0) = \underline{\underline{u}}^{(i)} .$$

Then

$$\underline{\underline{p}}(1) = -A_k \underline{\underline{u}}^{(i)} = -\lambda_i \underline{\underline{p}}(0)$$

hence

$$p^{(k-1)} = p^{(k-2)}(1) - p^{(k-2)}(0) = -(\lambda_i + 1) u_{k-2}^{(i)}$$

therefore, with (9),

$$p^{(k-2)}(0)p^{(k-1)}(0) < 0 < p^{(k-2)}(1)p^{(k-1)}(1)$$

and so, again with (9),

$$S^-(p(0), \ldots, p^{(k-1)}(0)) = i = S^+(p(1), \ldots, p^{(k-1)}(1)) .$$

Finally, these statements about the sign structure of p persist if we perturb p slightly because of (8) as long as we keep $p(0) = p(1) = 0$.

Corollary. *If $\underline{u}^{(i)}$ is a nonzero eigenvector belonging to the i-th eigenvalue of A_k (ordered as in the theorem), then, for any polynomial p of degree $< k$ vanishing at 0 and at 1 and with $\underline{p}(0)$ close to $\underline{u}^{(i)}$,*

$$S^-(p(0), \ldots, p^{(k-1)}(0)) = i = S^+(p(1), \ldots, p^{(k-1)}(1)) .$$

In particular, such a p has no zeros in $(0, 1)$, by the Budan-Fourier Theorem.

If now f is a cardinal spline function of order k with knots $\mathbb{Z}$ which vanishes at its knots, then

$$\underline{f}(m) = -A_k \underline{f}(m-1) \quad \text{for all } m \in \mathbb{Z}$$

hence

$$\underline{f}(m) = (-A_k)^m \underline{f}(0) \quad \text{for all } m \in \mathbb{Z} .$$

Further, expanding $\underline{f}(0)$ in terms of a complete eigenvector sequence $(\underline{u}^{(i)})_1^{k-2}$ for A_k, we have

$$\underline{\underline{f}}(m) = \sum_{i=1}^{k-2} (-\lambda_i)^m c_i \underline{\underline{u}}^{(i)} = \sum_{i=1}^{k-2} c_i \underline{\underline{U}}_i(m)$$

with U_i the unique cardinal spline of order k which vanishes at its knots $\mathbb{Z}$ and satisfies $\underline{\underline{U}}_i(0) = \underline{\underline{u}}^{(i)}$, hence satisfies

$$\underline{\underline{U}}_i(m) = (-\lambda_i)^m \underline{\underline{U}}_i(0), \quad \text{all } m \in \mathbb{Z},$$

i.e., U_i is the eigenspline belonging to the eigenvalue $-\lambda_i$. Therefore

$$f = \sum_{i=1}^{k-2} c_i U_i .$$

3. Cardinal nullsplines which vanish between knots. The analysis is slightly more complicated when the nullspline vanishes at $\tau + \mathbb{Z}$ with τ not a knot.

Let $\tau \in (0, 1)$ and let p be any spline of order k vanishing at τ and at $\tau+1$ and with simple knot at 1. Then

$$\sum_{j=1}^{k-1} (p^{(j)}(\tau)/j!)\binom{j}{i}(1-\tau)^{j-i} = p^{(i)}(1)/i!, \quad i=0,\ldots,k-2,$$

and

$$\sum_{j=0}^{k-2} (p^{(j)}(1)/j!)[\binom{j}{i} - \binom{k-1}{i}]\tau^{j-i} = p^{(i)}(\tau+1)/i! , \quad i=1,\ldots,k-1 .$$

Hence, with

$$B_k := (\binom{j}{i})_{i,j=0}^{k-1}$$

and B_k' the matrix derived from B_k by subtracting the last column from all other columns as before, we conclude that

$$-A_{k,\tau}\underline{\underline{p}}(\tau) = \underline{\underline{p}}(\tau+1)$$

where

$$(1)\quad \begin{aligned} A_{k,\tau} &:= -(H_\tau^{-1}B_k'H_\tau)\begin{pmatrix}1,\dots,k-1\\0,\dots,k-2\end{pmatrix}(H_{1-\tau}^{-1}B_kH_{1-\tau})\begin{pmatrix}0,\dots,k-2\\1,\dots,k-1\end{pmatrix} \\ &= -\frac{1-\tau}{\tau}H_\tau^{-1}B_k'\begin{pmatrix}1,\dots,k-1\\0,\dots,k-2\end{pmatrix}H_\tau H_{1-\tau}^{-1}B_k\begin{pmatrix}0,\dots,k-2\\1,\dots,k-1\end{pmatrix}H_{1-\tau}\end{aligned}$$

and now

$$\underline{\underline{p}}(x) := (p^{(1)}(x)/1!, \dots, p^{(k-1)}(x)/(k-1)!) .$$

By considering the spline q given by

$$q(1+x) = p(1-x), \quad \text{all } x,$$

which then vanishes at $1-\tau$ and at $1+(1-\tau)$ and has a simple knot at 1, we find that

$$-H_{-1}^{-1}A_{k,1-\tau}H_{-1}(-A_{k,\tau}) = 1,$$

hence $A_{k,\tau}$ is invertible and

$$A_{k,\tau}^{-1} = H_{-1}^{-1}A_{k,1-\tau}H_{-1} .$$

It follows that

(2) $$\text{spectrum}(A_{k,\tau}) = 1/\text{spectrum}(A_{k,1-\tau}) .$$

Both $B_k\begin{pmatrix}0,\dots,k-2\\1,\dots,k-1\end{pmatrix}$ and $-B_k'\begin{pmatrix}1,\dots,k-1\\0,\dots,k-2\end{pmatrix}$ are easily seen to be oscillation matrices with the aid of arguments used earlier to establish that $-B_k'\begin{pmatrix}1,\dots,k-2\\1,\dots,k-2\end{pmatrix}$ is an oscillation matrix. $A_{k,\tau}$ is therefore also an oscillation matrix.

Theorem 2. (i) *The spectrum of* $A_{k,\tau}$ *consists of* $k-1$ *(different) positive numbers which we will think of as ordered:*

$$0 < \lambda_{k-1}(\tau) < \dots < \lambda_1(\tau) .$$

Further, by (2),

(3) $$\lambda_i(\tau)\lambda_j(1-\tau) = 1 \quad \text{for} \quad i+j = k.$$

(ii) *If* $(\underline{\underline{u}}^{(i,\tau)})_{i=1}^{k-1}$ *is a corresponding (necessarily complete) sequence of eigenvectors for* $A_{k,\tau}$, *then*

$$S^-(\underline{\underline{u}}^{(i,\tau)}) = S^+(\underline{\underline{u}}^{(i,\tau)}) = i-1, \quad i=1, \dots, k-1.$$

(iii) *If* p *is a spline of degree* $< k$ *which vanishes at* τ *and at* $\tau+1$, *has a simple knot at* 1, *and for which* $\underline{p}(\tau)$ *is close to* $\underline{\underline{u}}^{(i,\tau)}$, *then*

$$S^-(p(\tau), \dots, p^{(k-1)}(\tau)) + 1 = i = S^+(p(\tau+1), \dots, p^{(k-1)}(\tau+1)) .$$

In particular, such a p *does not vanish on* $(\tau, \tau+1)$, *by the Budan-Fourier theorem for splines.*

Proof. In view of the preceding discussion, only assertion (iii) needs argument. For this, we note that $p(\tau) = p(\tau+1) = 0$ while, by (ii), $S(p^{(1)}(\tau), \ldots, p^{(k-1)}(\tau)) = i-1$. |||

Since $A_{k,\tau}$ is an analytic function of $\tau \in (0, 1)$, so is each $\lambda_i(\tau)$. Further,

$$\prod_{i=1}^{k-1} \lambda_i(\tau) = (-)^{k-1} \det A_{k,\tau} = (1-\tau)/\tau$$

so that

$$\lim_{\tau\to 0} \lambda_1(\tau) = +\infty, \qquad \lim_{\tau\to 1} \lambda_{k-1}(\tau) = 0 .$$

Also

$$\lim_{\tau\to 1} H_{1-\tau}^{-1} B_k H_{1-\tau} = 1, \qquad \lim_{\tau\to 1} H_\tau^{-1} B_k' H_\tau = B_k' ,$$

therefore

$$\lim_{\tau\to 1} A_{k,\tau} = A_{k-1} := B_k' \begin{pmatrix} 1, \ldots, k-1 \\ 0, \ldots, k-2 \end{pmatrix} \; 1 \begin{pmatrix} 0, \ldots, k-2 \\ 1, \ldots, k-1 \end{pmatrix} = \left(\begin{array}{ccc|c} & A_k & & 0 \\ \hline -1 & \ldots & -1 & 0 \end{array} \right) ,$$

with A_k the oscillation matrix discussed earlier. Hence

$$\lim_{\tau\to 1} \lambda_i(\tau) = \lambda_i(1) := \begin{cases} \lambda_i, & i=1, \ldots, k-2 \\ 0, & i=k-1 \end{cases}$$

where $\lambda_{k-2} < \ldots < \lambda_1$ are the eigenvalues of A_k. Consequently, from (3) and since $\lambda_i \lambda_j = 1$ for $i+j = k-1$ by (2.7) above,

$$\lim_{\tau\to 0} \lambda_i(\tau) = 1/\lim_{\tau\to 1} \lambda_{k-i}(\tau) = 1/\lambda_{k-i} = \lambda_{i-1} = \lim_{\tau\to 1} \lambda_{i-1}(\tau)$$

for $i=2,\ldots,k-1$.

Theorem 3. The function Λ_k defined on $(0, \infty)$ by

$$\Lambda_k(\tau) := \begin{cases} \lambda_i(\tau - i - 1), & i-1 \le \tau \le i, \quad i=1,\ldots,k-1 \\ 0 & , k-1 \le \tau \end{cases}$$

is continuous (in fact in $C^{(k-2)}(0, \infty)$) and maps $(0,k-1)$ to $(0, \infty)$. Also, Λ_k is strictly monotonely decreasing on $(0,k-1)$, and $\Lambda_k(\tau) = \Lambda_k(k-1-\tau)$.

Proof. We only have to establish the strict monotonicity of Λ_k on $(0,k-1)$. For this, it is sufficient to show:

if $\tau_1, \tau_2 \in (0,1)$ are such that $\lambda_{i_1}(\tau_1) = \lambda_{i_2}(\tau_2) = \lambda$, say, then $\tau_1 = \tau_2$ and $i_1 = i_2$.

For this, let q_r be a spline of order k vanishing at τ_r and at $1+\tau_r$, with a simple knot at 1 and such that $\underline{q}_r(\tau_r)$ is a nontrivial eigenvector for A_{k,τ_r} corresponding to $\lambda=\lambda_{i_r}(\tau_r)$. Then

$$q_r^{(j)}(1) = -\lambda q_r^{(j)}(0), = \quad j=0,\ldots,k-2 .$$

If now $q_r(0) = 0$, then $(q_r^{(1)}(0), \ldots, q_r^{(k-2)}(0)/(k-2)!)$ would be a nontrivial eigenvector of A_k, hence q_r would have to be nonzero on $(0, 1)$ by the Corollary to Theorem 1, a contradiction. Hence $q_r(0) \neq 0$, and (iii) of Theorem 2 implies that

(4) $$q_r(\tau)/q_r(0) \gtrless 0 \quad \text{for} \quad \tau \lessgtr \tau_r \ .$$

With this, the spline

$$q := q_1/q_1(0) - q_2/q_2(0)$$

is of degree $< k$, vanishes at 0 and at 1 and satisfies

$$q^{(j)}(1) = -\lambda q^{(j)}(0), \quad j=0,\ldots,k-2,$$

while

$$q(\tau_1)q(\tau_2) = -\Big(q_1(\tau_2)/q_1(0)\Big)\Big(q_2(\tau_1)/q_2(0)\Big) \leq 0$$

by (4). Hence q vanishes in $(0, 1)$, and therefore must vanish identically since otherwise $(q^{(1)}(0), \ldots, q^{(k-2)}(0)/(k-2)!)$ would be a nontrivial eigenvector for A_k, hence q couldn't vanish in $(0, 1)$. It follows that $q_1/q_1(0) = q_2/q_2(0)$, therefore $\tau_1 = \tau_2$ by (4), and $i_1 = i_2$ follows from (i) of Theorem 2. |||

A different proof of the theorem can be found in Schoenberg [18]. The theorem itself is due to C. Micchelli who proved it in [10] for the much more general case of cardinal $\mathcal{L}$-splines.

If we combine the earlier statement

(3) $$\lambda_i(\tau)\lambda_j(1-\tau) = 1 \quad \text{for} \quad i+j = k$$

with the strict decrease, just proven, of each $\lambda_i(\tau)$ as τ goes from 0 to 1, then we obtain the following corollary which will be helpful in the discussion of even degree cardinal spline interpolation in Part C.

Corollary. Let $p = p(\tau, k)$ be the smallest integer so that $\lambda_p(\tau) \leq 1$, and let q be the largest integer so that $\lambda_q(\tau) \geq 1$ and let $\tau \in (0, 1]$. Then, with $m := \lfloor k/2 \rfloor$,

$$(p,q) = \begin{cases} (m+1, m) & \text{for } k \text{ odd and } \tau \in (0, 1) \\ (m, m) & \text{for } k \text{ odd and } \tau = 1 \\ (m+1, m) & \text{for } k \text{ even and } \tau \in (0, \tfrac{1}{2}) \\ (m, m) & \text{for } k \text{ even and } \tau = \tfrac{1}{2} \\ (m, m-1) & \text{for } k \text{ even and } \tau \in (\tfrac{1}{2}, 1) \end{cases}$$

In particular, $\lambda_i(\tau) \neq 1$ unless $i = m$ and

$$\tau = \begin{cases} \tfrac{1}{2}, & k \text{ even} \\ 1, & k \text{ odd} \end{cases} .$$

It is now a simple matter to describe a cardinal spline of order k with knots $\mathbb{Z}$ which vanishes at $\tau + \mathbb{Z}$ for some $\tau \in (0, 1)$. If f is such a spline, then

$$f = \sum_{i=1}^{k-1} c_i U_{i,\tau}$$

with $U_{i,\tau}$ the eigenspline belonging to the eigenvalue $-\lambda_i(\tau)$ and satisfying

$$\underline{\underline{U}}_{i,\tau}(\tau) = \underline{\underline{u}}^{(i,\tau)} ,$$

and the c_i's so chosen that $\underline{\underline{f}}(\tau) = \sum_1^{k-1} c_i \underline{\underline{u}}^{(i,\tau)}$.

Part C. The sign structure of the fundamental function and the Peano kernel for cardinal spline interpolation

1. A theorem on cardinal Green's functions. The application of the Budan-Fourier theorem for splines to cardinal spline interpolation takes the form of the following theorem.

Theorem 1. *Let K be a real valued function on* $\mathbb{R}$ *having the following properties:*

(i) K *is a polynomial spline of order* $k > 2$ *not identically zero and has simple knots at all integers* n *with* $|n| \geq N$ *for some* N. K *has no more than* r *additional knots, all of which lie in* $(-N, N)$ *and are simple.*

(ii) *For some* $\tau \in (0, 1]$ *with*

$$\tau \neq k/2 - \lfloor k/2 \rfloor + 1/2 \tag{1}$$

and for all $n \geq N$, K *vanishes at* $\tau+n$ *and at* $\tau-n-1$. *In addition, there are at least* r *distinct points in* $(-n, N)$ *at which* K *vanishes and no more than one of these occurs in any interval* $(\alpha, \beta]$ *formed by neighboring knots of* K.

(iii) K *is of power growth at* $\pm\infty$.

Then, K *vanishes exactly* r *times in* $(-N, N]$ *and has simple zeros at the points* $\tau+n$ *and* $\tau-n-1$ *for* $n=N, N+1, \ldots,$ *and vanishes nowhere else, and its* $(k-1)$*st derivative changes sign strongly across each knot. Also, with* $(\lambda_i(\tau))$ *the strictly decreasing sequence of eigenvalues of*

$$(2) \qquad A := \begin{cases} A_{k,\tau}, & \tau < 1 \\ A_k, & \tau = 1 \end{cases}$$

as described in Part B, we have

$$(3) \qquad \begin{aligned} 0 &< \limsup_{x\to\infty} |K(x)|/|\lambda_p(\tau)|^x < \infty \\ 0 &< \limsup_{x\to-\infty} |K(x)|/|\lambda_q(\tau)|^x < \infty \end{aligned}$$

with

$$(4) \qquad (p,q) = \begin{cases} (m+1,\ m) & \text{if } k \text{ odd} \\ (m+1,\ m) & \text{if } k \text{ even and } \tau \in (0, \tfrac{1}{2}) \\ (m,\ m-1) & \text{if } k \text{ even and } \tau \in [\tfrac{1}{2}, 1] \end{cases}$$

and

$$m := \lfloor k/2 \rfloor .$$

In particular, $K(x)$ decays exponentially as $|x| \to \infty$.

Proof. We intend to apply the Budan-Fourier theorem to the spline K and therefore begin with the observation that K is necessarily of exact degree $k-1$ since otherwise K would be a polynomial (of degree $< k-1$) which vanishes infinitely often, hence would have to vanish identically.

We recall from Part B the abbreviation

$$\underline{\underline{f}}(x) := \begin{cases} (f(x), \ldots, f^{(k-1)}(x)/(k-1)!), & \tau < 1, \\ (f(x), \ldots, f^{(k-2)}(x)/(k-2)!), & \tau = 1. \end{cases}$$

Since A $(= A_{k,\tau}$ or $A_k)$ is diagonalizable, we can write $\underline{\underline{K}}(\tau+N)$ as

$$\underline{\underline{K}}(\tau+N) = \underline{\underline{u}}^{(p,\tau)} + \sum_{i>p} c_i \underline{\underline{u}}^{(i,\tau)}$$

with $(\underline{\underline{u}}^{(i,\tau)})$ an appropriate complete eigenvector sequence for A corresponding to the eigenvalue sequence $(\lambda_i(\tau))$. Then

$$\underline{\underline{K}}(\tau+N+n) = (-A)^n \underline{\underline{K}}(\tau+N)$$

$$= (-\lambda_p(\tau))^n \underline{\underline{u}}^{(p,\tau)} + \sum_{i>p} c_i(-\lambda_i(\tau))^n \underline{\underline{u}}^{(i,\tau)}$$

for $n=1,2,\ldots,$ hence

(5) $$\underline{\underline{K}}(\tau+N+n) = (-\lambda_p(\tau))^n \underline{\underline{u}}^{(p,\tau)} + o((\lambda_p(\tau))^n) \quad \text{as } n \to \infty .$$

Since K is of power growth at ∞, it now follows that

$$\lambda_p(\tau) \leq 1,$$

hence, by the Corollary to Theorem B3 we must have

$$(6) \qquad p \geq \begin{cases} m+1, & \text{if } k \text{ is odd} \\ m+1, & \text{if } k \text{ is even and } \tau \in (0, \tfrac{1}{2}) \\ m\,, & \text{if } k \text{ is even and } \tau \in (\tfrac{1}{2}, 1] \end{cases}$$

while, by (iii) of Theorem B2 (if $\tau < 1$) and by the Corollary to Theorem B1 (if $\tau = 1$),

$$(7) \qquad S(K^{(1)}(\tau+n), \ldots, K^{(k-1)}(\tau+n^-)) = p-1,$$

therefore

$$(8) \qquad S^+(K(\tau+n), \ldots, K^{(k-1)}(\tau+n^-)) = p$$

for all large $n \in \mathbb{N}$. One proves analogously that, for a possibly differently normalized eigenvector sequence for A,

$$(9) \qquad \underline{\underline{K}}(\tau - N-1-n) = (-\lambda_q(\tau))^{-n} \underline{\underline{u}}^{(q,\tau)} + o((\lambda_q(\tau))^{-n}) \text{ as } n \to \infty$$

with

$$(10) \qquad q \leq \begin{cases} m\,, & \text{if } k \text{ is odd} \\ m\,, & \text{if } k \text{ is even and } \tau \in (0, \tfrac{1}{2}) \\ m-1, & \text{if } k \text{ is even and } \tau \in (\tfrac{1}{2}, 1], \end{cases}$$

hence

(11) $$S^-(K(\tau-n-\varepsilon), \ldots, K^{(k-1)}(\tau-n-\varepsilon)) = q$$

for large $n \in \mathbb{N}$ and small $\varepsilon > 0$. Now take n large enough so that (8) and (11) hold and abbreviate

$$a := \tau-n-1-\varepsilon, \quad b := \tau+n .$$

Let $[\hat{a}, \hat{b}]$ be an interval in $[a, b]$ which is maximal with respect to the property that it contains only isolated zeros of K. Then either $\hat{a} = a$ or else $S^-(K(\hat{a}),\ldots,K^{(k-1)}(\hat{a}^+)) = 0$, and either $\hat{b} = b$ or else $S^+(K(\hat{b}),\ldots,K^{(k-1)}(\hat{b}^-)) = k-1$. Therefore, we obtain from assumption (ii), from the Budan-Fourier theorem for splines, from equations (8) and (11) and from inequalities (6) and (10) that

$$\begin{aligned}
Z_{K^{(k-1)}}(\hat{a},\hat{b}) &\le \text{number of active knots of } K \text{ in } (\hat{a},\hat{b}) \\
&\le \text{number of knots of } K \text{ in } (\hat{a},\hat{b}) \\
&\le Z_K(\hat{a},\hat{b}) + 1 \\
&\le Z_{K^{(k-1)}}(\hat{a},\hat{b}) + 1 + S^-(K(\hat{a}),\ldots,K^{(k-1)}(\hat{a}^+)) - S^+(K(\hat{b}),\ldots,K^{(k-1)}(\hat{b}^-)) \\
&\le Z_{K^{(k-1)}}(\hat{a},\hat{b}) + 1 + S^-(K(a),\ldots,K^{(k-1)}(a^+)) - S^+(K(b),\ldots,K^{(k-1)}(b^-)) \\
&\le Z_{K^{(k-1)}}(\hat{a},\hat{b}) + 1 + q - p \\
&\le Z_{K^{(k-1)}}(\hat{a},\hat{b}),
\end{aligned}$$

hence _equality must hold in all inequalities used to establish this string of inequalities_. In particular, $[\hat{a},\hat{b}] = [a,b]$ since $k > 2$, and (6) and (10) must be equalities, hence (4) holds and (3) follows from (5) and (9). |||

Remark. If (1) is violated, then the assumption (iii) of power growth at $\pm\infty$ is not sufficient to conclude the exponential decay of $K(x)$ as $|x| \to \infty$. For, then $\lambda_m(\tau) = 1$ by the Corollary to Theorem B3, and even boundedness of K would only imply that $q \leq m \leq p$, and would therefore not lead to equality in the Budan-Fourier inequality. Yet, replacing assumption (iii) by a stronger assumption such as that $K \in \mathbb{L}_s(\mathbb{R})$ for some $s < \infty$ would force $q \leq m-1 < m+1 \leq p$, and application of the Budan-Fourier theorem would lead to a contradiction unless we add additional freedom to K. We exploit this further in Part D.

Corollary. Under the theorem's assumptions, let (t_i) be the knot sequence for K, and let (τ_i) be the increasing sequence of its zeros, numbered so that

$$t_n \in (\tau_{n-1}, \tau_n] \quad \text{for all large } n,$$

as can be done by assumption. If also $t_1 \leq \tau_1 < t_2$, then

$$\operatorname{sign} K^{(1)}(\tau_1) K^{(k-1)}(\tau_1^-) = (-)^{p-1}.$$

Proof. Since all zeros of K are simple, and K changes sign strongly across each knot, we have

$$\operatorname{sign} K^{(1)}(\tau_1) K^{(1)}(\tau_n) = (-)^{1-n} = \operatorname{sign} K^{(k-1)}(t_1^-) K^{(k-1)}(t_n^-),$$

hence $\operatorname{sign} K^{(1)}(\tau_1)K^{(n-1)}(\tau_n^-) = \operatorname{sign} K^{(1)}(\tau_n)K^{(k-1)}(\tau_n^-)$ in case $t_1 \le \tau_1 < t_2$ and $t_n \le \tau_n < t_{n+1}$. On the other hand, $S(K^{(1)}(\tau_n),\ldots,K^{(k-1)}(\tau_n^-)) = p-1$ for all large n, by (7). |||

2. Cardinal spline interpolation. The k-th order cardinal spline interpolant $S_k f$ to a given function f on $\mathbb{R}$ of power growth at $\pm\infty$ is, by definition, the unique spline of order k with knots $\mathbb{Z}+k/2$ which agrees with f at all integers and is of power growth at $\pm\infty$. A detailed discussion of this interpolation process can be found in Schoenberg's monograph [15].

The fundamental function of the process is, by definition, the unique cardinal spline L_k of power growth at $\pm\infty$ with knots $\mathbb{Z}+k/2$ which satisfies

$$L_k(n) = \delta_{0n} \quad \text{for all } n \in \mathbb{Z} \tag{1}$$

and so allows one to write the interpolant as

$$(S_k f)(x) = \sum_{\nu \in \mathbb{Z}} f(\nu) L_k(x-\nu) .$$

L_k is a cardinal Green's function in that

$$L_k(x) = \begin{cases} L(x), & x \le -1 \\ R(x), & x \ge 1 \end{cases}$$

with both L and R cardinal nullsplines. For even k, $K := L_k$ satisfies the hypotheses of Theorem 1 with $N = 1$, $r = 1$ and $\tau = 1$. For odd k,

$K := L_k(\cdot - \frac{1}{2})$ satisfies these hypotheses with $N = 1$, $r = 1$ and $\tau = \frac{1}{2}$. It is therefore a consequence of Theorem 1 that L_k has a simple zero at every nonzero integer and vanishes nowhere else. Therefore

$$\text{(2)} \qquad \operatorname{sign} L_k(x) = \operatorname{sign} \frac{\sin \pi x}{\pi x} \quad \text{for all } x \in \mathbb{R},$$

a fact apparently known (see, e.g., F. Richards [13]) but not proved anywhere as far as we know. Further, $L_k^{(1)}(1)$ must be negative since L_k is positive on $(-1, 1)$, hence, by the Corollary to Theorem 1, $(-)^p L_k^{(k-1)}(1^-)$ must be positive, with p given by (1.4). In particular, $L_k^{(k-1)}(0^+) = L_k^{(k-1)}(1^-)$ and $p = k/2$ in case k is even, while $p = (k+1)/2$, and $L_k^{(k-1)}(0^+)$ and $L_k^{(k-1)}(1^-)$ have opposite sign in case k is odd. So,

$$\text{(3)} \qquad \operatorname{sign} L_k^{(k-1)}(0^+) = (-)^{\lfloor k/2 \rfloor}.$$

Theorem 1 also implies the known fact that

$$\text{(4)} \qquad 0 < \lim_{x \to \infty} |L_k(x)| / \gamma_k^{|x|} < \infty$$

with γ_k the largest eigenvalue less than one of A_k or of $A_{k,\frac{1}{2}}$ as k is even or odd.

Schoenberg [17] has obtained sharp estimates for the interpolation error $f - S_k f$ in terms of $\|f^{(k)}\|_\infty$ for even k. He uses the representation of the error

$$(5)\qquad f(x) - (S_k f)(x) = \int_{-\infty}^{\infty} K_k(x,t) f^{(k)}(t)\, dt/(k-1)!$$

with

$$(6)\qquad K_k(x,t) := (x-t)_+^{k-1} - \sum_\nu (\nu-t)_+^{k-1} L_k(x-\nu)$$

which he shows to be valid for $f \in \mathbb{L}_{1,loc}^{(k)}(\mathbb{R})$ with $f^{(k)}$ of power growth at $\pm\infty$, and for even k. He leaves to the present paper the proof of the following theorem.

Theorem 3 of [17]. For even k greater than 2 and for $t \in (0, 1)$, the function $K(x) := K_k(x,t)$ has simple zeros at all integer values of x and vanishes nowhere else.

Proof. For fixed t, $K_k(x,t)$ is the error in interpolating $(x-t)_+^{k-1}$ in x by cardinal splines of order k, hence $K_k(\cdot, t) = 0$ in case $t \in \mathbb{Z}$ since then $(\cdot - t)_+^{k-1}$ is its own cardinal spline interpolant. Further, for $t \in (0, 1)$, $K_k(\cdot, t)$ is of power growth at $\pm\infty$, vanishes at $\mathbb{Z}$ and has simple knots at $\mathbb{Z}$ and at t, and is of exact degree $k-1$ since its $(k-1)$st derivative has a nonzero jump at t. In short, $K := K_k(\cdot, t)$ satisfies the assumptions of Theorem 1 with $N = 1$, $r = 2$, and $\tau = 1$ and must, therefore, satisfy its conclusions. |||

3. Even degree cardinal spline interpolation. We discuss now in more detail the interpolation error for odd k, i.e., for even degree cardinal spline interpolation,

$$k = 2m+1,$$

say. Let K_k be as defined in (2.6). Then, as $(\nu-t)_+^{k-1}$ is a spline of order k in t with a simple knot at ν, and is of power growth, we can write

$$(\nu-t)_+^{k-1} = \sum_\mu (\nu - \mu - \tfrac{1}{2})_+^{k-1} L_k(t - \mu - \tfrac{1}{2}),$$

with the series converging uniformly on compact sets, by (2.4). Therefore

$$\begin{aligned}\sum_\nu (\nu-t)_+^{k-1} L_k(x-\nu) &= \sum_\nu \Big(\sum_\mu (\nu-\mu-\tfrac{1}{2})_+^{k-1} L_k(t-\mu-\tfrac{1}{2}) \Big) L_k(x-\nu) \\ &= \sum_\mu \Big(\sum_\nu (\nu-\mu-\tfrac{1}{2})_+^{k-1} L_k(x-\nu) \Big) L_k(t-\mu-\tfrac{1}{2}) \\ &= \sum_\mu (x-\mu-\tfrac{1}{2})_+^{k-1} L_k(t-\mu-\tfrac{1}{2})\end{aligned}$$

with the interchange permitted because of the absolute convergence of the series involved, and the last equality justified by the fact that $(x-\mu-\frac{1}{2})_+^{k-1}$ is a cardinal spline of order k in x. Since

$$(x-t)_+^{k-1} - (t-x)_+^{k-1} = (x-t)^{k-1}$$

for odd k and since cardinal spline interpolation of order k reproduces the right side of this identity, we conclude that

$$K_k(x, t) = -K_k(t-\tfrac{1}{2}, x-\tfrac{1}{2}) . \tag{1}$$

Further, we conclude that, for fixed $x \in (0, 1)$,

$$K_k(x, t+\tfrac{1}{2}) = (x-\tfrac{1}{2}-t)_+^{k-1} - \sum_\mu (x-\tfrac{1}{2}-\mu)_+^{k-1} L_k(t-\mu) ,$$

i.e., $K_k(x, t+\frac{1}{2})$ is the error in cardinal spline interpolation in t to $(x-\frac{1}{2}-t)_+^{k-1}$, hence is of power growth, vanishes at $\mathbb{Z}$, and has simple knots at $\mathbb{Z}-\frac{1}{2}$ and at $x-\frac{1}{2}$ and nowhere else. The function

$$K(t) := K_k(x, t)$$

therefore satisfies the hypotheses of Theorem 1 with $\tau = \frac{1}{2}$, $N = 1$, and $r = 2$, and must therefore also satisfy its conclusions. In particular, $K(t)$ has simple zeros at $\mathbb{Z}+\frac{1}{2}$ and vanishes nowhere else, i.e.,

$$\operatorname{sign} K_k(x, t) = \varepsilon_k(x)\omega(t-\tfrac{1}{2})$$

with

$$\omega(t) := \operatorname{sign} \sin \pi t$$

and $\varepsilon_k(x)$ equal to 1 or -1 or 0. In order to determine $\varepsilon_k(x)$, we observe that $\varepsilon_k(n) = 0$ for all $n \in \mathbb{Z}$. Further, for $x \in (0, 1)$,

$$\operatorname{jump}_x K^{(k-1)} = (k-1)!\,(-)^{k-1} = (k-1)!$$

since k is odd, therefore

$$\operatorname{sign} K^{(k-1)}(x^+) = 1 .$$

If now $x < \frac{1}{2}$, then the Corollary to Theorem 1 applies to K with $t_1 = x$ and $\tau_1 = \frac{1}{2}$, i.e., $\operatorname{sign} K^{(l)}(\tau_1)K^{(k-1)}(\tau_1^-) = (-)^{p-1}$ with $p=m+1$, hence, as $K^{(k-1)}(x^+) = K^{(k-1)}(\tau_1^-)$, we have

$$\operatorname{sign} K_k(x,x) = -\operatorname{sign} K^{(l)}(\tau_1) = -(-)^m ,$$

showing that $\varepsilon_k(x) = (-)^m$ in this case. If, on the other hand, $x \geq \frac{1}{2}$, then K satisfies the assumptions of the Corollary to Theorem 1 with $t_1 = 1$ and $\tau_1 = 3/2$, i.e., $\operatorname{sign} K^{(l)}(3/2)K^{(k-1)}(3/2^-) = (-)^m$, while $K^{(k-1)}(x^+)$ and $K^{(k-1)}(3/2^-)$ have opposite sign, therefore

$$\operatorname{sign} K_k(x, x) = -\operatorname{sign} K^{(l)}(\tau_1) = (-)^m$$

showing that $\varepsilon_k(x) = (-)^m$ also in this case. Since

$$K_k(x, t) = K_k(x+1, t+1)$$

trivially, this proves that

(2) $$\operatorname{sign} K_k(x, t) = (-)^m \omega(x)\omega(t-1/2) \text{ for all } x \text{ and } t.$$

We also obtain from Theorem 1 that, for each x, there exists a constant $a = a(x)$ so that

(3) $$|K_k(x,t)| \leq a(x) |\lambda_m(\tfrac{1}{2})|^{|t|} \text{ for all } t.$$

Here, $\lambda_m(\frac{1}{2})$ is the largest eigenvalue less than 1 of $A_{k,\frac{1}{2}}$. It follows that, for $j=1,\ldots,k-1$, $(d/dt)^j K_k(x,t)$ is (piecewise) continuous n t and decays exponentially as $|t| \to \infty$. Hence, if f has a locally integrable k-th derivative of power growth, then

$$E(x) := \int_{-\infty}^{\infty} K_k(x,t) f^{(k)}(t)dt/(k-1)!$$

defines a function E on $\mathbb{R}$ which vanishes at $\mathbb{Z}$. Further, for $x \notin \mathbb{Z}$, we can evaluate E(x) by repeated integration by parts, obtaining

$$E(x) = (-)^{k-1} \int_{-\infty}^{\infty} (d/dt)^{k-1} K_k(x,t) f^{(1)}(t)dt/(k-1)! \; .$$

But, since

$$(-)^{k-1}(d/dt)^{k-1} K_k(x,t)/(k-1)! = (x-t)_+^0 - \sum_\nu (\nu - t)_+^0 L_k(x-\nu),$$

we obtain that

$$E(x) = \int_{-\infty}^{\infty} [(x-t)_+^0 - \sum_\nu (\nu-t)_+^0 L_k(x-\nu) f^{(1)}(t)]dt$$

$$= f(x) - \sum_\nu f(\nu) L_k(x-\nu) = f(x) - (S_k f)(x) \; .$$

Theorem 2. *Let $k = 2m+1$ be odd. If f has k-1 locally absolutely continuous derivatives on $\mathbb{R}$ and $f^{(k)}$ is of power growth at $\pm\infty$, then*

$$f(x) = (S_k f)(x) + \int_{-\infty}^{\infty} K_k(x,t) f^{(k)}(t)dt/(k-1)! \tag{4}$$

with K_k given by (2.6). *Further*

$$|K_k(x,t)| \le a(x)\exp(-b_k |t|)$$

for some function a and some positive constant b_k, and, for $k \geq 3$,

$$\text{sign } K_k(x,t) = (-)^m \omega(x)\omega(t-\tfrac{1}{2})$$

with $\omega(r) := \text{sign} \sin r\pi$.

Specific choices for f in (4) give much information about K_k, much as in the discussion of K_{2m} in [17]. E.g., $f(x) := \sin \nu\pi x$ vanishes at $\mathbb{Z}$ and is bounded, hence $S_k f = 0$ and (4) gives

(5) $$\sin \nu\pi x = (-)^m (\nu\pi)^{2m+1} \int_{-\infty}^{\infty} K_{2m+1}(x,t) \cos \nu\pi t \, dt/(2m)! .$$

If we choose $f(x) := x^k/k!$, then $f^{(k)} = 1$ and $f - S_k f$ is known to be equal to the k-degree Bernoulli monospline $\overline{B}_k/k!$ (see [15, Lecture 4, §6C]), therefore

(6) $$\int_{-\infty}^{\infty} K_{2m+1}(x,t)dt/(2m)! = \overline{B}_{2m+1}(x)/(2m+1)!, \quad \text{all } x \in \mathbb{R} .$$

Finally, Hölder's inequality gives at once the following corollary.

Corollary. If $f \in \mathbb{L}_\infty^{(k)}(\mathbb{R})$ with $k = 2m+1$, then, for any particular x,

$$f(x) - S_k f(x) \leq \|f^{(k)}\|_\infty \int_{-\infty}^{\infty} |K_k(x,t)| dt/(k-1)!$$

with equality iff either $x \in \mathbb{Z}$ (in which case both sides vanish) or

(7) $$f^{(k)}(t) = (-)^m \omega(x) \|f^{(k)}\|_\infty \omega(t-\tfrac{1}{2}) .$$

One function f satisfying (7) is a shifted version of the k-th degree <u>Euler spline</u> (see [15, Lecture 4, §6B]). To recall, the k-th degree Euler spline $\mathcal{E}_k$ is a particular cardinal null spline, an eigenspline belonging to the eigenvalue -1, and normalized to satisfy

$$\mathcal{E}_k(\nu) = (-)^{\nu}, \quad \text{all } \nu \in \mathbb{Z} .$$

It has its knots at $\mathbb{Z}+(k+1)/2$, i.e., at $\mathbb{Z}$ since we took $k = 2m+1$, and, being an eigenspline with eigenvalue -1, must satisfy

$$\mathcal{E}_k(x+1) = -\mathcal{E}_k(x), \quad \text{all } x \in \mathbb{R},$$

therefore

$$\mathcal{E}_k(\nu+\tfrac{1}{2}) = 0, \quad \text{all } \nu \in \mathbb{Z}$$

$$\mathcal{E}_k^{(k)}(x) = (-)^m \|\mathcal{E}_k^{(k)}\|_\infty \ \omega(x) .$$

It follows that $f(x) := \mathcal{E}_k(x - \frac{1}{2})$ has 0 for its cardinal spline interpolant and, except for the factor $\omega(x)$, satisfies (7), hence

$$(8) \qquad \omega(x)\mathcal{E}_{2m+1}(x - \tfrac{1}{2}) = \|\mathcal{E}_{2m+1}^{(2m+1)}\|_\infty \int_{-\infty}^{\infty} |K_{2m+1}(x,t)|dt/(2m)! .$$

<u>Theorem</u> 3. <u>Let</u> $k = 2m+1$. <u>If</u> $f \in \mathbb{L}_\infty^{(k)}(\mathbb{R})$, <u>then</u>

$$|f(x) - S_k f(x)| \le \frac{|\mathcal{E}_k(x-\frac{1}{2})|}{\|\mathcal{E}_k^{(k)}\|_\infty} \ \|f^{(k)}\|_\infty$$

and this inequality is sharp since it becomes equality for $f = \mathcal{E}_k(\cdot - \frac{1}{2})$.
Moreover, if, for some $x \notin \mathbb{Z}$ *and for some* $f \in \mathbb{L}_\infty^{(k)}(\mathbb{R})$ *with* $\|f^{(k)}\|_\infty \leq 1$,

$$|f(x) - S_k f(x)| = |\mathcal{E}_k(x - \tfrac{1}{2})| / \|\mathcal{E}_k^{(k)}\|_\infty ,$$

then f *must be of the form*

$$f = \pm \mathcal{E}_k / \|\mathcal{E}_k^{(k)}\|_\infty + p$$

for some polynomial p *of degree* $< k$.

Part D. A proof of Kolmogorov's theorem

1. The Euler splines and statement of Kolmogorov's theorem.

We already discussed the Euler splines in Section 3 of Part C, referring the reader to Schoenberg [15, Lecture 4] for background and proofs. For $k = 0,1,2,\ldots,$ the Euler spline $\mathcal{E}_k$ is the unique spline function of degree k with simple knots at $\mathbb{Z}+(k+1)/2$ satisfying

(1) $$\mathcal{E}_k(\nu) = (-)^{\nu} \quad \text{for all } \nu \in \mathbb{Z},$$

(2) $$\|\mathcal{E}_k\|_\infty \le 1 .$$

Except for the name "Euler spline", these functions are very well known, their polynomial components being essentially the classical Euler polynomials. Our conditions normalize these functions in a way that is convenient for our purpose. In Schoenberg [16], the reader will find a direct recursive derivation of the $\mathcal{E}_k$.

The function $\mathcal{E}_k(x)$ is a kind of stylized version of $\cos \pi x$ to which it converges as $k \to \infty$. Its sign structure is described by the inequalities

(3) $$(-)^{\nu+\lfloor (j+1)/2 \rfloor}\mathcal{E}_k^{(j)}(x) > 0 \quad \text{in} \begin{cases} \nu - \frac{1}{2} < x < \nu + \frac{1}{2} & \text{if } j \text{ is even} \\ \nu < x < \nu + 1 & \text{if } j \text{ is odd.} \end{cases}$$

In particular, we find that for the supremum norm we have

$$\|\mathcal{E}_k^{(j)}\|_\infty = (-)^{\lfloor (j+1)/2 \rfloor} \begin{cases} \mathcal{E}_k^{(j)}(0) & \text{if } j \text{ is even,} \\ \mathcal{E}_k^{(j)}(\tfrac{1}{2}) & \text{if } j \text{ is odd,} \end{cases} \qquad j=0,\ldots,k . \tag{4}$$

For convenience, we write

$$\|\mathcal{E}_k^{(j)}\|_\infty =: \gamma_{k,j} \quad (j=0,\ldots,k) . \tag{5}$$

These are rational numbers expressible in terms of the Euler numbers, and in particular $\gamma_{k,0} = 1$ by (1) and (2).

We also need the class

$$\mathbb{L}_\infty^{(k)}(\mathbb{R}) := \{f \in C^{(k-1)}(\mathbb{R}) \mid f^{(k-1)} \text{ satisfies a Lipschitz cond. on } \mathbb{R}\} . \tag{6}$$

Evidently, $\mathcal{E}_k \in \mathbb{L}_\infty^{(k)}(\mathbb{R})$. By $\|f\|_\infty$ we mean the essential supremum of f on $\mathbb{R}$.

<u>Theorem of Kolmogorov</u> [8]. <u>If</u> $f \in \mathbb{L}_\infty^{(k)}(\mathbb{R})$ <u>is such that</u>

$$\|f\|_\infty \le \|\mathcal{E}_k\|_\infty \quad \underline{\text{and}} \quad \|f^{(k)}\| \le \|\mathcal{E}_k^{(k)}\|_\infty , \tag{7}$$

<u>then</u>

$$\|f^{(j)}\|_\infty \le \|\mathcal{E}_k^{(j)}\|_\infty \quad \underline{\text{for}} \quad j = 1, \ldots, k-1. \tag{8}$$

The right sides of (8) are the best constants for each j because the Euler spline $\mathcal{E}_k$ satisfies the conditions (7) and also the conclusions (8) with the equality signs. Note the Corollary 2 in Section 3 where it is shown that in a certain sense the Euler splines are the only functions for which we can have equality in (8), even for a single value of j.

In Section 2, wo derive certain approximate differentiation formulae. These are applied in Section 3 to establish Kolmogorov's theorem. In Section 4, we establish the needed properties of the formulae of Section 2. Finally, in Section 5, we give a characterization of these differentiation formulae.

2. Some approximate differentiation formulae. In this section, we consider a cardinal Green's function K of order k which fails to be a spline function only because we require

$$\text{jump}_{\alpha} K^{(k-j-1)} = (-)^{k-j} \tag{1}$$

for

$$\alpha := \begin{cases} 0 , & j \text{ even,} \\ \frac{1}{2}, & j \text{ odd .} \end{cases} \tag{2}$$

More explicitly, except for the jump condition (1), K is a spline function of order k with simple knots at $\mathbb{Z}$ and vanishes at $\mathbb{Z}+\tau$ where

$$\tau := \begin{cases} \frac{1}{2}, & k \text{ even,} \\ 0 , & k \text{ odd.} \end{cases} \tag{3}$$

This choice of τ was quite explicitly excluded in Theorem C1 since it would allow bounded eigensplines, viz the Euler splines $\mathcal{E}_k$ already used in Part C and again in the previous section. This exclusion was the subject of the remark following Theorem C1. The required "additional freedom" for K mentioned there is provided here by the condition (1).

Theorem 1. Let $k \geq 2$, and let $1 \leq j \leq k-1$. There exists a unique function K on $\mathbb{R}$ with the following three properties:

(i) K is a spline function of order k with simple knots at $\mathbb{Z}$ except that

$$\text{jump}_{\alpha} K^{(k-j-1)} = (-)^{k-j} ,$$

with $\alpha = 0$ or $\frac{1}{2}$ as j is even or odd.

(ii) K vanishes at $\mathbb{Z}+\tau$, with $\tau = \frac{1}{2}$ or 0 as k is even or odd, except that K is not required to vanish at τ in case $j+1 = k$ as then K is not continuous at τ, by (i).

(iii) $K \in \mathbb{L}_1(\mathbb{R})$.

Theorem 1 will be established in Section 4. Observe that the inequalities $1 \leq j \leq k-1$ imply that

$$0 \leq k-j-1 \leq k-2 .$$

Hence K fails to be in $C^{(k-2)}(\mathbb{R})$ and this is the reason why K is not a cardinal spline.

Let us now assume that Theorem 1 were established and use the function K as a kernel as follows: If $f \in \mathbb{L}_\infty^{(k)}(\mathbb{R})$ and if we integrate by parts repeatedly the integral

$$\int_{-\infty}^{\infty} K(x) f^{(k)}(x)\,dx$$

and use the jump condition (1), we obtain the following corollary.

Corollary 1. For $f \in \mathbb{L}_\infty^{(k)}(\mathbb{R})$,

(4) $$f^{(j)}(\alpha) = \sum_{\nu\in\mathbb{Z}} A_\nu f(\nu) + \int_{-\infty}^{\infty} K(x) f^{(k)}(x)\,dx$$

with

(5) $$A_\nu := (-)^{k-1} \operatorname{jump}_\nu K^{(k-1)}, \quad \text{all } \nu\in\mathbb{Z}.$$

The derivation of this differentiation formula by integration by parts requires the following remark: Our construction of K in Sec. 4 will prove that all derivatives $K^{(\nu)}(x)$ $(\nu=0,\ldots,k-1)$ decay exponentially as $x \to \pm\infty$. On the other hand, $f \in \mathbb{L}_\infty^{(k)}(\mathbb{R})$ implies that the derivatives $f^{(\nu)}$ $(\nu=0,\ldots,k)$ can be of power growth at most at $\pm\infty$. This explains the vanishing of all "finite parts" at $\pm\infty$ and also the convergence of the series in (4). Clearly, K and the A_ν depend also on k and j, but the values of k and j will be obvious from the context.

In our application of Corollary 1 to a proof of Kolmogorov's theorem, the following additional information on the A_ν and on K is vital.

Theorem 2. We assume that $k \geq 4$.

(i) For certain positive constants A and B,

(6) $$|K^{(r)}(x)| < Ae^{-B|x|} \quad \text{for all } x \in \mathbb{R} \text{ and for } r=0,\ldots,k-1.$$

(ii) The coefficients A_ν of the formula (4) satisfy

(7) $$(-)^{\nu+\lfloor (j+1)/2 \rfloor} A_\nu > 0$$

for all $\nu \in \mathbb{Z}$.

(iii) The kernel K of Theorem 1 satisfies the inequality

(8) $$(-)^{\nu+\lfloor (k+1)/2 \rfloor + \lfloor (j+1)/2 \rfloor} K(x) > 0 \quad \text{for } x \in (\nu-\tau,\ \nu+1-\tau)$$

for all $\nu \in \mathbb{Z}$.

(iv) The kernel K is symmetric around α. Specifically,

(9) $$K(\alpha + x) = (-)^{k-j} K(\alpha - x) \quad \text{for all } x \in \mathbb{R}\setminus\{0\}.$$

By (7), the coefficients A_ν alternate strictly in sign if $k \geq 4$. If $k = 2$ or 3, this is no longer the case, since then only a finite number of the A_ν are $\neq 0$ (see [16]). By (8), the kernel K vanishes only at the points $\mathbb{Z}+\tau$ (with the exception mentioned in (ii) of Theorem 1) and alternates strictly in sign as we cross from one unit interval to the next.

In the next section, we use Corollary 1 and Theorem 2 to establish Kolmogorov's theorem and to describe its extremizing functions. In Section 4, we establish Theorem 1 and Theorem 2 jointly by constructing K and then applying the Budan-Fourier theorem to its two pieces.

3. A proof of Kolmogorov's theorem. We retain the definitions

$$\alpha := \left\{\begin{matrix} 0 \\ \frac{1}{2} \end{matrix}\right\} \text{ if } j \text{ is } \left\{\begin{matrix} \text{even} \\ \text{odd} \end{matrix}\right\}, \quad \tau := \left\{\begin{matrix} \frac{1}{2} \\ 0 \end{matrix}\right\} \text{ if } k \text{ is } \left\{\begin{matrix} \text{even} \\ \text{odd} \end{matrix}\right\}$$

introduced in the preceding section. Our earlier description (1.3) - (1.4) of certain properties of the Euler spline then give that

$$(1) \qquad (-)^{\nu+\lfloor (k+1)/2 \rfloor} \mathcal{E}_k^{(k)}(x) > 0 \text{ on } (\nu-\tau, \nu+1-\tau), \text{ for all } \nu \in \mathbb{Z}$$

and

$$(2) \qquad (-)^{\lfloor (j+1)/2 \rfloor} \mathcal{E}_k^{(j)}(\alpha) = \|\mathcal{E}_k^{(j)}\|_\infty =: \gamma_{k,j} \quad \text{for } j=0, \ldots, k.$$

We apply the differentiation formula (2.4) to the special function

$$f_0 := (-)^{\lfloor (j+1)/2 \rfloor} \mathcal{E}_k .$$

By (2),

$$(3) \qquad f_0^{(j)}(\alpha) = (-)^{\lfloor (j+1)/2 \rfloor} \mathcal{E}_k^{(j)}(\alpha) = \gamma_{k,j} > 0 .$$

The interpolation property (1.1) shows that $f_0(\nu) = (-)^{\nu+\lfloor (j+1)/2 \rfloor}$ and Theorem 2.(ii) shows then that

(4) $$A_\nu f_0(\nu) = |A_\nu| > 0 \quad \text{for all} \quad \nu \in \mathbb{Z} .$$

From (1) and (2), we find that

$$(-)^{\nu + \lfloor (k+1)/2 \rfloor + \lfloor (j+1)/2 \rfloor} f_0^{(k)}(x) = \gamma_{k,k} > 0 \quad \text{on } (\nu - \tau, \nu + 1 - \tau)$$

for all $\nu \in \mathbb{Z}$, and Theorem 2.(iii) then shows that

(5) $$K(x) f_0^{(k)}(x) = |K(x)| \gamma_{k,k} > 0 \quad \text{for} \quad x \in \mathbb{R} \setminus (\mathbb{Z} + \tau) .$$

By (3), (4) and (5), the relation (2.4) becomes

(6) $$f_0^{(j)}(\alpha) = \sum_{\nu=-\infty}^{\infty} |A_\nu| + \gamma_{k,k} \int_{-\infty}^{\infty} |K(x)| dx .$$

If f is an arbitrary function in $\mathbb{L}_\infty^{(k)}(\mathbb{R})$ satisfying the assumptions (1.7), we may also assume that $f^{(j)}(\alpha) \geq 0$, for if not we consider $-f$ instead, which also satisfies all assumptions. Applying the differentiation formula (2.4) to our f, we obtain from (1.7) that

(7) $$\begin{aligned} 0 \leq f^{(j)}(\alpha) &= \sum_{\nu=-\infty}^{\infty} A_\nu f(\nu) + \int_{-\infty}^{\infty} K(x) f^{(k)}(x) dx \\ &\leq \sum_\nu |A_\nu| + \gamma_{k,k} \int_{-\infty}^{\infty} |K(x)| dx = f_0^{(j)}(\alpha) , \end{aligned}$$

the last equality following from (6). We have just shown that

(8) $$|f^{(j)}(\alpha)| \leq |f_0^{(j)}(\alpha)| = \gamma_{k,j} .$$

If x_0 is real and if we apply our result to $f(x + x_0 - \alpha)$, then we obtain that $|f^{(j)}(x_0)| \leq \gamma_{k,j}$ and the conclusion (1.8) is established.

Let us now assume that in (7) we have $f^{(j)}(\alpha) = f_0^{(j)}(\alpha)$. This implies the equality of the two middle terms of (7), and this we may write as

(9) $$\sum_{\nu} (|A_\nu| - A_\nu f(\nu)) + \int_{-\infty}^{\infty} \{\gamma_{k,k} - \frac{K(x)}{|K(x)|} f^{(k)}(x)\} |K(x)| dx = 0.$$

By (1.7), we see that all terms of the series are nonnegative, and so is the integrand almost everywhere. From (4), we conclude that

$$A_\nu f(\nu) = A_\nu f_0(\nu) \quad \text{for all} \quad \nu \in \mathbb{Z}$$

and (5) shows that

$$K(x) f^{(k)}(x) = K(x) f_0^{(k)}(x) \quad \text{for almost all} \quad x.$$

Therefore

(10) $$f(\nu) = f_0(\nu) \quad \text{for all} \quad \nu, \quad f^{(k)} = f_0^{(k)} \quad \text{a.e.}.$$

Integrating the last relation k times, we conclude that f and f_0 may differ only by a polynomial of degree $< k$, and the first relations (10) show that this polynomial is identically zero. Hence

(11) $$f = (-)^{\lfloor (j+1)/2 \rfloor} \mathcal{E}_k .$$

This completes our proof of Kolmogorov's theorem and also a proof of

Corollary 2. Let f satisfy the assumptions (1.7) and therefore also the conclusions (1.8). If, for some value of $j \in [1, k-1]$, the equality sign holds in (1.8), and if the extremum $\|f^{(j)}\|_\infty = \gamma_{k,j}$ is attained in the sense that

$$(12) \qquad |f^{(j)}(x_0)| = \|f^{(j)}\|_\infty \text{ for some } x_0,$$

then f is necessarily of the form

$$(13) \qquad f(x) = \pm \mathcal{E}_k(x - c) .$$

Remarks. (i) Since we have used Theorem 2 in our proof, we have also implicitly assumed that $k \geq 4$. For the elementary cases when $k = 2$ and 3 see [16]. Corollary 2 is valid also for $k = 3$, but requires a special proof given in [16, §8]. For a discussion of the extremizing functions in the weak sense (when there is no x_0 satisfying (12)) see [16, §8].

(ii) For the case j=1, Kolmogorov proved the stronger result: If $f \in \mathbb{L}_\infty^{(k)}(\mathbb{R})$, and a, b, and c are such that the function $g(x) := a\mathcal{E}_k(bx+c)$ satisfies $\|g\|_\infty = \|f\|_\infty$, $\|g^{(k)}\|_\infty = \|f^{(k)}\|_\infty$, and $g(x_0) = f(x_0)$, then $|g^{(1)}(x_0)| \geq |f^{(1)}(x_0)|$.

(iii) For the general formulation of Kolmogorov's theorem see, e.g., [16, §6].

(iv) For the special case k=4, j=1, we find that $\alpha = \frac{1}{2}$ and $\text{jump}_\alpha K^{(2)} = -1$. This shows that the right hand side of (1.13) in Lecture 9 of [15] has the wrong sign as does the first line of the right hand side of (1.21) in the same section.

4. A proof of Theorems 1 and 2. The proof is divided into six parts.

(α) A proof of the unicity of K. If we have two functions, K_1 and K_2 say, satisfying the conditions of Theorem 1 for the same k and j, then, in forming their difference

$$S := K_1 - K_2 ,$$

the discontinuity in the $(k-j-1)$st derivative at α cancels out, and we conclude that S is a spline of order k with simple knots at $\mathbb{Z}$, and vanishes at $\mathbb{Z}+\tau$; in short, S is a cardinal nullspline. Hence, with

$$\underline{\underline{S}}(x) := \begin{cases} (S(x), \ldots, S^{(k-2)}(x)/(k-2)!), & \text{if } \tau = 0, \\ (S(x), \ldots, S^{(k-1)}(x)/(k-1)!), & \text{if } \tau = \frac{1}{2}, \end{cases}$$

we infer from Part B that

$$\underline{\underline{S}}(\nu+\tau) = \sum_{i=p}^{q} c_i(-\lambda_i(\tau))^{\nu}\,\underline{\underline{u}}^{(i,\tau)}, \quad \text{all } \nu\in\mathbb{Z},$$

for a certain basis $(\underline{\underline{u}}^{(i,\tau)})_i$ and a certain strictly decreasing positive sequence $(\lambda_i(\tau))_i$. On the other hand, S must be in $\mathbb{L}_1(\mathbb{R})$ since both

K_1 and K_2 are, by assumption (iii). Therefore, letting $\nu \to \infty$, we see that $c_i = 0$ for all i with $\lambda_i(\tau) \geq 1$, while, letting $\nu \to -\infty$, we see that $c_i = 0$ for all i with $\lambda_i(\tau) \leq 1$. In short, $S = 0$.

(β) A proof of the symmetry relation (iv) of Theorem 2. We are to show that the unique K of Theorem 1, if it exists, must satisfy

(1) $$K(\alpha + x) = (-)^{k-j} K(\alpha - x) \quad \text{for all positive } x.$$

For this, define K_1 by

$$K_1(\alpha + x) := (-)^{k-j} K(\alpha - x) \quad \text{for all } x \neq 0 .$$

Then K_1 is in $\mathbb{L}_1(\mathbb{R})$ since K is, K_1 vanishes at $\mathbb{Z}+\tau$ since K does, and K_1 is a spline of order k with knots at $\mathbb{Z}$, K having this property, except for the discontinuity in the $(k-j-1)$st derivative at α. But,

$$\begin{aligned} \text{jump}_\alpha K_1^{(k-j-1)} &= K_1^{(k-j-1)}(\alpha^+) - K_1^{(k-j-1)}(\alpha^-) \\ &= (-)^{k-j}(-)^{k-j-1}\{K^{(k-j-1)}(\alpha^-) - K^{(k-j-1)}(\alpha^+)\} \\ &= \text{jump}_\alpha K^{(k-j-1)} = (-)^{k-j}, \end{aligned}$$

so that K_1 also satisfies the jump condition (2.1). But now (1) follows from the unicity of K just proven.

(γ) A proof of existence of K when k is odd. Motivated by the symmetry (1), we actually construct K by determining a spline function

S in $\mathbb{I}_1[\alpha, \infty)$ of order k with simple knots at the positive integers which vanishes at $\nu + \tau$ for all nonnegative integers ν, and which satisfies

(2) $$S^{(r)}(\alpha^+) = (-)^{k-j}\delta_{r,k-j-1}/2 \text{ for } r = \begin{cases} 0, 2, \dots, k-3 & \text{if } j \text{ even} \\ 1, 3, \dots, k-2 & \text{if } j \text{ odd} \end{cases}$$

Any such S will give rise to a K of the required sort by

(3) $$K(x) := \begin{cases} S(x) & x > \alpha \\ (-)^{k-j} S(2\alpha - x), & x < \alpha \end{cases}$$

with the conditions (2) guaranteeing that

$$\text{jump}_\alpha K^{(r)} = (-)^{k-j}\delta_{r,k-j-1} \quad \text{for } r=0, \dots, k-\begin{Bmatrix} 2 \\ 1 \end{Bmatrix} \text{ if } \alpha = \begin{Bmatrix} 0 \\ \frac{1}{2} \end{Bmatrix} .$$

Since k is odd, $\tau = 0$, i.e., S is to vanish at its knots 1, 2, ... With

$$\underline{\underline{S}}(x) := (S(x), \dots, S^{(k-2)}(x)/(k-2)!)$$

as in Part B, the condition that S have simple knots at the positive integers and vanish at these implies that

(4) $$\underline{\underline{S}}(\nu+1) = (-A_k)^{\nu}\underline{\underline{S}}(1) \quad \text{for } \nu=1, 2, 3, \dots,$$

where A_k is the matrix described in Section B2. Hence, S is determined on $[1, \infty)$ once we choose $\underline{\underline{S}}(1)$. In particular, with $(\underline{\underline{u}}^{(i)})_1^{k-2}$ a complete eigenvector sequence for A_k corresponding to the decreasing eigenvalue sequence $(\lambda_i)_1^{k-2}$, any $\underline{\underline{S}}(1)$ of the form

(5) $$\underline{\underline{S}}(1) = \sum_{i=m+1}^{k-2} c_i \underline{\underline{u}}^{(i)}$$

will give rise to an S in $\mathbb{L}_1[1, \infty)$ since $\lambda_i < 1$ for $i > m := \lfloor k/2 \rfloor$ (see Theorem B1). On the other hand,

(6) $$S^{(r)}(\alpha^+) = \sum_{i=r}^{k-2} S^{(i)}(1)(\alpha-1)^{i-r}/(i-r)! + S^{(k-1)}(1^-)(\alpha-1)^{k-1-r}/(k-1-r)!$$

for $r=0,\ldots,k-1$, so that, with the choice (5) for $\underline{\underline{S}}(1)$, (2) constitutes an inhomogeneous linear system of $(k-1)/2$ equations in the

$$k-2-m+1 = (k-1)/2$$

unknowns $c_{m+1}, \ldots, c_{k-2}$ and $S^{(k-1)}(1^-)$.

We are therefore assured of the existence of exactly one solution (necessarily nontrivial) in case the corresponding homogeneous equations have only the trivial solution. But that is certainly so here, since a nontrivial solution would give rise via (3) to a cardinal nullspline in $\mathbb{L}_1(\mathbb{R})$, a possibility already rejected when proving unicity.

We conclude that (2), considered via (5) and (6) as a linear system for $c_{m+1}, \ldots, c_{k-2}$ and $S^{(k-1)}(1^-)$, has exactly one solution, proving the existence of K for this case.

(δ) A proof of Theorem 2 when k is odd. We obtain the exponential decay as described in (i) of Theorem 2 at once from (4) and (5) above with $e^{-B} = \lambda_{m+1}$. As to (ii) and (iii), we begin with the observation that roughly half the numbers $S(\alpha), \ldots, S^{(k-1)}(\alpha^+)$ vanish. Precisely, as we saw in the existence proof, (2) is comprised of $(k-1)/2$ equations all but one being homogeneous, hence $(k-3)/2$ of the k numbers $S(\alpha), \ldots, S^{(k-1)}(\alpha^+)$ must be zero. Therefore,

$$S^-(S(\alpha), \ldots, S^{(k-1)}(\alpha^+)) \leq k-1 - (k-3)/2 = m+1, \tag{7}$$

with $m := \lfloor k/2 \rfloor = (k-1)/2$ as before. Let now $p+1$ be the smallest integer $\geq m+1$ for which $c_{p+1} \neq 0$ in (5). Then

$$\underline{\underline{S}}(\nu) = (-\lambda_{p+1})^{\nu-1} c_{p+1} \underline{\underline{u}}^{(p+1)} + o((\lambda_{p+1})^{\nu-1}) \text{ as } \nu \to \infty,$$

therefore, by the Corollary to Theorem B1,

$$S^+(S(\nu), \ldots, S^{(k-1)}(\nu^-)) = p+1 \geq m+1 \text{ for } \nu \text{ near } \infty . \tag{8}$$

But, on (α, ν), S is a spline of order k with simple knots only, and with at least as many zeros as knots, and all these zeros must be isolated since, by (ii) of Theorem B1 and by (5),

$$S^-(S^{(1)}(\nu), \ldots, S^{(k-2)}(\nu)) \geq p \geq m > 0 \text{ for } \nu = 1, 2, \ldots,$$

hence S cannot vanish identically on a positive interval. Also, S is not just a polynomial of degree $< k-1$ since $S \neq 0$. Therefore, from the Budan-Fourier theorem for splines, and from (7) and (8) we have for ν near ∞ that

$$Z_{S^{(k-1)}}(\alpha,\nu) \le \text{number of active knots of } S \text{ in } (\alpha,\nu)$$

$$\le \text{number of knots of } S \text{ in } (\alpha,\nu)$$

$$\le Z_S(\alpha,\nu)$$

(9) $$\le Z_{S^{(k-1)}}(\alpha,\nu) + S^-(S(\alpha), \ldots, S^{(k-1)}(\alpha^+))$$

$$- S^+(S(\nu), \ldots, S^{(k-1)}(\nu^-))$$

$$\le Z_{S^{(k-1)}}(\alpha,\nu) + (m+1) - (m+1),$$

$$= Z_{S^{(k-1)}}(\alpha,\nu),$$

showing that equality must hold in all inequalities used to establish this string of inequalities.

We harvest the fruits of this statement one at a time. Equality in (7) implies that all entries of the sequence $S(\alpha), \ldots, S^{(k-1)}(\alpha^+)$ not explicitly set to zero by (2) must be nonzero and alternate in sign. Since we know that

$$S^{(k-j-1)}(\alpha^+) = (-)^{k-j}/2,$$

we therefore know that

(10a) $$(-)^{k-j-r} S^{(k-j-2r)}(\alpha) > 0 \quad \text{for } r = 1, 2, \ldots$$

and

(10b) $$(-)^{k-j+r}S^{(k-j+2r)}(\alpha^+) < 0 \quad \text{for } r = 0,1,2,\ldots .$$

If now j is odd, then $\alpha = \frac{1}{2}$ and $k-j$ is even and (10a) implies that

(11) $$(-)^{(k-j)/2}S(\tfrac{1}{2}) > 0 \quad \text{for } j \text{ odd.}$$

Further, $k-1 = k-j + 2r$ with $r = (j-1)/2$, hence (10b) gives that

(12) $$(-)^{(j-1)/2}S^{(k-1)}(\tfrac{1}{2}) < 0 \quad \text{if } j \text{ is odd.}$$

If, on the other hand, j is even, then $\alpha = 0$ and $k-j$ is odd, and $1 = k-j-2r$ with $r = (k-j-1)/2$, so (10a) implies (for $r \geq 1$) that $-(-)^{(k-j-1)/2}S^{(1)}(0) > 0$, therefore

(13) $$(-)^{(k-j-1)/2}S(\tfrac{1}{2}) > 0 \quad \text{for } j \text{ even,}$$

since this follows directly in case $j = k-1$. Also, $k-2 = k-j+2r$ with $r = (j-2)/2$, so $-(-)^{(j-2)/2}S^{(k-2)}(0) < 0$ by (10b), hence

(14) $$(-)^{j/2}S^{(k-1)}(0^+) > 0 \quad \text{for } j \text{ even.}$$

Further, since the number of active knots of S in (α, ν) must equal the number of zeros of $S^{(k-1)}$ there, it follows that $S^{(k-1)}$ <u>changes sign strongly across each integer</u> $1, 2, 3, \ldots$. But then $K^{(k-1)}$ must change sign strongly across each $\nu \in \mathbb{Z}$: This is obvious in case $\alpha = \frac{1}{2}$; but it is also true in case $\alpha = 0$ for then j is even, hence $k-j$ is odd, and therefore all even derivatives of K are odd around $\alpha = 0$, hence $K^{(k-1)}$ is odd around $\alpha = 0$, showing that $K^{(k-1)}$ changes sign strongly also across $\alpha = 0$. It follows that

$$A_\nu A_{\nu+1} < 0 \quad \text{for all} \quad \nu \in \mathbb{Z}$$

and it remains only to show (2.7) for a particular value of ν, say for $\nu = 1$, in which case (2.7) asserts that

$$(-)^{\lfloor (j+1)/2 \rfloor} \operatorname{jump}_1 K^{(k-1)} < 0 .$$

But that is now a consequence of the fact that, by (12) and (14),

$$(-)^{\lfloor (j+1)/2 \rfloor} S^{(k-1)} > 0 \quad \text{on} \quad (0, 1).$$

Finally, even counting multiplicities, S must have exactly as many zeros in (α, ν) as it has knots, hence S changes sign strongly at all positive integers and nowhere else in (α, ∞). K therefore changes sign at the integers and nowhere else: This is clear for $\alpha = \frac{1}{2}$. But it is also true for $\alpha = 0$, since then, as we just said, K must be odd around 0, hence must change sign strongly across 0. It remains to verify (2.8) for some ν, say for $\nu = 0$, in which case (2.8) asserts that

(15) $$(-)^{\lfloor k+1)/2 \rfloor - \lfloor (j+1)/2 \rfloor} K > 0 \quad \text{on} \quad (0,1) .$$

But $\lfloor (k+1)/2 \rfloor = (k+1)/2$. Further, for odd j, $\lfloor (j+1)/2 \rfloor = (j+1)/2$ while, by (11), $(-)^{(k-j)/2} K > 0$ on $(0, 1)$, proving (15) for this case. If, on the other hand, j is even, then $\lfloor (j+1)/2 \rfloor = j/2$ while, by (13), $(-)^{(k+1-j)/2} K > 0$ on $(0,1)$, thus proving (15) for this case, too.

This proves all assertions about K made in Theorem 2, for odd k.

(ε) <u>A proof of existence of K when k is even.</u> In this case, it becomes convenient (and perhaps more diverting) to construct K in the form

(16) $$K(x) := \begin{cases} (-)^{k-j} S(2\alpha - x), & x > \alpha \\ S(x) & , \quad x < \alpha \end{cases}$$

with S a spline of order k in $\mathbb{L}_1(-\infty, \alpha]$ with simple knots at the non-positive integers and which vanishes at $\nu + \frac{1}{2}$ for all negative integers ν and satisfies

(17) $$S^{(r)}(\alpha^-) = -(-)^{k-j} \delta_{r,k-j-1}/2 \quad \text{for } r = \begin{cases} 1,3,\ldots,k-3 & \text{if } j \text{ even} \\ 0,2,\ldots,k-2 & \text{if } j \text{ odd} \end{cases} .$$

As S is to vanish at $\nu + \tau$ for $-\nu \in \mathbb{N}$ and $\tau = \frac{1}{2}$, we recall from Sec. B3 the abbreviation

$$\underline{\underline{S}}(x) := (S(x), \ldots, S^{(k-1)}(x)/(k-1)!) ,$$

in terms of which then

(18) $$\underset{=}{S}(\nu - \tfrac{1}{2}) = (-A_{k,\tau})^{\nu}\ \underset{=}{S}(-\tfrac{1}{2}), \qquad \nu = -1, -2, \ldots,$$

where $A_{k,\tau}$ is the matrix described in Section B3. Hence, S is determined on $(-\infty, -\frac{1}{2}]$ once we have chosen $\underset{=}{S}(-\frac{1}{2})$. In particular, with $(\underset{=}{u}^{(i,\tau)})_1^{k-1}$ a complete eigenvector sequence for $A_{k,\tau}$ corresponding to the decreasing eigenvalue sequence $(\lambda_i(\tau))_1^{k-1}$, any $\underset{=}{S}(-\frac{1}{2})$ of the form

(19) $$\underset{=}{S}(-\tfrac{1}{2}) = \sum_{i=1}^{m-1} c_i\ \underset{=}{u}^{(i,\tau)}$$

gives rise to an S in $\mathbb{L}_1(-\infty, -\frac{1}{2}]$, since $\lambda_i(\tau) > 1$ for $i < m := \lfloor k/2 \rfloor$ and $\tau = \frac{1}{2}$, by Theorem B2 or by the Corollary to Theorem B3. On the other hand,

(20) $$S^{(r)}(\alpha^-) = \sum_{i=r}^{k-1} S^{(i)}(-\tfrac{1}{2})(\alpha+\tfrac{1}{2})^{i-r}/(i-r)! + S^{(k-1)}(0^+)\alpha^{k-1-r}/(k-1-r)!$$

for $r = 0, \ldots, k$, so that, with the choice (19) for $\underset{=}{S}(-\frac{1}{2})$, (17) constitutes an inhomogeneous linear system of $\begin{Bmatrix} m-1 \\ m \end{Bmatrix}$ equations for $\begin{Bmatrix} \text{even} \\ \text{odd} \end{Bmatrix}$ j in the unknowns $c_1, \ldots, c_{m-1}$, and also in $S^{(k-1)}(0^+)$ in case $\alpha \neq 0$. Hence, in terms of (19) and (20), (17) constitutes an inhomogeneous linear system in as many unknowns as equations and is therefore uniquely solvable (since a nontrivial solution to the homogeneous system would give rise to a nontrivial null spline in $\mathbb{L}_1(\mathbb{R})$, an impossibility). This proves the existence of K when k is even.

(ζ) A proof of Theorem 2 when k is even. The argument parallels closely that given when k is odd. The exponential decay is again obvious from the construction. Further, Equations (17) set to zero $\begin{Bmatrix} m-2 \\ m-1 \end{Bmatrix}$ terms in the sequence $S(\alpha), \ldots, S^{(k-1)}(\alpha^-)$ for $\begin{Bmatrix} \text{even} \\ \text{odd} \end{Bmatrix}$ j. Hence, choosing the sign of these zeros to alternate in conjunction with the nonzero term $S^{(k-j-1)}(\alpha^-)$, we see that

(21) $$S^+(S(\alpha), \ldots, S^{(k-1)}(\alpha^-)) \geq \begin{Bmatrix} m-2 \\ m-1 \end{Bmatrix} \text{ for } j \begin{Bmatrix} \text{even} \\ \text{odd} \end{Bmatrix} .$$

Also, with $p-1$ the largest integer $\leq m-1$ for which $c_{p-1} \neq 0$ in (19), we have

$$\underline{\underline{S}}(\nu-\tfrac{1}{2}) = (-\lambda_{p-1}(\tau))^{\nu} \underline{\underline{u}}^{(p-1,\tau)} + o((\lambda_{p-1}(\tau))^{\nu}) \text{ as } \nu \to -\infty .$$

Therefore, by Theorem B2.(iii),

(22) $$S^-(S(\nu-\tau), \ldots, S^{(k-1)}(\nu-\tau^+)) = p-2 \leq m-2$$

for all integers ν near $-\infty$. Further, on $(\nu-\tau,\alpha)$, S is a spline of order k (and certainly not just a polynomial of degree $< k-1$) with simple knots at $\nu-1, \nu-2, \ldots, -1$, and also at 0 in case $\alpha = \frac{1}{2}$, i.e., when j is odd, and nowhere else, while S vanishes in $(\nu-\tau,\alpha)$ at $\nu-1-\frac{1}{2}, \ldots, -\frac{1}{2}$. Since these zeros are necessarily isolated, we have

$$\text{number of knots of } S \text{ in } (\nu-\tau,\alpha) \leq Z_S(\nu-\tau,\alpha) + \begin{Bmatrix} 0 \\ 1 \end{Bmatrix} \text{ for } j \begin{Bmatrix} \text{even} \\ \text{odd} \end{Bmatrix} .$$

The Budan-Fourier theorem for splines, and the inequalities (21) and (22) now give, with $\beta := \nu-\tau$ and ν near $-\infty$,

$$Z_{S^{(k-1)}}(\beta,\alpha) \le \text{number of active knots of } S \text{ in } (\beta,\alpha)$$

$$\le \text{number of knots of } S \text{ in } (\beta,\alpha)$$

$$\le Z_S(\beta,\alpha) + \begin{Bmatrix} 0 \\ 1 \end{Bmatrix}$$

(23) $$\le Z_{S^{(k-1)}}(\beta,\alpha) + S^-(S(\beta),\ \ldots,\ S^{(k-1)}(\beta^+)) - S^+(S(\alpha),\ \ldots,\ S^{(k-1)}(\alpha^-))$$

$$\le Z_{S^{(k-1)}}(\beta,\alpha) + m-2 - \begin{Bmatrix} m-2 \\ m-1 \end{Bmatrix} + \begin{Bmatrix} 0 \\ 1 \end{Bmatrix}$$

$$= Z_{S^{(k-1)}}(\beta,\alpha)$$

showing that <u>equality must hold in all inequalities used to establish this string of inequalities</u>.

In particular, S changes sign strongly across $\nu+\frac{1}{2}$ for each negative integer ν, and changes sign nowhere else in $(-\infty,\alpha)$. K must therefore change sign strongly across $\nu+\frac{1}{2}$ for each $\nu \in \mathbb{Z}$ and nowhere else, which reduces the proof of (iii) to checking (2.8) for $\nu = 0$. Also, $S^{(k-1)}$, and therefore $K^{(k-1)}$, changes sign strongly across each knot, which reduces the proof of (ii) to checking (2.7) for $\nu = 0$. But since there must be equality in (21), and since $K(\alpha+x) = (-)^{k-j}\ K(\alpha-x)$, we have with A.1.(1) that

$$S^-(K(\alpha), \ldots, K^{(k-1)}(\alpha^+)) = k/2 + \begin{Bmatrix} 1 \\ 0 \end{Bmatrix} \text{ for } j \begin{Bmatrix} \text{even} \\ \text{odd} \end{Bmatrix} .$$

This forces the $k-m+\begin{Bmatrix} 2 \\ 1 \end{Bmatrix}$ of the k terms $K(\alpha), \ldots, K^{(k-1)}(\alpha^+)$ not explicitly set to zero by (17) to be in fact nonzero and to alternate in sign, and the verification of (ii) and (iii) proceeds from this and from the fact that

$$(-)^{k-j}K^{(k-j-1)}(\alpha^+) > 0$$

much as in the case k odd.

5. A characterization of the differentiation formulae of §2.

If α and τ are defined by (2.2) and (2.3), $1 \leq j \leq k-1$, and $f \in \mathbb{L}_\infty^{(k)}(\mathbb{R})$, then we know that

(1) $$f^{(j)}(\alpha) = \sum_{\nu \in \mathbb{Z}} A_\nu f(\nu) + \int_{-\infty}^{\infty} K(x) f^{(k)}(x) dx .$$

Here $K(x)$ is the kernel of Theorem 1, having the properties

(2) $$K(\tau+\nu) = 0 \text{ if } \nu \in \mathbb{Z} ,$$

and

(3) $$\text{jump}_\alpha K^{(k-j-1)} = (+)^{k-j} .$$

Let us first assume that

(4) $$j \leq k-2 .$$

Since $K \in C(\mathbb{R})$, it follows that we can write (1) in the form

(5) $$f^{(j)}(\alpha) = \sum_{\nu \in \mathbb{Z}} A_\nu f(\nu) + \int_{-\infty}^{\infty} K(x) df^{(k-1)}(x)$$

where we interpret the remainder as a Stieltjes integral. Then (5) is surely valid if we only assume that $f \in L_\infty^{(k-1)}(\mathbb{R})$, where $f^{(k-1)}$ is uniformly locally of bounded variation, meaning that the total variation of $f^{(k-1)}$ on the interval $[a, a+\ell]$ is bounded for every fixed ℓ and all a. This is surely the case if

(6) $$f \in \mathbb{S}_{k,\tau+\mathbb{Z}} \cap \mathbb{L}_\infty^{(k-1)}(\mathbb{R}) .$$

In this case, $f^{(k-1)}$ is a step-function with jumps at $\tau+\mathbb{Z}$, and (2) shows that the remainder term of (5) vanishes <u>because of</u> (4).

<u>Lemma 1.</u> <u>If</u> (6) <u>holds then</u>

(7) $$f^{(j)}(\alpha) = \sum_{\nu \in \mathbb{Z}} A_\nu f(\nu),$$

<u>where in case that</u>

(8) $$j = k-1, \ \underline{\text{hence}} \ \alpha = \tau ,$$

<u>we interpret</u> $f^{(k-1)}(\alpha) = f^{(k-1)}(\tau)$ <u>to mean</u>

(9) $$f^{(k-1)}(\alpha) = \Box f^{(k-1)}(\alpha) := (f^{(k-1)}(\alpha^+) + f^{(k-1)}(\alpha^-))/2 .$$

<u>Proof</u>: Since the case when (4) holds has already been established before stating the lemma, we may assume (8) to hold, and are to show that

(10) $$\Box f^{(k-1)}(\alpha) = \sum_{\nu\in\mathbb{Z}} A_\nu f(\nu) .$$

The only difficulty is that by (3), or $\text{jump}_\alpha K = -1$, the Stieltjes integral in (5) is not defined. However, it is defined if

(11) <u>the point</u> $x = \alpha$ $(= \tau)$ <u>is not an active, or actual, knot of the spline</u>.

In this case again (10) holds.

Let us remove the restriction (11). Observe that, whatever the parity of k may be, the Euler spline $\mathcal{E}_{k-1}(x-\frac{1}{2})$ has its knots at $\tau+\mathbb{Z}$, and $\text{jump}_\tau \mathcal{E}_{k-1}^{(k-1)} \neq 0$. It follows that for some appropriate constant c the spline

$$f_0(x) := f(x) - c\mathcal{E}_{k-1}(x-\tfrac{1}{2})$$

will satisfy the condition (11). Moreover, $\mathcal{E}_{k-1}(\nu-\frac{1}{2}) = 0$ $(\nu\in\mathbb{Z})$ and therefore $f(\nu) = f_0(\nu)$ for all ν. It follows that $f(x) = f_0(x) + c\mathcal{E}_{k-1}(x-\frac{1}{2})$ has the property that

$$\Box f^{(k-1)}(\alpha) = f_0^{(k-1)}(\alpha) + \Box c\mathcal{E}_{k-1}^{(k-1)}(\alpha-\tfrac{1}{2})$$

$$= \sum_{\nu\in\mathbb{Z}} A_\nu f_0(\nu) + 0 = \sum_{\nu\in\mathbb{Z}} A_\nu f(\nu) ,$$

which proves (10). |||

We may now establish the

Theorem 3. The differentiation formula (7) is the unique diff. formula having absolutely summable coefficients A_ν and which is valid for all splines $f(x)$ satisfying (6).

Proof: Suppose that also the formula

$$f^{(j)}(\alpha) = \sum_{\nu\in\mathbb{Z}} A'_\nu f(\nu) \tag{12}$$

shares all these properties with (7). Subtracting them and setting $\tilde{A}_\nu = A_\nu - A'_\nu$, we conclude that

$$\sum_\nu \tilde{A}_\nu f(\nu) = 0 \tag{13}$$

for all f satisfying (6). If we apply (13) to the sequence of B-splines (see, e.g. [15, p. 11])

$$f(x) = Q_k(n-x+\tau) \quad (n\in\mathbb{Z})$$

we obtain that

$$\sum_{\nu\in\mathbb{Z}} \tilde{A}_\nu Q_k(n-\nu+\tau) = 0 \quad (n\in\mathbb{Z}). \tag{14}$$

This shows that the cardinal spline $g := \sum_{\nu\in\mathbb{Z}} \tilde{A}_\nu Q_k(\cdot-\nu)$ of order k vanishes at $\mathbb{Z}+\tau$ while also, by assumption on the $\tilde{A}_\nu$, being in $\mathbb{L}_1(\mathbb{R})$. But this implies, as in the proof of unicity of K (see Section 4.(α) above) that $g = 0$, therefore $\tilde{A}_\nu = 0$, for all ν. |||

REFERENCES

1. G. Birkhoff and C. de Boor, Error bounds for spline interpolation, J. Math. Mech. 13 (1964) 827-836.

2. C. de Boor, On cubic spline functions which vanish at all knots, MRC TSR 1424, 1974; Adv. Math. (1975).

3. H. G. Burchard, Extremal positive splines with applications to interpolation and approximation by generalized convex functions, Bull. Amer. Math. Soc. 79 (1973) 959-963.

4. A. Cavaretta, Jr., An elementary proof of Kolmogorov's theorem, Amer. Math. Monthly, 81 (1974), 480-486.

5. F. R. Gantmacher and M. G. Krein, "Oszillationsmatrizen, Oszillationskerne und kleine Schwingungen mechanischer Systeme," (transl. from 2nd Russian ed. of 1950), Akademie Verlag, Berlin, 1960.

6. Charles A. Hall and W. Weston Meyer, Optimal error bounds for cubic spline interpolation, GMR-1556, Gen. Motors Res. Labs., Warren, Michigan, Mar. 1974, iii+25pp.

7. S. Karlin and C. Micchelli, The fundamental theorem of algebra for monosplines satisfying boundary conditions, Israel J. Math. 11 (1972) 405-451.

8. А. Н. Колмогоров, О неравенствах между верхними гранями последовательных производных функций на бесконечном интервале, Учен. зап. МГУ, вып. 30, "Математика", 30 (1939) 3-13; a translation into English has appeared as:

A. N. Kolmogorov, On inequalities between upper bounds of the successive derivatives of an arbitrary function on an infinite interval, in Amer. Mathem. Soc. Translations 4 (1949) 233-243.

9. E. Landau, Einige Ungleichungen für zweimal differentiirbare Funktionen, Proc. London Math. Soc. (2) 13 (1913) 43-49.

10. C. A. Micchelli, Cardinal $\mathcal{L}$-splines, in "Studies in splines and approximation theory", S. Karlin, C. A. Micchelli, A. Pinkus and I. J. Schoenberg, Academic Press, New York, 1975.

11. C. A. Micchelli, Oscillation matrices and cardinal spline interpolation, in "Studies in splines and approximation theory", S. Karlin et al., Academic Press, New York, 1975.

12. E. N. Nilson, Polynomial splines and a fundamental eigenvalue problem for polynomials, J. Approx. Theory 6 (1972) 439-465.

13. F. Richards, Best bounds for the uniform periodic spline interpolation operator, J. Approx. Theory 7 (1973) 302-317.

14. I. J. Schoenberg, Zur Abzählung der reellen Wurzeln algebraischer Gleichungen, Math. Zeit. 38 (1934) 546-564.

15. I. J. Schoenberg, "Cardinal Spline Interpolation", CBMS Vol. 12, SIAM, Philadelphia, 1973.

16. I. J. Schoenberg, The elementary cases of Landau's problem of inequalities between derivatives, Amer. Math. Monthly 80 (1973) 121-148.

7. I. J. Schoenberg, On remainders and the convergence of cardinal spline interpolation for almost periodic functions, MRC TSR 1514, Dec. 1974; in "Studies in splines and approximation theory", S. Karlin et al., Academic Press, New York, 1975.

8. I. J. Schoenberg, On Charles Micchelli's theory of cardinal $\mathcal{L}$-splines, MRC TSR 1511, Dec. 1974; in "Studies in splines and approximation theory", S. Karlin et al., Academic Press, New York, 1975.

Appendix on periodic spline interpolation

In his talk entitled "Periodische Splines" at this conference, Professor G. Meinardus raised the question of the existence and uniqueness of a periodic spline interpolant of order k with simple knots at the $(b-a)$-periodic extension $(x_i)_{-\infty}^{\infty}$ of the sequence $x_1 < \ldots < x_n$ in (a, b) which agrees at all its knots with a given $(b-a)$-periodic function g on $\mathbb{R}$. Specifically he conjectured that, for odd $k > 1$, this interpolation problem has exactly one solution in case n is odd.

This is indeed the case, as has been proved with some effort by Friedrich Krinzessa in his doctoral dissertation entitled "Zur periodischen Spline-interpolation,", Bochum, 1969, a fact pointed out to us by Larry Schumaker.

But the conjecture can also be established directly from the Budan-Fourier theorem for splines: For, suppose that f is a $(b-a)$-periodic spline of order k with knot sequence (x_i) and that f vanishes at all its knots but is not identically zero. Then f cannot be just a polynomial of degree $< k-1$, since periodicity would otherwise force it to be constant, hence to vanish identically since it vanishes at x_1 (we do assume that $n \geq 1$). Further, f can have only isolated zeros, since otherwise we could find an interval $(\hat{a}, \hat{b})$ in which f has only isolated zeros but so that f vanishes identically in a left neighborhood of $\hat{a}$ and in a right neighborhood of $\hat{b}$; but then

$$S^-(f(\hat{a}), \ldots, f^{(k-1)}(\hat{a}^+)) = 0, \quad S^+(f(\hat{b}), \ldots, f^{(k-1)}(\hat{b}^-)) = k-1,$$

and so, by the Budan-Fourier theorem for splines,

$$\text{number of knots in } (\hat{a}, \hat{b}) \le Z_f(\hat{a}, \hat{b})$$

$$\le Z_{f^{(k-1)}}(\hat{a}, \hat{b}) - (k-1)$$

$$\le \text{number of knots in } (\hat{a}, \hat{b}) - (k-1)$$

a contradiction for $k > 1$. Therefore

$$n \le Z_f(a, b) \le Z_{f^{(k-1)}}(a, b) \le n,$$

the second inequality by the Budan-Fourier theorem for splines since

$$S^-(f(a), \ldots, f^{(k-1)}(a)) - S^+(f(b), \ldots, f^{(k-1)}(b)) \le 0$$

by periodicity. It follows that f has simple zeros at its knots and vanishes nowhere else, therefore

$$f^{(1)}(x_i)f^{(1)}(x_j)(-)^{i-j} > 0, \quad \text{all } i, j,$$

and so, since $f^{(1)}(x_{n+1}) = f^{(1)}(x_1)$ by periodicity,

$$(f^{(1)}(x_1))^2(-)^n > 0.$$

In short, if the interpolation problem fails to have exactly one solution, then n is even. Q.E.D.

AN EXTENSION OF SARD'S METHOD

Franz-Jürgen Delvos and Walter Schempp

In several papers concerning the theory of optimal approximation Sard has developed a method for the construction of spline approximants in an abstract setting [8,9,10]. In particular, the minimal quotient theorem provides precise error bounds in optimal approximation formulae [9]. On the other hand, in concrete spline approximation improved error bounds have been derived using the so-called second integral relation [1] [14].

It is the purpose of this paper to present an extension of Sard's method as described in [3] which allows to derive improved error bounds in the framework of Sard's method. For the case of optimal interpolation in spaces of continuous functions, the existence of Green kernel corresponding to a self-adjoint operator will be proved. Thus, the construction of optimal interpolants with the aid of Green kernels as proposed by Karlin [5] for L-splines can be performed in the abstract setting of Sard's method.

1. Sequences of Sard systems

The theory of optimal interpolation as developed by Sard [8,9,10] is based on the "Sard system" [3] :

$$(X,Y,Z_o;U,F_o) \quad . \tag{1.1}$$

Here X, Y, Z_o are (complex, separable) Hilbert spaces, and

$$U : X \to Y \quad , \quad F_o : X \to Z_o$$

are continuous linear mappings. It is supposed that the completeness condition holds [9]. Thus

$$((x,y)) = (Ux,Uy) + (F_o x,F_o y) \tag{1.2}$$

is a scalar product on X which induces the original topology of X. For simplicity assume that

$$Im(U) = Y \quad , \quad Im(F_o) = Z_o \quad .$$

The corresponding spline projector P_o is the orthogonal projector of $(X;((.,.)))$ defined by

$$Im(P_o) = Ker(F_o)^{\perp} \, .$$

Consider now a sequence of Sard systems

$$(X,Y,Z_n;U,F_n) \quad (n=1,2,\ldots) \tag{1.3}$$

such that

$$Ker(F_{n+1}) \subset Ker(F_n) \quad (n=0,1,\ldots) \quad . \tag{1.4}$$

Because of (1.4) the corresponding spline projectors P_n are also spline projectors of $(X;((.,.)))$ such that

$$Im(P_n) = Ker(F_n)^{\perp} \tag{1.5}$$

(see [3]).

Suppose that X is a subspace of Y such that the imbedding mapping is continuous:

$$X \hookrightarrow Y \, . \tag{1.6}$$

Defining ($||x|| = ||x||_Y$)

$$C_n = \sup_{F_n x=0, Ux\neq 0} \frac{||x||}{||Ux||} \quad (n=1,2,\ldots) \; , \tag{1.7}$$

an application of the minimal quotient theorem yields the appraisal [10] :

$$||x - P_n x|| \leq C_n ||Ux|| \tag{1.8}$$

($x \in X$). Because of (1.4) and (1.6) we have

$$0 \leq C_{n+1} \leq C_n < \infty \quad (n=1,2,..) .$$

We have restricted our considerations to " a priori " error bounds (1.8) of "L^2-type". This seems quite natural in the abstract setting of Sard's method. Other error bounds are easily obtained by Sard's theory of admissible approximation [9].

2. The extended Sard system

Suppose that the assumptions of the preceding section hold. Put

$$X_o = \mathrm{Ker}(F_o) \quad . \tag{2.1}$$

Then

$$(X,Y,Z_o;U,F_o)$$

is called an _extended Sard system_ iff

$$\overline{X}_o = Y \quad . \tag{2.2}$$

LEMMA 2.1

The operator U_o _defined by_

$$\mathrm{Dom}(U_o) = X_o \quad , \quad U_o x = Ux \tag{2.3}$$

is closed in Y.

Proof: Consider a sequence (x_n) in X_o such that

$$\lim_{n\to\infty} x_n = x_o \in Y \quad , \quad \lim_{n\to\infty} U_o x_n = y_o \in Y \quad .$$

Since $(U_o x_n)$ is a Cauchy sequence in Y, the sequence

(x_n) is also a Cauchy sequence in the subspace X_o of $(X;((.,.)))$ (see (1.2)). Hence, there is a unique element $z_o \in X$ with

$$\lim_{n\to\infty} ||U(x_n - z_o)|| = 0 \quad .$$

Taking into account (1.6), there exists a positive constant B such that

$$||U(x_n - z_o)|| \geq B||x_n - z_o|| \; .$$

Thus

$$z_o = x_o \in \mathrm{Dom}(U_o) \quad , \quad y_o = U_o x_o \quad ,$$

i.e. U_o is closed.

LEMMA 2.2

<u>The</u> <u>self-adjoint</u> <u>operator</u>

$$A = U_o^* U_o \tag{2.4}$$

<u>is</u> <u>positive</u> <u>definite</u> <u>in</u> Y.

Proof. U_o^* denotes the adjoint operator of U_o. It is a well-known fact in operator theory that A is self-adjoint andpositive [15]. Since

$$\mathrm{Dom}(A) \subset X_o \quad ,$$

for any $x \in \mathrm{Dom}(A)$:

$$(Ax,x) = (U_o x, U_o x) = ((x,x)) \geq B(x,x)$$

(see (1.6)). Thus, A is positive definite.

THEOREM 2.3

<u>The operator</u> $U_o^*U_o$ <u>is the unique self-adjoint operator</u> A <u>in</u> Y <u>such that</u>

$$\mathrm{Dom}(A) \subset X_o$$

$$(Ux,Uy) = (x,Ay) \tag{2.5}$$

($x \in X_o$; $y \in \mathrm{Dom}(A)$).

Proof. We have to apply the Friedrichs extension theorem [7]. First, remark that X_o is a Hilbert space having the scalar product

$$((x,y)) = (Ux,Uy) \quad . \tag{2.6}$$

This scalar product induces a densely defined , closed symmetric form bounded from below. Therefore, by the Friedrichs extension theorem, there is a unique positive definite and self-adjoint operator A satifying the hypotheses of the theorem. On the other hand, Lemma 2.2 implies

$$\mathrm{Dom}(U_o^*U_o) \subset X_o$$

and

$$(Ux,Uy) = (x,U_o^*U_oY)$$

($x \in X_o$; $y \in \mathrm{Dom}(U_o^*U_o)$). Since A is unique, we obtain

$$A = U_o^*U_o \quad .$$

REMARK 2.4

Because of Theorem 2.3 the operator $A = U_o^*U_o$ is called the <u>operator associated with the extended Sard system</u> $(X,Y,Z_o;U,F_o)$.

3. Improved error bounds

It is obvious from the definition of the spline projector that the "best approximation property" is equivalent to the " first integral relation "

$$(UP_n x, U(x - P_n x)) = 0 \qquad (3.1)$$

($x \in X$) .

Replacing in (1.8) x by $x-P_n x$ and taking into account the idempotence of P_n we obtain

LEMMA 3.1

For each $x \in X$ the inequality

$$||x - P_n x|| \leq C_n ||U(x - P_n x)||$$

is true.

The following Lemma 3.2 represents an abstract version of the "second integral relation" .

LEMMA 3.2

Suppose that $x \in \mathrm{Dom}(A)$. Then the following relation is valid:

$$||U(x - P_n x)||^2 = (x - P_n x, Ax) \quad . \qquad (3.3)$$

Proof. Because of (1.4), we have

$$x - P_n x \in X_o \quad .$$

In the terminology of Mikhlin [6] (see also [16]) the Hilbert space X_o with the scalar product (2.6) is just the energy space of A :

$$H_A = X_o \quad . \tag{3.4}$$

Now

$$(Ux,Uy) = (x,Ay) \tag{3.5}$$

if $x \in H_A$ and $y \in \text{Dom}(A)$ [6]. Therefore, the second integral relation (3.3) is an immediate consequence of (3.5).

THEOREM 3.3

Suppose that $x \in \text{Dom}(A)$. Then the following estimate is valid:

$$||x - P_n x|| \leq C_n^2 ||Ax|| \ . \tag{3.6}$$

Proof. By Lemma 3.1, we have

$$||x - P_n x|| \leq C_n ||U(x - P_n x)|| \quad . \tag{3.7}$$

An application of the Cauchy-Schwarz inequality in (3.3) yields

$$\begin{aligned} ||U(x - P_n x)||^2 &\leq ||x - P_n x|| \; ||Ax|| \\ &\leq C_n ||U(x - P_n x)|| \; ||Ax|| \end{aligned}$$

whence

$$||U(x - P_n x)|| \leq C_n ||Ax|| \quad . \tag{3.8}$$

Combining (3.8) and (3.7), this proves (3.6).

4. The associated kernel

Let D be a relative compact set of R^m. It is supposed

that X is a vector space of continuous functions on D such that X possesses a reproducing kernel K(s,t).
For each t ∈ D, put

$$K_t(s) = K(s,t) \qquad (4.1)$$

(s ∈ D) . Then we have for every x ∈ X

$$x(t) = ((x,K_t)) \qquad (4.2)$$

(t ∈ D). The connections between the theory of reproducing kernels and optimal interpolation have been studied by several authors (see for instance [2] [13]).

THEOREM 4.1

Put

$$G(s,t) = K(s,t) - P_o(K_t)(s) \qquad (4.3)$$

(s,t ∈ D). Then for every x ∈ Dom(A) we have:

$$x(t) = (Ax,G_t) \qquad (4.4)$$

(t ∈ D).

Proof. First remark that

$$G_t = K_t - P_o K_t$$

whence

$$G_t \in X_o \quad .$$

Combining (3.5) and (4.2) we obtain

$$\begin{aligned} x(t) &= ((x,K_t)) \\ &= ((x,G_t)) \\ &= (Ux,UG_t) \\ &= (Ax,G_t) \quad . \end{aligned}$$

REMARK 4.2

Because of (4.4), the function $G(s,t)$ is called the Green kernel of A associated with the extended Sard system $(X,Y,Z_o;U,F_o)$. Obviously, $G(s,t)$ is the reproducing kernel of the energy space $H_A = X_o$.

Consider now the points

$$t_1,\dots,t_n \in D$$

such that the Dirac-measures

$$\varepsilon_{t_1},\dots,\varepsilon_{t_n}$$

are linear independent. Put

$$F_n = F_o \times \varepsilon_{t_1} \times\dots\times \varepsilon_{t_n} \quad (n=1,2,..) \quad .$$

THEOREM 4.2

For every $x \in X_o$ we have

$$P_n x(s) = \sum_{i=1}^{n} a_i G(s,t_i) \quad (s \in D) \tag{4.5}$$

with

$$\sum_{i=1}^{n} a_i G(t_k,t_i) = x(t_k) \quad (k=1,..,n) \quad . \tag{4.6}$$

Proof. The relations (4.5), (4.6) follow from the fact that the determination of $P_n x$ can be considered as an interpolation problem with minimal norm in the Hilbert space X_o with reproducing kernel $G(s,t)$.

Finally, we remark that the representation (4.8), (4.9), is closely related to the results of Karlin [5] (see also [11, 12]). Sometimes, the spectral properties of A can be used for the computation of G(s,t) (see [4]). We conclude with the determination of the operator A corresponding to L-splines of type I.

Let

$$Lx = \sum_{i=o}^{n} p_i D^i x$$

be a linear differential operator with smooth coefficients $p_i \in C^n(I)$ ($i = o,\dots,n$; $I = [a,b]$) and leading coefficient $p_n(t) \neq o$ ($t \in I$). Obvioulsy, L induces a continuous linear map U from $X = W_2^n(I)$ onto $Y = L_2(I)$. L-splines of type I are characterized by the boundary conditions

$$F_o x = (x(a),..,D^{n-1}x(a),x(b),..,D^{n-1}x(b)).$$

Thus the operator U_o is given by

$$U_o x = Lx \ , \quad Dom(U_o) = \mathring{W}_2^n(I) \quad .$$

Because of the smoothness of L the adjoint operator L* exists. An application of Lagrange's identity yields that the operator U_o^* is defined by

$$U_o^* x = L^* x \ , \ Dom(U_o^*) = W_2^n(I) \quad .$$

Hence the operator A corresponding to L-splines of type I is given by

$$Ax = L^*Lx \quad , \quad Dom(A) = W_2^{2n}(I) \cap \mathring{W}_2^n(I) \quad .$$

REFERENCES

1. J. H. AHLBERG, E. N. NILSON, and J.L. WALSH, "The theory of splines and their applications", Academic Press, New York, 1967.

2. C. DE BOOR and R. E. LYNCH, On splines and their minimum properties, J. Math. Mech. 15 (1966), 953 - 989.

3. F. J. DELVOS and W. SCHEMPP, Sard's method and the theory of spline systems, J. Approximation Theory 14 (1975), 230-243.

4. F. J. DELVOS and W. SCHEMPP, On optimal periodic spline interpolation, J. Math. Analysis Appl., to appear.

5. S. KARLIN, "Total positivity", Stanford University Press, Stanford, California, 1968.

6. S. G. MIKHLIN, "The problem of the minimum of a quadratic functional", Holden Day, San Francisco-London-Amsterdam, 1965.

7. F. RIESZ and B. SZ. NAGY, "Vorlesungen über Funktionalanalysis", Deutscher Verlag der Wissenschaften, Berlin, 1956.

8. A. SARD, Optimal approximation, J. Functional Analysis 1 (1967), 222-244; 2 (1968), 368-369.

9. A. SARD, Approximation based on nonscalar observations, J. Approximation Theory 8 (1973), 315-334.

10. A. SARD, Instances of generalized splines, in "Spline-Funktionen" (eds.: K. Böhmer, G. Meinardus, W. Schempp), BI-Wissenschaftsverlag, Mannheim-Wien-Zürich, 1974.

11. R. SCHABACK, Konstruktion and algebraische Eigenschaften von M-Spline-Interpolierenden, Numer. Math. 21 (1973), 166-180.

12. R. SCHABACK, Konstruktion von Spline-Interpolierenden und Peano-Kerne, in "Spline-Funktionen" (ed.: K. Böhmer, G. Meinardus, W. Schempp), BI-Wissenschaftsverlag, Mannheim-Wien-Zürich, 1974.

13. W. SCHEMPP und U. TIPPENHAUER, Reprokerne zu Spline-Grundräumen, Math. Z. 136 (1974), 357-369.

14. M. H. SCHULTZ, "Spline Analysis, Prentice-Hall, Englewood Cliffs, New Jersey, 1973.

15. W. I. SMIRNOW, "Lehrgang der höheren Mathematik V" Deutscher Verlag der Wissenschaften, Berlin, 1967.

16. H. TRIEBEL, "Höhere Analysis", Deutscher Verlag der Wissenschaften, Berlin, 1972.

Dr. F. J. Delvos
Prof. Dr. W. Schempp
Lehrstuhl für Mathematik I
Universität Siegen
D-59 Siegen 21
Hölderlinstr. 3

ZUR NUMERISCHEN BEHANDLUNG VON HOMOGENEN FREDHOLMSCHEN INTEGRALGLEICHUNGEN 2. ART MIT SPLINES

Günther Hämmerlin

In order to approximate the kernel of a Fredholm integral equation of second kind, two - dimensional splines over a net of square meshes are constructed by using B - splines. The approximating kernel defines a substituting equation which is equivalent to a linear system of equations. Solving this linear system, we get approximating eigenvalues and eigenfunctions of the integral equation. The procedure is based upon a method published in [1] which already contains the case of two - dimensional splines of zeroth and first degree. With the aid of new estimates for the quality of multidimensional splines, error bounds for the approximations of eigenvalues are given.

In dieser Abhandlung soll gezeigt werden, wie in Anlehnung an frühere Untersuchungen Ersatzkernverfahren zur numerischen Behandlung linearer Integralgleichungen mit Hilfe von Splines definiert werden können, die die Eigenschaften der bequemen Lösbarkeit und der Möglichkeit einer Fehlerabschätzung miteinander verbinden.

1) Sei $\varphi \in L^2[a,b]$, K ein linearer beschränkter Integraloperator, $K: L^2[a,b] \longrightarrow L^2[a,b]$. Wir betrachten die homogene Fredholmsche Integralgleichung 2. Art

$$(1.1) \qquad \varkappa\varphi = K\varphi \quad \text{bzw.} \quad \bigwedge_{s\in[a,b]} \varkappa\varphi(s) = \int_a^b K(s,t)\varphi(t)dt \; .$$

Sei $\tilde{K}$ ein approximierender Operator an K, dessen Kern die ausgeartete Darstellung

$$(1.2) \qquad \tilde{K}(s,t) = \sum_{j,k=0}^{n} c_{jk}y_j(s)y_k(t) \; ,$$

$$c_{jk} \in \mathbb{R} \; , \; y_l \in L^2[a,b], \; (\; l = 0,1,\dots,n \;),$$

mit Hilfe geeigneter Elemente y_l erlaube. Jede Lösung $(\tilde{\varkappa}, \tilde{\varphi})$ der Ersatzgleichung

$$(1.3) \qquad \tilde{\varkappa}\tilde{\varphi} = \tilde{K}\tilde{\varphi}$$

hat dann die Gestalt

$$\tilde{\varphi} = \sum_{j=0}^{n} d_j y_j \; ,$$

wie man durch Einsetzen erkennt:

$$(1.4) \qquad \tilde{\varkappa}\tilde{\varphi}(s) = \int_a^b \sum_{j,k=0}^{n} c_{jk}y_j(s)y_k(t)\tilde{\varphi}(t)dt.$$

Zur Bestimmung einer Lösung von (1.4) ergibt sich über

$$\tilde{\varkappa}\sum_{j=0}^{n} d_j y_j(s) = \sum_{j,k=0}^{n} y_j(s)c_{jk} \sum_{l=0}^{n}\Big(\int_a^b y_k(t)y_l(t)dt\Big)d_l$$

das lineare Gleichungssystem

$$(1.5) \qquad \tilde{\varkappa}\underline{d} = C\,Y\,\underline{d} \quad ,$$

$$\underline{d} := \begin{pmatrix} d_0 \\ \vdots \\ d_n \end{pmatrix}, \quad C := (c_{jk})_{j,k=0}^{n}, \quad Y := ((y_j, y_k))_{j,k=0}^{n}$$

$$\text{mit} \quad (y_j, y_k) := \int_a^b y_j(t) y_k(t) dt.$$

Die Lösung von (1.5) liefert die exakte Lösung von (1.4) bzw. (1.3) und damit eine Näherung für die Lösung von (1.1), sofern nur $\tilde{K}$ den Integraloperator K hinreichend gut approximiert.

In [1] wurde das folgende spezielle Ersatzkernverfahren zur Konstruktion eines geeigneten approximierenden Kerns $\tilde{K}(s,t)$ der Gestalt (1.2) entwickelt. Dazu nehmen wir jetzt an, daß $K(s,t)$ in $[a,b]\times[a,b]$ stetig sei. Nach Einteilung von $[a,b]\times[a,b]$ in n^2 gleichgroße, quadratische Maschen wird $\tilde{K}(s,t)$ durch Interpolation von $K(s,t)$ mit Polynomen über den einzelnen Maschen definiert, die in s und t von jeweils gleichem Höchstgrad sind. Bei Interpolation nullten Grades entsteht so ein maschenweise konstanter, insgesamt aber nicht stetiger, bei Interpolation ersten Grades ein insgesamt stetiger Ersatzkern, der maschenweise in s und t jeweils linear ist. Heute ist es üblich, in diesem Fall von der Tensorproduktdarstellung eines zweidimensionalen Splines nullten bzw. ersten Grades zu sprechen.

Daran anknüpfend soll nun das Ersatzkernverfahren auf allgemeinere Splines erweitert werden, und mit Hilfe neuer Fehlerschranken für den Approximationsfehler zweidimensionaler Splines kann man zu Fehlerabschätzungen für die Genauigkeit der Näherungen an die Eigenwerte des Problems kommen.

2) Eine lokale Darstellung des zweidimensionalen interpolierenden Splines an $K(s,t)$ bezüglich der Stützstellen (s_j,t_k), $a = s_0 = t_0$, $s_j := s_0 + jh$, $t_k := t_0 + kh$, $h := (b-a)/n$, $(j,k = 1,2,\ldots,n)$, gestattet die Wahl von B - Splines als Basis.

Sei b_j der B-Spline des jeweils betrachteten Grades bezüglich der Stützstelle s_j. Dann gilt

$$(2.1) \qquad \tilde{K}(s,t) = \sum_{j,k=o}^{n} c_{jk} b_j(s) b_k(t) .$$

Besonders einfach läßt sich der Fall des linearen Splines behandeln: Hier ist $c_{jk} := K(s_j, t_k)$, ($j,k = 0,1,\ldots,n$), falls die Normierung $b_j(s_j) := 1$ getroffen wird.

Nach (1.5) entsteht die Ersatzgleichung

$$(2.2) \qquad \tilde{\varkappa}\, \underline{d} = C B \underline{d} \quad ,$$

die die Eigenwerte $\tilde{\varkappa}_1, \ldots, \tilde{\varkappa}_n$ und die zugehörigen Eigenvektoren liefert. Mit dem Eigenvektor $\underline{d}_j$ ist dann die Eigenlösung

$$(2.3) \qquad \tilde{\varphi} = \sum_{k=o}^{n} d_{jk} b_k \ , \quad \underline{d}_j = : \begin{pmatrix} d_{jo} \\ \vdots \\ d_{jn} \end{pmatrix} ,$$

von (1.3) gegeben. Vom Kern K(s,t) hängt dabei allein die Matrix C ab. Sie ergibt sich bei Berechnung des Splines an K(s,t) in der Darstellung durch B-Splines. Die Matrix

$$B := ((b_j, b_k))_{j,k=o}^{n}$$

dagegen hängt allein vom gewählten Grad der Spline-Approximation ab; so erhält man etwa für den bilinearen Spline mit

$$b_o(s) := \begin{cases} \frac{1}{h}(s_1 - s) & s \in [s_o, s_1] \\ 0 & s \in [s_1, s_n] \end{cases} ,$$

$$b_j(s) := \begin{cases} 0 & s \in [s_o, s_{j-1}] \\ \frac{1}{h}(s - s_{j-1}) & s \in [s_{j-1}, s_j] \\ \frac{1}{h}(s_{j+1} - s) & s \in [s_j, s_{j+1}] \\ 0 & s \in [s_{j+1}, s_n] \end{cases} ,$$

($j = 1,\ldots,n-1$),

$$b_n(s) := \begin{cases} 0 & s \in [s_0, s_{n-1}] \\ \frac{1}{h}(s - s_{n-1}) & s \in [s_{n-1}, s_n] \end{cases}$$

die Bandmatrix

$$B = \frac{h}{6} \begin{pmatrix} 2 & 1 & 0 & \cdot & \cdot & \cdot & 0 \\ 1 & 4 & 1 & 0 & \cdot & & \cdot \\ 0 & 1 & 4 & 1 & 0 & \cdot & \cdot \\ \cdot & \cdot & \cdot & \cdot & \cdot & \cdot & \cdot \\ \cdot & \cdot & \cdot & \cdot & \cdot & \cdot & \cdot \\ \cdot & \cdot & 0 & 1 & 4 & 1 & 0 \\ \cdot & \cdot & \cdot & 0 & 1 & 4 & 1 \\ 0 & \cdot & \cdot & \cdot & 0 & 1 & 2 \end{pmatrix} ,$$

während der bikubische Spline naturgemäß eine symmetrische Bandmatrix B mit insgesamt sieben besetzten Diagonalen erzeugt.

3) Eine Abschätzung für die Abweichung der Näherungen $\tilde{\varkappa}_j$ von den wahren Eigenwerten $\varkappa_j$ läßt sich mit Hilfe eines Satzes von H. Weyl [3] gewinnen (s. auch [1], S. 446). Wir benutzen die folgende Aussage, die aus dem Satz von H. Weyl fließt; für die Anwendung dieser Abschätzung schränken wir die zugelassenen Kerne auf (reelle) symmetrische ein:

Sei $K(s,t) = \tilde{K}(s,t) + K^*(s,t)$ eine Zerlegung des symmetrischen Kerns $K(s,t)$, $\tilde{K}(s,t)$ und damit $K^*(s,t)$ ebenfalls symmetrisch. Seien $\varkappa_1 \geqq \varkappa_2 \geqq \ldots \geqq 0$ die Eigenwerte von K, entsprechend $\tilde{\varkappa}_j$ und $\varkappa_j^*$, (j=1,2,...), die nach der Größe des Betrags geordneten Eigenwerte von $\tilde{K}$ und von K^*. Dann gilt die Abschätzung

$$(3.1) \qquad |\varkappa_j - \tilde{\varkappa}_j| \leqq |\varkappa_1^*| , \quad (j=1,2,\ldots) .$$

Beachten wir gleichzeitig die Tatsache, daß

$$(3.2) \qquad \bigwedge_{j=1,2,\ldots} |\varkappa_j^*| \leqq \| K^* \|$$

bei beliebiger Norm gilt, die nicht Hilbertraumnorm zu sein braucht, so reduziert sich die Frage der Genauigkeit der Näherungseigenwerte $\tilde{\varkappa}_j$ nach (3.1) und (3.2) auf die Abschätzung

$$(3.3) \qquad |\varkappa_j - \tilde{\varkappa}_j| \leqq \| K - \tilde{K} \| ,$$

$$(j = 1,2,\dots) ,$$

und damit auf Abschätzungen der Approximationsgenauigkeit von Splines. Die Abschätzung (3.3) ist gleichmäßig bezüglich j und damit am besten für die betragsgrößten der Eigenwerte von $\tilde{K}$ bzw. K.

4) Zur Anwendung von (3.3) dienen die folgenden Abschätzungen, die sich auf den Fall hinreichend oft differenzierbarer Kerne beziehen:

Bilinearer Spline (nach [3] , S. 19)

$$(4.1 \qquad \| K - \tilde{K} \|_2 \leqq \frac{h^2}{\pi^2} \left[\| K_{ss} \|_2 + \| K_{st} \|_2 + \| K_{tt} \|_2 \right]$$

oder (nach [3] , S. 20)

$$(4.2) \qquad \| K - \tilde{K} \|_\infty \leqq \frac{h^2}{8} \left[\| K_{ss} \|_\infty + \| K_{tt} \|_\infty \right] .$$

Bikubischer Spline (nach [3] , S. 60)

$$(4.3) \qquad \| K - \tilde{K} \|_2 \leqq \frac{4}{\pi^4} h^4 \left[\| K_{ssss} \|_2 + \| K_{sstt} \|_2 + \| K_{tttt} \|_2 \right]$$

oder (nach [3] , S. 60)

$$(4.4) \qquad \| K - \tilde{K} \|_\infty \leqq h^4 \left[\frac{5}{384} \| K_{ssss} \|_\infty + \frac{4}{9} \| K_{sstt} \|_\infty + \frac{5}{384} \| K_{tttt} \|_\infty \right] .$$

5) Die Ausführungen in dieser Abhandlung sind keineswegs vollständig, sondern in mehreren Richtungen erweiterungs- und verallgemeinerungsfähig. Dabei ist etwa an die

Frage der Genauigkeit der Näherungen an die Eigenfunktionen nach (2.3) zu denken, die wie in [1] zu behandeln ist, oder an die Verwendung anderer approximierender, nicht notwendig interpolierender Splines; schließlich ist auch die Anwendung auf andere Typen von Integralgleichungen zu untersuchen, um nur die nächstliegenden Möglichkeiten zu nennen.

Literatur:

[1] G. Hämmerlin: Ein Ersatzkernverfahren zur numerischen Behandlung von Integralgleichungen 2. Art. Z. Angew. Math. Mech. 42 (1962), 439 - 463.

[2] M. H. Schultz: Spline Analysis. Englewood Cliffs,N.J., Prentice - Hall, Inc., 1973.

[3] H. Weyl: Das asymptotische Verteilungsgesetz der Eigenwerte linearer partieller Differentialgleichungen (mit einer Anwendung auf die Hohlraumstrahlung),Math. Ann. 71 (1912), 441 - 479.

Prof. Dr. G. Hämmerlin
Mathematisches Institut der
Ludwig - Maximilians - Universität
D 8000 München 2
Theresienstr. 39

ANWENDUNG DER SPLINE-FUNKTIONEN ZUR BEARBEITUNG GEOPHYSIKALISCHER MESSREIHEN

Gerhard Jentzsch, Gerald Lange, Otto Rosenbach

Applications of natural spline functions to geophysical time series are discussed. These functions may be used to interpolate unavoidable gaps in the observations, and also to smooth data where commonly used methods are not applicable.

Empirically determined features of spline functions with different smoothing factors (weights) are given. An exemple of the use of bicubic splines in processing potential field data is also presented.

1. Einleitung und Problemstellung

In der Geophysik muß i.a. unter Bedingungen registriert werden, die nicht optimal der Meßgröße angepaßt sind: Bei Feldmessungen ist die Stationsauswahl nicht nur abhängig von dem zu erwartenden Signal, sondern vor allem auch von der Zeitdauer des Einsatzes (z.B. Kosten) und dem Meßaufwand (Zugänglichkeit des Ortes). Bei Dauerregistrierungen z.B. von geodynamischen Effekten über mehrere Jahre kommen noch Störungen durch äußere Einflüsse wie Stromausfall oder Geräteeffekte wie Driften der Elektronik und Bereichsüberschreitungen hinzu. Deshalb muß der Vorbearbeitung der Meßreihen vor der eigentlichen Analyse besondere Beachtung geschenkt werden.

Abgesehen von individuellen Korrekturen einzelner Meßwerte soll die Vorbearbeitung die digitale Meßreihe im Hinblick auf das erwünschte Analysenergebnis aufbereiten. Dies geschieht i.a. durch numerische Filterung, die in zweierlei Hinsicht eingesetzt wird:

- Aliasfilterung:
  Der zeitliche oder lokale Stützstellenabstand der Meßreihe muß der Meßgröße und der Charakteristik des Meßverfahrens angepaßt werden, um auch alle hochfrequenten Anteile zu erhalten. Nach der numerischen Tiefpaßfilterung kann anschließend der Stützstellenabstand entsprechend dem zu erwartenden Resultat vergrößert werden. Ohne vorherige Tiefpaßfilterung jedoch werden durch hochfrequente Anteile, die durch den größeren Stützstellenabstand nicht mehr aufgelöst werden können, tieffrequente Schwingungen vorgetäuscht. Diese haben die "Alias"-frequenzen ν, $2\nu_N \pm \nu$; $4\nu_N \pm \nu$;...; mit ν_N als Nyquist-(Abtast-)Frequenz und $\nu > \nu_N$.
- Bandpaßfilterung:
  Ist das Signal beschränkt auf ein bestimmtes Frequenzband, dann werden durch diesen Prozeß die übrigen in der Registrierung enthaltenen Frequenzanteile unterdrückt.

Bei diesen Filterprozessen werden Operatoren im Zeitbereich mit der Meßreihe gefaltet. Die Länge dieser Operatoren wird aufgrund eines Kompromisses zwischen der gewünschten Filterwirkung und der Zahl der vorhandenen Daten bestimmt: Einerseits soll der Operator möglichst lang sein, andererseits gehen i.a. am Anfang und Ende der Meßreihe Datenabschnitte der halben Operatorlänge bei der Faltung verloren. Weiterhin fordern diese Filteroperatoren Vollständigkeit (keine Lücken) und Stationarität der Daten. Ist beispielsweise ein Nadelimpuls enthalten, so wird dieser durch die Filterung auf die benachbarten Werte verteilt und so der Mittelwert der Meßkurve verfälscht.

Werden diese Forderungen an die Daten nicht erfüllt, muß eine Methode gefunden werden, die mit Hilfe möglichst "glatter" Funktionen derartige Fehlstellen überbrückt. Die Nachteile der bisher meist angewandten Verfahren - Polynominterpolation, stückweise Polynominterpolation - können durch die Verwendung von natürlichen Spline-Funktionen überwunden werden.

2. Natürliche Spline-Funktionen

Die Menge der natürlichen Spline-Funktionen vom Grad (2k-1) mit n Knoten x_i (i=1,...,n) ist definiert als

$$S_{2k-1}(x_1,\dots,x_n) = \left\{ s \in C^{2k-2}[a,b]: \begin{array}{l} s \in p_{k-1} \text{ in } (a,x_1) \text{ und } (x_n,b) \\ s \in p_{2k-1} \text{ in } (x_i,x_{i+1}),\ i=1,\dots,n-1 \end{array} \right\}$$

Dabei ist p_k die Menge der Polynome vom Grad $\leq$ k.

Die natürlichen Spline-Funktionen werden z.B. dargestellt durch

$$(1) \quad s(x) = p_{k-1}(x) + \sum_{i=1}^{n} c_i (x-x_i)_+^{2k-1}$$

wobei $x_+^k := \begin{cases} x^k & \text{für } x > 0 \\ 0 & \text{für } x \leq 0 \end{cases}$

Für die c_i gelten die Bedingngen

$$\sum_{i=1}^{n} c_i x_i^j = 0 \qquad (j=0,\dots,k-1)$$

Sind n Meßwerte y_i (i=1,...,n) an den Stützstellen x_i (i=1,...,n) vorgegeben, dann gibt es genau eine $s_o \in S_{2k-1}$ mit den Knoten x_i, für die gilt $s_o(x_i) = y_i$.

Diese Spline-Funktion hat folgende Eigenschaft:

Unter allen Funktionen f aus

$$H_2^k[a,b] = \left\{ f \in C^{k-1}[a,b]: f^{(k-1)} \text{ absolut stetig, } f^{(k)} \in L_2[a,b] \right\}$$

mit $f(x_i) = y_i$ ist s_o die eindeutig bestimmte Lösung des Minimumproblems

$$\min_{f \in H_2^k} F(f) := \min_{f \in H_2^k} \int_a^b \left[f^{(k)}(x) \right]^2 dx = \int_a^b \left[s_o^{(k)}(x) \right]^2 dx$$

Betrachtet man F als Maß für die Glätte einer Funktion, so ist s_o die in diesem Sinne glatteste Funktion durch n vorgebene Meßwerte.

Meist sind Meßwerte nur genau innerhalb gewisser Schranken. Somit ist es oft gar nicht wünschenswert, die Beziehung $s_o(x_i) = y_i$ (i=1,...,n) exakt zu erfüllen. Vorteilhafter ist vielmehr, Abweichungen zugunsten der Glättung zuzulassen.

Die Lösung des Minimumproblems

$$(2) \quad \min_{f \in H_2^k} F^+(f) := \min_{f \in H_2^k} \left[F(f) + \sum_{i=1}^{n} w_i \, (f(x_i) - y_i)^2 \right]$$

mit vorgegebenen Gewichten $w_i > 0$ ist dann ebenfalls ein eindeutig bestimmtes Element aus $S_{2k-1}(x_1,\ldots,x_n)$.

Ausführliche Darstellungen dieser Zusammenhänge finden sich bei Anselone und Laurent (1968), Greville (1969), Lyche and Schumaker (1973), Böhmer (1974).

3. Beispiele zur Aliasfilterung

3.1 Filterung einer gravimetrischen Kurve

Schweremessungen auf Profilen werden in hohem Maße von Topographie und oberflächennahen Dichteinhomogenitäten am Meßort beeinflußt. Man bringt deshalb anhand der aufgenommenen Umgebung Korrekturen an, die durch Annäherung der Topographie an geometrische Körper konstanter Dichte berechnet werden. Die Dichten werden durch Messung an Handstücken oder geeignet angelegte Feldmessungen ermittelt.

Trotz dieser Korrekturen streuen die Meßwerte noch sehr stark. In Abb. 1 ist ein derartiges Meßprofil aufgetragen (Krimmler Achental, Alpen): Die negativen Schwerewerte geben die Differenz an zu einem Anschlußpunkt im Grundschwerenetz 1. Ordnung. Deutlich zu erkennen ist die Variation der Punktdichte, die auf Schwierigkeiten im Gelände zurückzuführen ist.

Die Glättung dieser Meßkurve dient der Elimination von Meßfehlern und lokalen topographischen Effekten sowie kleinräumigen geologischen Störungen. Eine Filterung nach Kap. 1 ist nicht möglich, da

- der Stützstellenabstand variabel ist und
- der Operator zu lang sein müßte, um die eingezeichnete Kurve zu erhalten.

Zur Berechnung der natürlichen Spline-Funktion, die das Minimumproblem (1) löst, wurde eine ALGOL-Prozedur nach Reinsch (1967) angewandt. Dort wird die Interpolationsbedingung ersetzt durch

$$(2) \qquad \sum_{i=0}^{n} \left(\frac{f(x_i) - y_i}{dy_i}\right)^2 \leq S$$

wobei $dy_i \neq 0$ und $S \geq 0$ wählbare Zahlen sind; dy_i kann sich nach der Standardabweichung der Meßwerte richten, und S ist ein Maß für die gewünschte Glätte der Ausgleichskurve (über den Zusammenhang zwischen dieser Aufgabe und (2) siehe z.B. Reinsch (1971)).

Im vorliegenden Beispiel wurde $dy_i = 1$ gewählt, da Aussagen über die Meßfehler dieser Größenordnung nicht möglich sind. Die Variation von S ergab für S = 20 das "beste" Ergebnis.

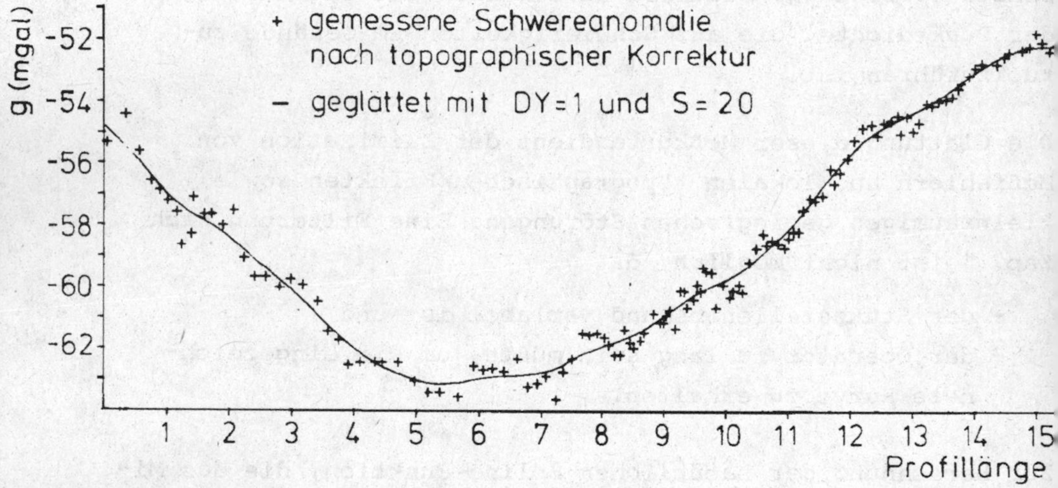

Abb. 1: Schwereanomalie gemessen im Krimmler Achental (Alpen); geglättet mit einem Programm nach Reinsch (1967)

Zur weiteren Bearbeitung werden die Punkte dieser Ausgleichskurve im Abstand von 250 m berechnet, wodurch die anschließenden Modellrechnungen zur Ermittlung der Struktur des Untergrundes vereinfacht werden.

3.2 Glättung und Interpolation von Erdgezeitenregistrierungen

In Abb. 2 ist in der oberen Spur eine 30 Tage lange Registrierung von Erdgezeiten dargestellt, die mit einem Askania-Bohrlochneigungsmesser durchgeführt wurde. Abgesehen von langperiodischen Anteilen sind in dieser Kurve im wesentlichen etwa halb- und gangtägige Wellen enthalten, die den Einfluß von Mond und Sonne auf die Änderung der Neigung der Erdoberfläche (hier in Skt) angeben. Diese Registrierung ist gestört durch Lücken und "Ausreißer".

Wegen der elektronischen Ausgangsfilter des Meßgerätes hat die Meßreihe einen Stützstellenabstand von 12 Min; für die Gezeitenanalyse sind aber 60 Min ausreichend. Eine übliche Aliasfilterung ist wegen der Lücken und Spitzen nicht möglich.

Zur Konstruktion der geglätteten Spline-Funktion konnte das Verfahren von Reinsch (1967) nicht übernommen werden, weil die Anzahl der Daten sehr groß ist, so daß die Schnelligkeit und Speicherkapazität der zur Verfügung stehenden Rechenanlage TR 4 nicht ausreichen.

Grundlage für das 1971/72 am Institut für Geophysik der Technischen Universität Clausthal entwickelte Programm wurde deshalb die Arbeit von Anselone and Laurent (1968). Dieser Ansatz wurde gleichzeitig auch von Reinsch (1971) veröffentlicht.

Zur Berechnung der Koeffizienten der Spline-Funktion in (1) muß das lineare Gleichungssystem

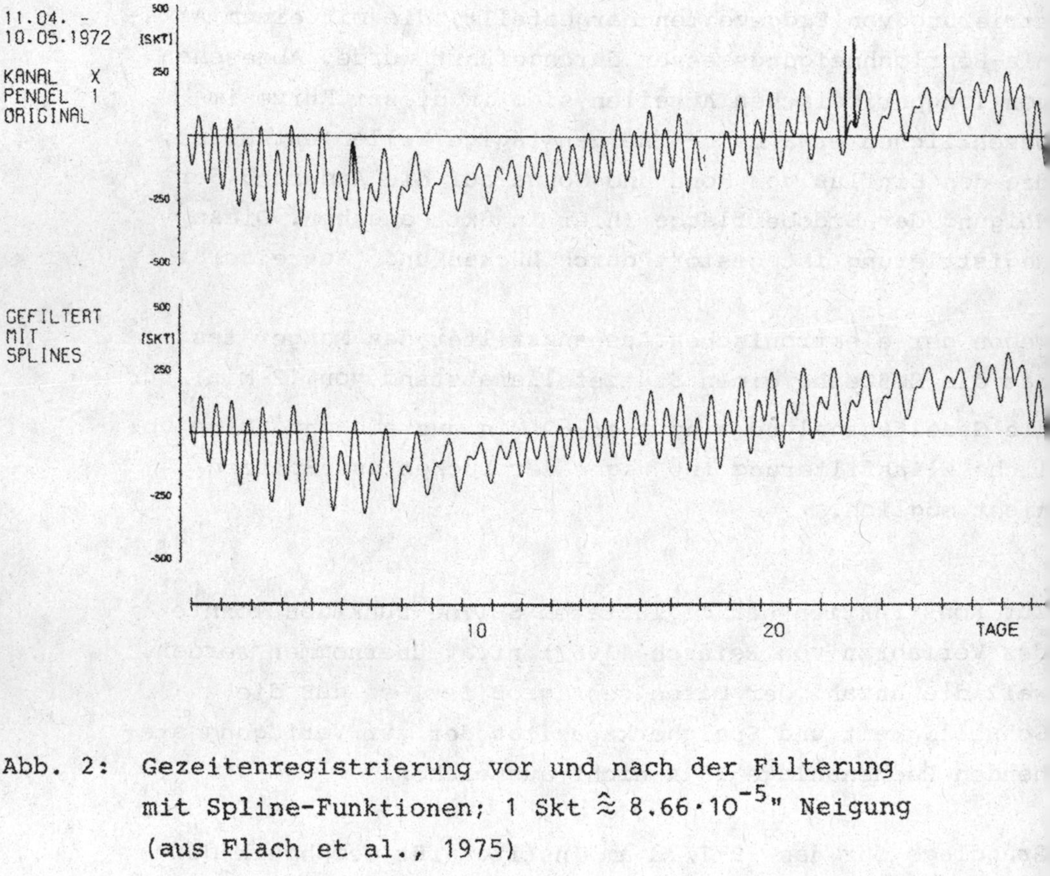

Abb. 2: Gezeitenregistrierung vor und nach der Filterung mit Spline-Funktionen; 1 Skt $\hat{\approx} 8.66\cdot10^{-5}$" Neigung (aus Flach et al., 1975)

$$(3.1) \quad \left\{ DMD^T + DG^{-1} D^T \right\} \vec{r} = D\vec{y}$$

gelöst werden. Die c_i ergeben sich aus

$$(3.2) \quad \vec{c} = \left[(-1)^k/(2k-1)!\right] D^T \vec{v}$$

Dabei bedeuten:

$\vec{y}$ = Vektor der Meßwerte y_i an den Knoten x_i

G = Diagonalmatrix mit $g_{ii} = w_i$ (Gewichte)

M = $\{m_{ij}\} = \left[(-1)^k/(2k-1)!\right] (x_i-x_j)_+^{2k-1}$

D = $\{d_{ij}\} = 1/l'(x_{i+j})$ mit $l(x) = (x-x_i)(x-x_{i+1})\ldots$
$\ldots (x-x_{i+k})$

$i=1,\ldots,n-k;\ j=0,\ldots,k$

(Koeffizienten der k-ten dividierten Differenz)

D^T = D transponiert

Die Matrix $\{DMD^T + DG^{-1} D^T\}$ ist eine symmetrische Bandmatrix der Bandbreite (2k+1).

Die Koeffizienten des Polynoms $\vec{p}_{k-1}(x)$ werden bestimmt aus

$$\vec{y} - (-1)^k (2k-1)!\, G^{-1} \vec{c} = M\vec{c} + \vec{p}_{k-1} = \vec{s}$$

mit

$\vec{p}_{k-1}(x) := (p_{k-1}(x_1),\ldots,p_{k-1}(x_n))$

$\vec{s} := (s(x_1),\ldots,s(x_n))$

Besondere Erwähnung verdient der Fall, wo die Stützstellen äquidistant verteilt sind. Dies ist im vorliegenden Problem der Fall. Die praktische Rechnung wird dann einfacher, weil die Matrizen D und DMD^T durch eine Zeile bereits vollständig bestimmt sind. Das zugehörige Programm ist in FORTRAN IV geschrieben und wird an der TR 4 des Rechenzentrums der TU Clausthal eingesetzt.

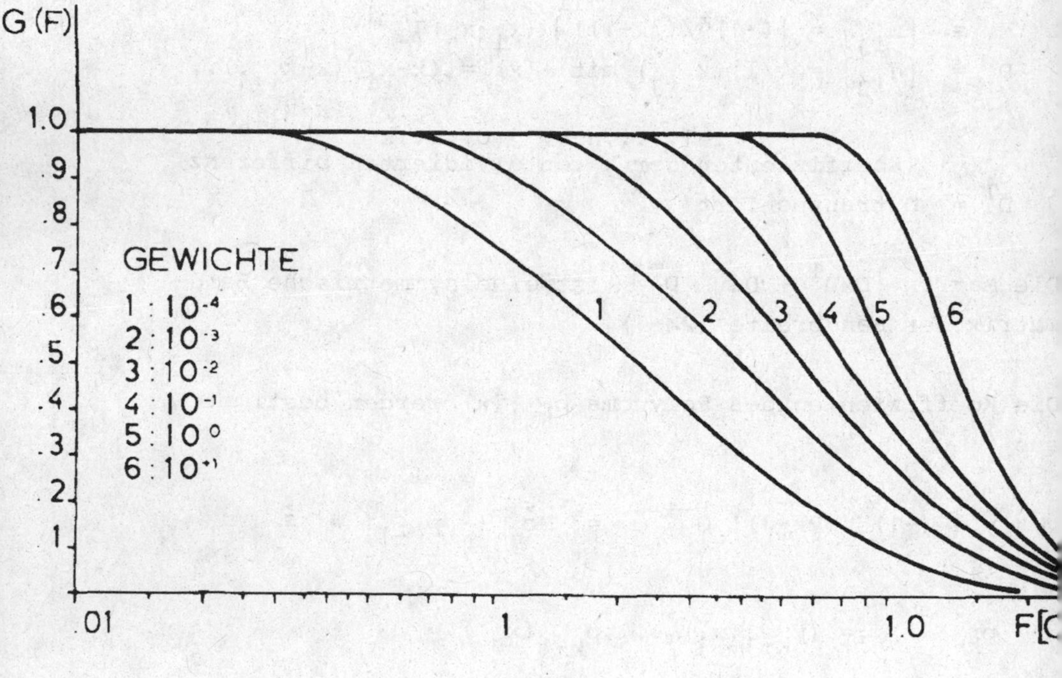

Abb. 3: Filtercharakteristik der Spline-Funktion in Abhängigkeit vom Gewicht (empirisch ermittelt) (aus Flach et al., 1975)

Für die Anwendung des Rechenprogramms werden in der Meßkurve die Fehlstellen mit einem sehr kleinen Gewicht belegt, so daß an dieser Stelle in Anlehnung an die benachbarten Stützstellen interpoliert wird. Das Resultat ist in der unteren Spur von Abb. 2 dargestellt: Die Lücken und Spitzen sind verschwunden. Zur Verdeutlichung der Glättung ist in Abb. 3 die Abhängigkeit der Filtercharakteristik vom Gewicht der Spline-Funktion gegeben: Je kleiner das Gewicht, desto flacher die Flanke. Im vorliegenden Fall wurde 10^{-2} gewählt, um gerade noch über den sehr kleinen dritteltägigen Gezeiten zu liegen.

Neben dem Herausfiltern von kurzperiodischen Frequenzanteilen wird eine weitere Stelle hinter dem Komma zugelassen, um auf diese Weise den Rundungseffekt des Digitalvoltmeters ($\pm$ 0.5 Skt) auszugleichen. Weitere Einzelheiten sind der Arbeit Flach, Jentzsch, Rosenbach (1975) zu entnehmen.

4. Beispiele für die Interpolation

4.1 Eichung von Gezeitenregistrierungen

Die vorliegenden Gezeitenregistrierungen werden i.a. dadurch geeicht, daß automatisch im Meßgerät eine wohldefinierte Veränderung vorgenommen wird, die eine entsprechende Verschiebung der Meßkurve zur Folge hat. Neben der Ermittlung dieser Differenz, des Eichfaktors, ist die Untersuchung der Langzeitstabilität dieses Wertes von besonderer Bedeutung. Der Eichfaktor soll genauer als $\pm$ 0.2 % sein. In Abb. 4 ist eine schematische Darstellung des Eichvorgangs gegeben.

Durch die Einlaufzeit der elektronischen Filter von 45 Min ist keine quasi-synchrone Ausschlagsmessung möglich: Der Punktabstand auf der ausgelenkten wie auch auf der verbliebenen Kurve ist jetzt 120 Min. Die Interpolation des jeweils "fehlenden" Wertes ermöglicht die Differenzbildung und somit die Ermittlung des Eichfaktors und seines Fehlers.

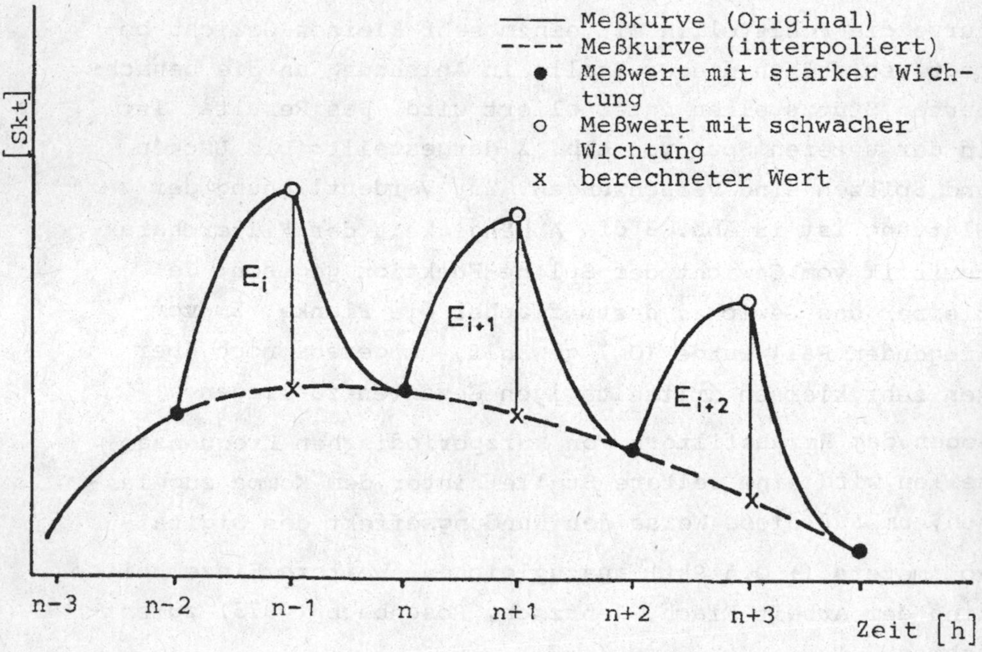

Abb. 4: Schematische Darstellung der Auswertung der Eichung, E_i: berechnete Eichsprünge (aus Flach et al., 1975)

Versuche, dieses Problem mit Ausgleichspolynomen verschiedener Grade zu lösen, zeigten, daß in diesem Falle die Güte des Eichfaktors mehr von der verwendeten Methode als vom Meßgerät abhängt. Dies ist verständlich, da bei den vorherrschenden halbtägigen Wellen ein Polynom 4-ten Grades schon die Werte von fast einer ganzen Schwingung (also zwei Wendepunkte) approximieren und gleichzeitig auf kürzerperiodische Anteile Rücksicht nehmen soll. Es mußten deshalb z.T. mehr als die Hälfte der berechneten Differenzen aufgrund mehr oder weniger subjektiver Kriterien aussortiert werden (Flach et al., 1975).

Die Lösung dieses Problems durch Interpolation nur der nichtausgelenkten Kurve mit Hilfe desselben Programms, wie in 3.2 beschrieben, führt zu einem überzeugenden Ergebnis:

über 90 % aller Differenzen können verwendet werden; nur offensichtliche "Ausreißer" fallen weg. Die Genauigkeit des Eichfaktors verbessert sich noch (Flach et al., 1975).

Damit ist insgesamt die Vertrauenswürdigkeit des Ergebnisses gestiegen.

4.2 Berechnung des gravimetrischen Regionalfeldes

In diesem Beispiel werden auch Werte interpolativ gewonnen, die durch Messung nicht zu erhalten sind: Es geht um die Ermittlung des gravimetrischen Regionalfeldes im Bereich der Alpen.

Dieses Regionalfeld ist ein Maß für die großräumige Struktur der Erdkruste; Masseninhomogenitäten bis zur Tiefe von 60 - 100 km werden erfaßt. Ist das Regionalfeld bekannt,so können interessierende lokale, flachere Anomalien aus der Schweremessung durch Differenzenbildung ermittelt werden.

In Abb. 5 ist ein Meßprofil über den Hauptkamm der Alpen (hier: Hohe Tauern) dargestellt: In den Tälern nördlich (Krimmler Achental, Obersulzbachtal) und südlich (Dorfertal) wurde gemessen, dazwischen befindet sich eine größere Lücke wegen des dortigen unzugänglichen Gletschergebietes und Hochgebirges.

Es galt nun, ein Regionalfeld so zu bestimmen, daß es bekannte geologische Fakten und die erhaltenen Meßpunkte berücksichtigt: Das Schwereminimum muß direkt unter dem Hauptkamm liegen.

Diese Bestimmung wurde wieder mit dem Programm nach Reinsch (1967) vorgenommen. Variationen von DY und S führten schließlich zu der gewünschten Kurve.

Durch den Vorteil der beliebigen Anpassung ist eine völlig neue Interpretationsmöglichkeit der Messungen gegeben,die mit Hilfe einfacher Ausgleichspolynome nicht vorhanden war.

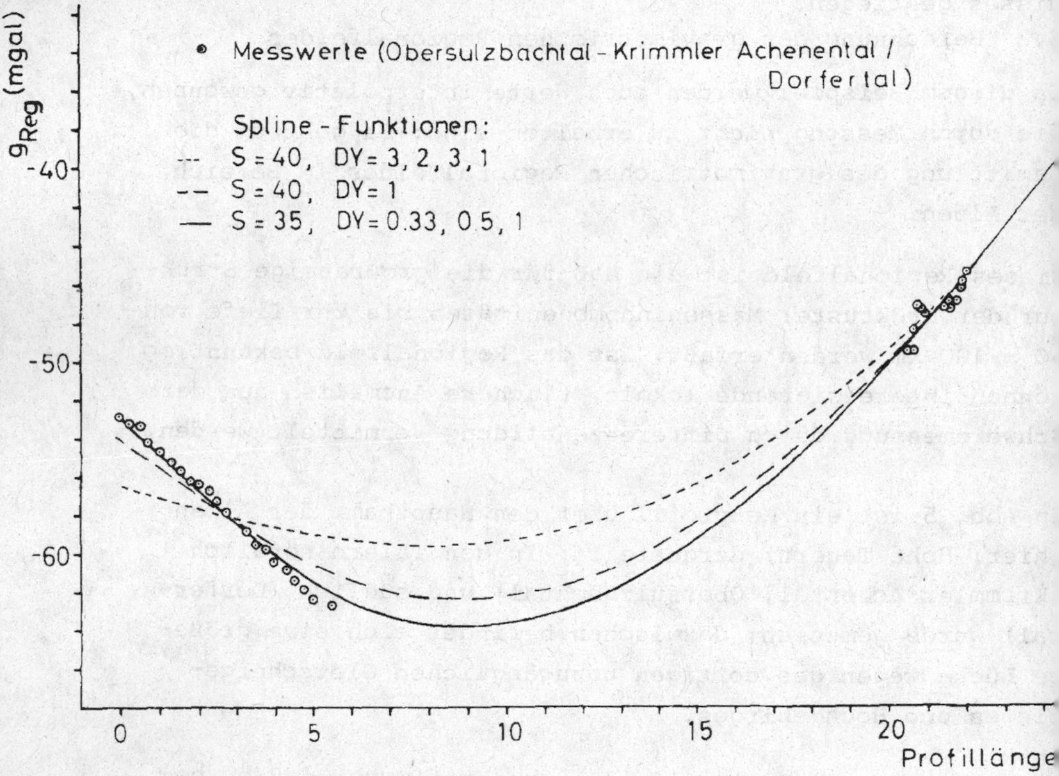

Abb. 5: Gravimetrisches Meßprofil angenähert durch verschiedene Spline-Funktionen (die DY wurden entlang der Meßkurve variiert, z.B. bei S=35: 0-2.5 km DY=0.33, 2.5-5.5 km DY=0.5, Punkte über 20 km: DY=1.0)

5. Bearbeitung flächenhafter Meßdaten

Zum Schluß ein Beispiel aus der Magnetik, in dem flächenhafte Meßdaten verarbeitet werden:

Bhattacharyya (1969) hat Auswertungen von magnetischen Messungen im nördlichen Kanada mit Hilfe von zweidimensionalen kubischen Splines vorgenommen. Großräumige, flächenhafte magnetische Messungen werden vom Flugzeug aus durch Überfliegen des Geländes auf nebeneinanderliegenden Profilen durchgeführt. Die Meßpunkte sind aber trotz Kurs- und Geschwindigkeitskorrekturen i.a. nicht gleichabständig.

Ziel der Arbeiten ist die Berechnung von Amplituden und Phasenspektren magnetischer Anomalien in vertikaler und horizontaler Richtung. Dies erfordert demnach eine Interpolationsmethode, die numerisch leicht zu handhaben ist und genaue horizontale und vertikale Ableitungen ermöglicht.

Zunächst berechnete Bhattacharyya mit Hilfe eindimensionaler kubischer Spline-Funktionen Projektionen der Meßkurven auf ein rechtwinkeliges Netz gleichabständiger Meßpunkte. Für diese so berechneten Daten wurde eine zweidimensionale kubische Spline-Funktion ermittelt und die Ableitungen in die gewünschten Richtungen berechnet.

Die Funktionen in den einzelnen Richtungen wurden nach Fourier transformiert. Dabei zeigten Tests, daß das Amplitudenspektrum aus den Splines dem wahren Spektrum bemerkenswert ähnlich ist, besonders bei relativ großen Wellenlängen. Dies erklärt auch die guten Eigenschaften der Ableitungen.

6. Zusammenfassung

Die Spline-Funktionen haben in den letzten Jahren in der Geophysik eine immer größere Bedeutung erlangt, und es ist zu erwarten, daß die erfolgreiche Anwendung gerade der zweidimensionalen Splines weitere Anwendungsgebiete finden wird, z.B. etwa zur Darstellung der Meeresgezeiten der Nordsee. Der Vorteil der Spline-Funktionen liegt hier darin, daß zur Untersuchung eines bestimmten Bereiches auch nur Daten dieser Region benötigt werden, und nicht, wie bei der Verwendung von Kugelflächenfunktionen, die Pegelstände der Gezeiten der ganze Erde bekannt sein müssen.

Für die Glättung und Interpolation von Meßkurven werden im Institut für Geophysik der Technischen Universität Clausthal seit 1971/72 Spline-Funktionen verwendet. Im Fall geringer Datenmengen nicht gleichabständiger Meßdaten wird eine Prozedur nach Reinsch (1967) erfolgreich angewendet. Die diesbezüglichen Abbildungen 1,5 stammen aus der Abteilung Gravimetrie und wurden von Dipl.-Geophys. H.J. Götze und W. Schöler zur Verfügung gestellt. Für die Bearbeitung gleichabständiger, großer Datenfelder wurde ein spezieller Formalismus entwickelt, so daß die Anwendung einer Alias-Filterung entspricht.

7. Literatur

Anselone, P.M. und P.J. Laurent: A general method for the construction of interpolating or smoothing spline functions. Num. Math. 12 (1968), 66-82

Bhattacharyya, B.K.: Bicubic spline interpolation as a method for treatment of potential field data. Geophysics, 34, 3 (1969), 402-423

Böhmer, K.: Spline-Funktionen. Stuttgart, Teubner 1974

Flach, D., G. Jentzsch und O. Rosenbach: Interpolation and smoothing of tidal records by spline functions. Deutsche Geodät. Komm., Reihe B, 1975

Greville, T.N.E.: Theory and application of spline functions. New York-London, Academic Press, 1969

Lyche, T. and L.L. Schumaker: Computation of smoothing and interpolating natural splines via local bases. SIAM J. Num. Anal. 10, No. 6 (1973), 1027-1038

Reinsch, C.H.: Smoothing by spline functions. Numer. Math. 10 (1967), 177-183

Reinsch, C.H.: Smoothing by spline functions II. Numer. Math. 16 (1971), 451-454

Dipl.-Geophys. G. Jentzsch
Prof. Dr.-Ing. O. Rosenbach
Institut für Geophysik der
Technischen Universität
3392 Clausthal-Zellerfeld
Postfach 230

Dipl.-Math. G. Lange
Rechenzentrum der Technischen Universität
3392 Clausthal-Zellerfeld
Erzstraße 51

Direct and Inverse Theorems for Best Approximation by Λ-Splines

by H. Johnen and K. Scherer

Introduction. Let g be a realvalued function on the intervall $[a,b]$, $Dg(x)$ its derivative evaluated at x, $C[a,b] = C^0[a,b]$ the space of continuous functions on $[a,b]$, and for a positive integer j $C^j[a,b]$ the space of functions $g \in C^{j-1}[a,b]$ such that $Dg \in C^{j-1}[a,b]$. For a nonnegative integer r and $1 \leq p \leq \infty$ the Sobolev space $W_p^r(a,b)$ provided with the norm $||f||_p^r = \sum_{j=0}^{r} ||D^j f||_p$ consists of functions g having absolutely continuous derivatives up to the order r-1 with $D^r g \in L_p(a,b)$. The linear differential operator $\Lambda : W_p^r(a,b) \to$ $\to L_p(a,b)$ defined by

$$(1) \qquad \Lambda = D^n + \sum_{j=0}^{n-1} a_j D^j$$

with $a_j \in C^j[a,b]$, $i = 0,1,\dots,n-1$, has the n-dimensional kernel $N_\Lambda \subset C^n[a,b]$.

Let be given a partition

$$(2) \qquad \Delta: \quad a = x_0 < x_1 < \dots x_{N-1} < x_N = b$$

of $[a,b]$ with $\overline{\Delta} = \max_{1 \leq i \leq N} (x_i - x_{i-1})$, $\underline{\Delta} = \min_{1 \leq i \leq N} (x_i - x_{i-1})$.

Denoting by $N_\Lambda([a,b]\setminus\Delta)$ the set of all functions g such that $\Lambda g(x) = 0$ for all $x \in [a,b]\setminus\Delta$, the set of Λ-splines $Sp(\Lambda,\Delta,m)$ for $0 \leq m \leq n-1$ is defined by

$$(3) \qquad Sp(\Lambda,\Delta,m) = W_\infty^m(a,b) \cap N_\Lambda([a,b]\setminus\Delta).$$

This paper is concerned with the best approximation

$$(4) \qquad E_\Delta^{(\Lambda,m)}(f;p) = \inf \{||f-s||_p;\ s \in Sp(\Lambda,\Delta,m)\}$$

of $f \in L_p(a,b)$ by Λ-splines of the class $Sp(\Lambda,\Delta,m)$, in particular its behaviour when $\overline{\Delta}$ or $\underline{\Delta}$ tends to zero. In this connection we regard direct theorems, i.e. estimates of $E_\Delta^{\Lambda,m}(f;p)$ from above in terms of smoothness conditions of f, and corresponding inverse theorems where smoothness conditions of f result from the behaviour of the best approximation when the partitions are getting finer. In the first case we base on results in [4],[8], in the second case we generalize those in [6],[7].

Stating the main results of this paper in this section, we collect in the second section the Lemmas needed for the proofs of the main theorems in the third seetion. A fourth section is devoted to several conclusions concerning approximation by Λ-splines. To every differential operator Λ of type (1) we associate the one-parameter family of continuous semi-norms on $L_p(a,b)$ by

(5) $K(t,f;p,\Lambda) = \inf \{ ||f-g||_p + t^n ||\Lambda g||_p ; g \in W_p^n(a,b) \}$

for $0 < t < \infty$, $f \in L_p(a,b)$.

Then one has

<u>Theorem 1</u>: For every partition Δ of $[a,b]$ there exists a constant $c > 0$, depending only on Λ,p, and the quotient $\overline{\Delta}/\underline{\Delta}$, and indepent of the individual partition Δ, such that for each $f \in L_p(a,b)$, $1 \leq p \leq \infty$,

(6) $E_\Delta^{(\Lambda,n-1)}(f;p) \leq c\ K(\overline{\Delta}, f;p,\Lambda)$.

There connot hold an inverse estimate of the form $K(\underline{\Delta},f;p,\Lambda) \leq c\ E_\Delta^{(\Lambda,n-1)}(f;p)$ for all $f \in L_p(a,b)$ where $c > 0$ is independent of f. Otherwise, by Lemma 3 below, $f \in Sp(\Lambda,\Delta,n-1)$ would imply $f \in N_\Lambda$, an obvious contradiction. For this reason one needs at least two different best approximations $E_\Delta^{(\Lambda,m)}(f;p)$ with different partitions to estimate $K(t,f;p,\Lambda)$ from above. Moreover the connection between these partitions will be of importance. In particular we introduce the following definitions (cf.[6]) for a sequence $\{\Delta_k\}_{k=1}^\infty$ of partitions of $[a,b]$: We call these partitions mixed (stongly mixed, respectively) if there exists a constant $d > 0$

and an integer k_o (as well as an integer r) such that for every $k \geq k_o$ there holds $\sup_{l \geq k} \operatorname{dist}(t, \Delta_l) \geq d\, \underline{\Delta}_k$ ($\max_{1 \leq s \leq r} \operatorname{dist}(t, \Delta_{l_s})$ $\geq d\, \underline{\Delta}_k$ for some integers $l_1, \ldots, l_r \geq k$, respectively).

An equivalent formulation for $\{\Delta_k\}_{k=1}^{\infty}$ being mixed is

(7) $\liminf_k (1/\underline{\Delta}_k) \min_{t \in \Delta_k} \sup_{l \geq k} \operatorname{dist}(t, \Delta_l) \geq d > 0.$

One now has

<u>Theorem 2:</u> Let $\{\Delta_k\}_{k=1}^{\infty}$ be a strongly mixed sequence of partitions of $[a,b]$. Then there exists a constant $c > 0$ such that for all $f \in L_p(a,$ and all $k \geq k_o$

(8) $K(\overline{\Delta}_k, f; p, \Lambda) \leq c \sup_{l \geq k} E^{(\Lambda,0)}_{\Delta_l}(f;p).$

In case $p = \infty$ the assumption "strongly mixed" can be weakened to "mixed".

One example for a sequence of strongly mixed partitions is given by the sequence $\{\Delta_k^*\}_{k=1}^{\infty}$ where

$$\Delta^*_{2k-1} = \{x_i : x_i = a + i(b-a)/k \ 0 \leq i \leq k\}$$
$$\Delta^*_{2k} = \{\overline{x}_i : \overline{x}_o = a,\ \overline{x}_{k+1} = b,\ \overline{x}_i = x_i - (b-a)/2k,\ 1 \leq i \leq k\}.$$

Concerning the subsequence $\{\Delta^*_{2k-1}\}_{k=1}^{\infty}$ of equidistant partitions we could only show that it is mixed so that in this simplest case Theorem 2 holds only for $p = \infty$. However, if in addition the coefficients of Λ are constant, by a somewhat different argument (compare [7]) one can establish even the stronger result

(9) $K(\overline{\Delta}_{2k-1}, f; p, \Lambda) \leq c\,[E^{(\Lambda,0)}_{\Delta^*_{2k-1}}(f;p) + E^{(\Lambda,0)}_{\Delta^*_{2k+1}}(f;p]$

The inequalities (8), (9) together with (6) imply saturation for the best approximation by Λ-sylines with (strongly) mixed partitions. This phenomenon cannot occur in the classical case of best approximation by increasing sequences of linear subspaces in view of a familiar theorem of Bernstein (see [10]). In contrast to mixed partitions, taking a sequence of nested partitions $\{\Delta_k\}_{k=1}^{\infty}$, i.e. $\Delta_k \subset \Delta_{k+1}$, one obtains an increasing sequence of subspaces $Sp(\Lambda, \Delta_k, m)$.

Then best approximation by splines fits in the classical scheme and one has in contrast to Theorem 2

<u>Theorem 3</u>: Let $\{\Delta_K\}_{K=1}^{\infty}$ a sequence of nested partitions of the intervall $[a,b]$ with $\sup \underline{\Delta}_K/\underline{\Delta}_{K+1} < \infty$. To every $1 \leq p \leq \infty$ there exists a constant $c > 0$ such that for all $f \in L_p(a,b)$ and all $0 < t < \infty$

$$(10) \qquad K(t,f;p,\Lambda) \leq c \max\,(1,t^n\underline{\Delta}_K^{-n}) \sum_{j=m}^{n-1} \underline{\Delta}_K^j \,[\,||f||_p + \sum_{l=1}^{K} \underline{\Delta}_l^{-j}\, E_{\Delta_l}^{(\Lambda,m)}(f;p)].$$

<u>Auxilary Results.</u> In this section we prove a series of auxiliary lemmas necessary for the proof of the above inequalities.

<u>Lemma 1</u>: Let Λ be the differential operator (1) and $g \in W_p^n(a,b)$, $1 \leq p \leq \infty$.

a) There exists a constand $c > 0$, depending only on Λ and b-a such that

$$(11) \qquad ||\Lambda g||_p \leq c\,[\,||g||_p + ||D^n g||_p].$$

b) There exists a constant $c > 0$, depending only on Λ and n, such that for $0 < h \leq \delta$, where $\delta = \min\,(1,(2\sum_{j=1}^{n-1}||a_j||_\infty)^{-1})$,

$$(12) \qquad ||D^n g||_p \leq 2\,||\Lambda g||_p + c\,h^{-n+1}||g||_p$$

Proof: By familiar arguments (compare [2, p.19]) one first establishes the inequality

$$(13) \qquad ||D^j g||_p \leq c\,h^{-j}||g||_p + h^{n-j}||D^n g||_p \quad (0 \leq j \leq n-1)$$

for all h with $0 < h \leq b-a$, where the constant c depends only on j and n. From this, taking h = b-a, part a) immediately follows in view of

$$||\Lambda g||_p \leq ||D^n g||_p + \sum_{j=1}^{n-1} ||a_j||_\infty ||D^j g||_p$$

To prove part b) observe that by (13) for $0 < h \leq \delta \leq 1$

$$\begin{aligned} &||D^n g||_p \\ &\leq ||\Lambda g||_p + \sum_{j=1}^{n-1}||a_j||_\infty\{c\,h^{-j}||g||_p + h^{n-j}||D^n g||_p\} \\ &\leq ||\Lambda g||_p + h^{-n+1}||g||_p\; c \sum_{j=1}^{n-1} ||a_j||_\infty + \tfrac{1}{2}||D^n g||_p. \end{aligned}$$

An immediate consequence of Lemma 1 is

Lemma 2: There exist constants $c_1 > 0$, $c_2 > 0$ depending only on Λ and b-a such that for all $0 < t \leq 1$ and $f \in L_p(a,b)$

(14) $K(t,f;p,D^n) \leq c_1[K(t,f;p,\Lambda) + t^n||f||_p]$,

(15) $K(t,f;p,\Lambda) \leq c_2[K(t,f;p,D^n) + t^n||f||_p]$.

Another property of the K-functional now can be proved.

Lemma 3: Let Λ be the differential operator in (1) and $f \in L_p(a,b)$, $1 \leq p \leq \infty$. The function equals almost everywhere to a function in N_Λ if and only if

(16) $\liminf_{t\to 0} t^{-n} K(t,f;p,\Lambda) = 0.$

Proof: For arbitrary $g \in W_p^n(a,b)$ and $0 < t \leq 1$ holds

$$K(1,f;p,\Lambda) \leq t^{-n}[||f-g||_p + t^n||\Lambda g||_p],$$

and hence

$$K(1,f;p,\Lambda) \leq t^{-n} K(t,f;p,\Lambda).$$

Therefore (16) implies $K(1,f;p,\Lambda) = 0$ and thus there exists a sequence $\{g_n\}_{n=1}^{\infty} \subset W_p^n(a,b)$ with

(17) $\lim_n ||f-g_n||_p = \lim_n ||\Lambda g_n||_p = 0.$

By (12) $||g||_p + ||\Lambda g||_p$ is an equivalent norm on $W_p^n(a,b)$, so that (17) tells us that $\{g_n\}_{n=1}^{\infty}$ is a Cauchy sequence in $W_p^n(a,b)$. Hence it converges to a $g \in W_p^n(a,b)$ which by (17) satisfies $||f-g||_p = 0$ and $||\Lambda g||_p = 0$. This concludes the "only if" part of the proof. The "if" part is trivial.

For $x \in [a,b]$, $h > 0$ such that $x + h \in [a,b]$ and $f \in L_p(a,b)$, $1 \leq p \leq \infty$, let $||f||_{x,h,p}$ denote the L_p-norm of f on the subinterval $[x,x+h]$ of $[a,b]$. With the aid of Lemma 1 one gets also

Lemma 4. Let Λ be the differential operator in (1) and δ as in part b) of Lemma 1. There exists a constant $c > 0$, depending only on Λ and j,n with $0 \leq j \leq n$ such that for $1 \leq p,q \leq \infty$, $0 < h \leq \delta$ and all $x \in [a,b]$ with $x + h \in [a,b]$ there holds

$$(18) \qquad ||D^j u||_{x,h,p} \leq c\, h^{-j+(1/p)-(1/q)} ||u||_{x,h,q} \qquad (u \in N_\Lambda)$$

Proof: We first consider the case $p = q$, $1 \leq p \leq \infty$. Then in view of $\Lambda u = 0$ by (13) and then (12) it follows that

$$(19) \qquad ||D^j u||_{x,h,p} \leq c_0 h^{-j} ||u||_p + h^{n-j} ||D^n u||_p \leq c\, h^{-j} ||u||_p .$$

In case $1 \leq p < q \leq \infty$ we use in addition Hölder's inequality giving

$$(20) \qquad ||u|| x,h,p \leq h^{-(1/q)+1/p} ||u||_{x,h,q} .$$

In case $1 \leq q < p \leq \infty$ one defines the function $\overline{u}$ for $a \leq t \leq b$ by

$$\overline{u}(t) = u(t) - \text{sign}\, (u(\xi) h^{-1/q} ||u||_{x,h,q}$$

where $\xi \in [x,x+h]$ is determined by the mean value theorem such that $|u(\xi)| = h^{-1/q} ||u||_{x,h,q}$. Since $\overline{u}(\xi) = 0$ one concludes by Hölder's inequality and (19) in case $j = 1$ that

$$||\overline{u}||_{x,h,q} = || \int_x^\xi u'(t)dt ||_{x,h,p} \leq || h^{1-1/q} ||u'||_{x,h,q} ||_{x,h,p} \leq c\, h^{-(1/q)+(1/p)} ||u||_{x,h,q} .$$

This together with

$$||u||_{x,h,p} \leq ||\overline{u}||_{x,h,p} + || h^{-1/q} ||u||_{x,h.q} ||_{x,h.p}$$

shows (18) also in the remaining case $1 \leq q < p \leq \infty$.

Though we need only inequality (18) in the following we formulate here as an easy consequence of Lemma 4 a Bernstein-type inequality for Λ-splines.

Corollary: Let Λ, j, n, p, q and δ be as in Lemma 4. There exists a constant $c > 0$ depending only on Λ, n, and j such that for every partition $\overline{\Delta} \leq \delta$ there holds for $s \in Sp(\Lambda, \Delta, 0)$

(21) $$||D^j s||_p \leq c\, \underline{\Delta}^{-j+\min(0,1/p-1/q)} ||s||_q$$

For the proof note that $D^j s$ is defined for all $x \in [a,b] \setminus \Delta$ which gives a natural meaning for $||D^j s||_p$. Setting $h_i = x_{i+1} - x_i$ it follows from (18) that

$$||D^j s||_p = \{ \sum_{i=0}^{N-1} ||D^j s||^p_{x_i, h_i, p} \}^{1/p}$$

$$\leq \{ \sum_{i=0}^{N-1} h_i^{-j + (1/p) - (1/q)} ||s||^p_{x_i, h_i, q} \}^{1/p}$$

$$\leq \underline{\Delta}^{-j+(1/p)-(1/q)} \{ \sum_{i=0}^{N-1} ||s||^p_{x_i, h_i, q} \}^{1/p}$$

Then it is easy to see that the last sum can be bounded by $||s||_q$ in case $q \leq p$ and by $N^{(1/p)-(1/q)} ||s||_q \leq [(b-a)\underline{\Delta}]^{(1/p)-(1/q)} ||s||_q$ in case $q \geq p$. This establishes (21).

Now let P_{2n-1} be the polynome of order $2n-1$ with $D^j P_{2n-1}(0) = a_j$, $D^j P_{2n-1}(h) = b_j$ $0 \leq j \leq n-1$, where $h > 0$. One concludes directly by the explicit representation of P_{2n-1} that there exists a constant $c = c(n)$, depending only on the integer n such that $||P_{2n-1}||_{0,h,\infty} \leq$ $\leq c \max\limits_{0 \leq k \leq n-1} (h^k \max(|a_k|, |b_k|))$. Hence a consequence of Lemma 4 is (compare [9])

Lemma 5. Let $h > 0$ and P_{2n-1} the above interpolation polynomial with $D^j P_{2n-1}(a) = a_j$ and $D^j P_{2n-1}(h) = b_j$, $0 \leq j \leq n$. There exists a constant $c > 0$, depending only on j and n such that for all $1 \leq p \leq \infty$ and all $h > 0$

(22) $$||D^j P_{2n-1}||_{0,h,p} \leq c\, h^{-j+\frac{1}{p}} \max_{0 \leq k \leq n-1} (h^k \max(|a_k|, |b_k|)).$$

The last Lemma allows to give an estimate of the K-functional applied to a spline $s \in Sp(\Lambda, \Delta, 0)$ in terms of the jumps of s and its derivatives. With the notation $[s]^j_i = D^j s\,(x_i+) - D^j s(x_i-)$,

$0 \le j \le n$, where $x_i \in (a,b), 1 \le i \le N-1$, are the knots of the partition Δ one obtains

Lemma 6. Let Λ be as in (1) and Δ be a partition of $[a,b]$ with $\underline{\Delta} \le 1$. There exists a constant c, depending only on Λ and n such that for all $s \in Sp(\Lambda,\Delta,0)$ and $1 \le p \le \infty$

$$(23)\quad K(t,s;p,\Lambda) \le c \max(\Lambda, t^n \underline{\Delta}^{-n})\, \underline{\Delta}^{1/p} \cdot \left(\sum_{i=1}^{N-1} \max_{0\le j\le n-1} |\underline{\Delta}^j [s]_i^j|^p\right)^{1/p}$$

Proof. To the spline $s \in Sp(\Lambda,\Delta,0)$ we assosiate the function s^*, defined by

$$s^*(x) = \begin{cases} s(x), & x \in [x_{i-1}, x_i - \frac{1}{2}\underline{\Delta}),\ 1 \le i \le N-1 \\ s(x) + P_{2n-1,i}(x), & x \in [x_i - \frac{1}{2}\underline{\Delta}, x_i),\ 1 \le i \le N-1 \\ s(x) & x \in [x_{N-1}, b] \end{cases}$$

where $P_{2n-1,i}$ is the interpolation polynomial satisfying

$$D^j P(x_i - \tfrac{1}{2}\underline{\Delta}) = 0\ ,\ D^j P(x_i) = [s]_i^j\ ,\ 0 \le j \le n-1.$$

Clearly s^* is a function in $W_p^n(a,b)$ and

$$(24)\quad K(t,s;p,\Lambda) \le ||s-s^*||_p + t^n ||\Lambda s^*||_p$$

Now for the rest of the proof let $1 \le p < \infty$. For $p = \infty$ the proof requires the usual modification. Because of the fact that $s(x) - s^*(x) = -P_{2n-1,i}(x)$ on the invalls $(x_i - \frac{1}{2}\underline{\Delta}, x_i), 1 \le i \le N-1$, and the difference $s(x)-s^*(x)$ vanishes outside of these, one obtains by inequality (22) with $x - \frac{1}{2}\underline{\Delta} = y_i$

$$(25)\quad ||s-s^*||_r = \left(\sum_{i=1}^{N-1} ||P_{2n-1,i}||^p_{y_i,\frac{1}{2}\underline{\Delta},p}\right)^{1/p}$$

$$\le c'(n)\, \underline{\Delta}^{1/p} \left(\sum_{i=1}^{N-1} \max_{0\le k\le n-1} |\underline{\Delta}^k [s]_i^k|^p\right)^{1/p} .$$

Since $\Lambda s^*(x)$ vanishes outside of the intervalls (y_i, x_i), $1 \le i \le N-1$, with exception of the knots x_i, and since $\Lambda s^*(x) = \Lambda P_{2n-1,i}(x)$ for $x \in (y_i, x_i)$, one deduces by Minkowski's inequality similarly from (22) ($a_n = 1$)

$$||\Lambda s^*||_p = \Big(\sum_{i=1}^{N-1} ||\Lambda p_{2n-1,i}||^p_{y_i,\underline{\Delta}/2,p}\Big)^{1/p}$$

$$\leq \sum_{l=0}^{n} ||a_l||_\infty \Big(\sum_{i=1}^{N-1} ||D^l p_{2n-1,i}||^p_{y_i,\underline{\Delta}/2,p}\Big)^{1/p}$$

$$\leq \sum_{l=0}^{n} ||a_l||_\infty \underline{\Delta}^{-l+1/p}\Big(\sum_{i=1}^{N-1} \max_{0\leq j\leq n-1} |\underline{\Delta}^j[s]^j_i|^p\Big)^{1/p}$$

$$\leq c\, \underline{\Delta}^{-n+1/p}\Big(\sum_{i=1}^{N-1} \max_{0\leq j\leq n-1} |\underline{\Delta}^j[s]^j_i|^p\Big)^{1/p}$$

Combining this with (25) one obtains the desired inequality (23).

Proof of the main theorems

Proof of Theorem 1: In case Λ is a so-called Polya-operator this theorem has already been established in [7]. Since the argument used there carries immediately over to the general case considered, we only give a brief account of these ideas.

Let $\{u_i\}_{i=1}^n$ be a basis of N_Λ and $\{u_j^*\}_{j=1}^n$ a such one of N_{Λ^*} where Λ^* is the (formally) adjoint of Λ. For each $g \in W_p^n(a,b)$, $1 \leq p \leq \infty$ we have the formula

$$g(x) = u_o(x) + \int_a^b \hat{v}(x,\xi)\, \Lambda g(\xi) d\xi$$

where $u_o \in N_\Lambda$ is determined by $D^j u_o(a) = D^j g(a)$, $0 \leq j \leq n-1$, and $\hat{v}(x,\xi) = \sum_{k=1}^{n} u_k^*(\xi) u_k(x)$ for $a \leq \xi \leq x \leq b$ and $\hat{v}(x,\xi) = 0$ for $a \leq x < \xi < b$. Following [4] a mapping $s_\Delta : W_p^n(a,b) \to Sp(\Lambda,\Delta,n-1)$ is constructed by setting

$$s_\Delta(g;x) = u_o(x) + \int_a^b \gamma(x,\xi) \Lambda g(\xi) d\xi$$

where

$$\gamma(x,\xi) = \hat{v}(x,\xi) - \sum_{k=0}^{n} \beta_k\, \hat{v}(x,t_k).$$

The numbers t_k $(k=1,\dots,n)$ are taken as those points of the partition Δ which are next to $\xi =: t_o$. In [4] it is shown that then for each ξ the numbers β_k can be chosen such that the ratios (β_k/β_o) as functions of ξ are bounded on (a,b) by a constant only depending on the mesh ratio $\overline{\Delta}/\underline{\Delta}$ and such that the function $\gamma(x,\xi) - \hat{v}(x,\xi)$ has support in $|x-\xi| \leq (n+1)\bar{\Delta}$. Since by Taylor's formula $|\hat{v}(x,y)|$ is seen to be bounded by $C'(x-y)_+^{n-1}$ one now easily concludes that

$$||g - S_\Delta(g)||_\infty \leq C\,\bar{\Delta}^n ||\Lambda g||_\infty ,$$
$$||g - S_\Delta(g)||_1 \leq C\,\bar{\Delta}^n ||\Lambda g||_1 .$$

By an argument of the theory of interpolation of Banach spaces from these inequalities follows also the validity for the norms $||\cdot||_p$, $1 \leq p \leq \infty$. Passing over to the best approximation $E_\Delta^{(\Lambda,n-1)}(f;p)$ again an "interpolation" argument establishes the desired result.

Proof of Theorem 2:

For $k \geq k_0$ let $s_k \in Sp\ (\Lambda,\Delta,0)$ be such that $||f-s_k||_p = E_{\Delta_k}^{(\Lambda,0)}(f;p)$. By Lemma 6 one obtains

$$K(\underline{\Delta}_k,f;p,\Lambda) \leq ||f-s_k||_p + K(\underline{\Delta}_k,s_k;p,\Lambda)$$
$$\leq E_{\Delta_k}^{(\Lambda,0)}(f;p) + c\,\underline{\Delta}_k^{1/p}\,(\sum_{i=1}^{N-1} \max_{0\leq j\leq n-1} |\underline{\Delta}_k^j[s_k]_i^j|^p)^{1/p}$$

where $[s_k]_i^j$ denotes the (possible) jump of s_k at the knots x_i of Δ_k. Now suppose $1 \leq p < \infty$ and let $\{\Delta_k\}_{k=1}^\infty$ be a sequence of strongly mixed partitions. Then for each x_i we find an integer $l=l(i) \geq k$ such that dist $(x_i,\Delta_l) \geq c\,\underline{\Delta}_k$. Let I_i be the intervall with center x_i and length $c\,\underline{\Delta}_k$. Then one has by Lemma 4

$$[s_k]_i^j \leq ||D^j(s_k-s_l)||_{I_i,\infty} \leq c\,\underline{\Delta}_k^{-j-1/p}||s_k-s_l||_{I_i,p}$$
$$\leq c\,\underline{\Delta}_k^{-j-1/p}\{||s_k-f||_{I_i,\infty} + ||s_l-f||_{I_i,p}\}$$

Summing up it follows that

$$(26)\qquad \underline{\Delta}_k^{1/p}(\sum_{i=1}^{N-1} \max_{0\leq j\leq n-1} |\underline{\Delta}_k^j\,[s_k]_i^j|^p)^{1/p} \leq$$
$$c\{E_{\Delta_k}^{(\Lambda,0)}(f;p) + (\sum_{i=1}^{N-1} ||s_{l(i)} - f||_{I_i,p}^p\} .$$

Now $l(i)$ can take only a fixed number r of different values l_s, $1 \leq s \leq r$. Therefore

$$(\sum_{i=1}^{N-1} ||s_{l(i)} - f||_{I_i,p}^p)^{1/p} \leq (\sum_{i=1}^{N-1} \sum_{s=1}^{r} ||S_{l_s} - f||_{I_i,p}^p)^{1/p}$$

$$\leq r \sup_{l\geq k} E^{(\Lambda,0)}_{\Delta_l}(f;p)$$

follows, giving together with (26) the assertion (8).

In case $p = \infty$ we simply proceed

$$\sup_i \|s_{l(i)} - f\|_{I_i,\infty} \leq \sup_i E^{(\Lambda,0)}_{\Delta_{l(i)}}(f;\infty)$$

$$\leq \sup_{l\geq k} E^{(\Lambda,0)}_{\Delta_l}(f;\infty),$$

using only the "mixed" property of the partitions Δ_l.

Proof of Theorem 3.

As in the proof of Theorem 2 let s_k be an element of best approximation of $Sp(\Lambda,\Delta_k,m)$ to $f \in L_p(a,b)$. Setting $\Delta_o = \{a,b\}$ one has $s_k = \sum_{l=1}^{k}(s_l - s_{l-1}) + s_o$. With this observation and Lemma 6 we deduce

$$(27)\qquad K(t,f;p,\Lambda) \leq \|f - s_k\|_p + c\max(1,t^n\underline{\Delta}_k^{-n})\underline{\Delta}_k^{1/p}\cdot$$

$$\cdot\Big\{\sum_{i=1}^{N-1}\max_{m\leq j\leq n-1}|\underline{\Delta}_k^j[s_k]_i^j|^p\Big\}^{1/p},$$

and

$$(28)\qquad \Big\{\sum_{i=1}^{N-1}\max_{m\leq j\leq n-1}|\underline{\Delta}_k^j[s_k]_i^j|^p\Big\}^{1/p}$$

$$\leq \sum_{j=m}^{n-1}\underline{\Delta}_k^j\sum_{l=1}^{k}\Big\{\sum_{i=1}^{N-1}|[s_l - s_{l-1}]_i^j|^p\Big\}^{1/p}$$

Since Δ_{l-1} is contained in Δ_l it follows as in the proof of the preceeding theorem that ($h = \underline{\Delta}_l/2$)

$$|[s_l-s_{l-1}]_i^j| \leq \|D^j(s_l-s_{l-1})\|_{x_i-h,2h,\infty}$$

$$\leq c\,\underline{\Delta}_l^{-j-\frac{1}{p}}\{\|s_l-f\|_{x_i-h,2h,p} + \|s_{l-1}-f\|_{x_i-h,2h,p}\}$$

Summing up, inserting the result in (28) and then in (27) gives

$$K(t,f;p,\Lambda) \leq E^{(\Lambda,m)}_{\Delta_k}(f;p)$$

$$+ c\max(1,t^n\underline{\Delta}_k^{-n})\underline{\Delta}_k^{1/p}\sum_{j=m}^{n-1}\underline{\Delta}_k^j\sum_{l=1}^{k}\underline{\Delta}_l^{-j-1/p}\{E^{(\Lambda,m)}_{\Delta_l}(f;p) + E^{(\Lambda,m)}_{\Delta_{l-1}}(f;p)\}$$

rom this inequality (10) of Theorem 3 immediately follows by bserving $E^{(\Lambda,m)}_{\Delta_o}(f;p) \leq ||f||_p$.

e remark that it is not possible to improve or to simplify the stimate (10) by dropping the index j = m or still some further ndices. A simple counterexample can be constructed in the case hen the sequence of partitions is obtained by successive bisection. et

$$f^*(x) = \begin{cases} u_1 \in N_\Lambda, & x \in [a(a+b)/2] \\ u_2 \in N_\Lambda, & x \in [(a+b)/2,b] \end{cases}$$

uch that f* is a function in $W^m_\infty(a,b)$, $0 \leq m \leq n-1$. Then the best pproximation of f* by splines of dess $Sp(\Lambda,\Delta,m)$ would be zero for ll partitions Δ of this sequence. An improved estimate (10) in the bove sense would then yield $K(t,f^*;p,\Lambda) = O(t^{m+1})$, $t \to 0$, thus by he remark following the corollary below $f^* \in Lip(m+1,n;p)$, which n turn, for $1 < p \leq \infty$, would imply that $f^{*(m)}$ is continuous(cf.[3]), iving a contradiction.

Equivalence Theorems

e now combine the previous results in order to characterize the rate f convergence of the best approximation by Λ-splines in terms of ecessary and sufficient. According to Theorems 2 and 3 we have to istinguish between sequences of (strongly) mixed and nested partiions.

Theorem 5: a) Suppose $1 \leq p \leq \infty$, $0 \leq m \leq n-1$, and let $\{\Delta_k\}_{k=1}^\infty$ be a sequence of strongly mixed (in case $p = \infty$ only mixed) partitions atisfying $\bar{\Delta}_k \geq \bar{\Delta}_{k+1}$, $\lim_{k\to\infty} \bar{\Delta}_k = 0$ and $\bar{\Delta}_k/\underline{\Delta}_k \leq M < \infty$. Then for each on-decreasing function $\varphi(t)$ on (0,1) with $\lim_{t\to 0} \varphi(t) = 0$ the folloing assertions are equivalent:

i) $E^{(\Lambda,m)}_{\Delta_k}(f;p) = O[\varphi(\bar{\Delta}_k)]$, $k \to \infty$,

ii) $K(t,f;p,\Lambda) = O[\varphi(t)]$, $t \to 0$.

b) Let $\{\Delta_k\}_{k=1}^{\infty}$ be a sequence of nested partitions where $\overline{\Delta}_k$, $\underline{\Delta}_k$ satisfies in addition to the properties of part a) $\overline{\Delta}_k/\underline{\Delta}_{k+1} \leq M^* < \infty$. Let $\varphi(t)$ be such as in part a) satisfying in addition

$$(29) \qquad \int_t^1 u^{-m-1}\varphi(u)du = O\,[t^{-m}\varphi(t)], \quad t \to 0.$$

Then the equivalence of assertions i), ii) remains true.

Concerning the proof we note that the implication ii) → i) is immediate by Theorem 1. The converse follows in part a) directly from Theorem 2, whereas in part b) we use Theorem 3, insert assertion i), transform the sum (over 1) into an integral and estimate this integral with the help of (29). Here we should remark that (29) implies the same relation for any j replacing m, where $j \geq m$.

Examples of sequences of partitions satisfying the conditions of part a) have bean given after Theorem 2; examples satisfying those of part b) are obtained by successive dividing of intervals. Concerning condition (29) we note that it implies $t^m = O[\varphi(t)]$, $t \to 0$, which yields the restriction $0 < \alpha < m$ in the particular case $\varphi(t) = t^{\alpha}$, $\alpha > 0$.

Using the results of Section 1 assertion ii) of Theorem 5 can be brought into connection with more familiar structural properties of f. We state this in the following

Corollary: a) If $t^n/\varphi(t) = O(1)$, $t \to 0$, assertion ii) of Theorem 5 is equivalent to

iii) $f \in \mathrm{Lip}(\varphi,n;p)$, i.e. $\omega_n(t;f)_p = O[\varphi(t)]$, $t \to 0$,

where

$$\omega_n(t;f)_p = \sup_{0<|h|<t} \Big\{ \int_{x,x+nh\in[a,b]} |\Delta_h^n f(x)|^p dx \Big\}^{1/p}$$

for $1 \leq p < \infty$, with the obious modification for $p = \infty$.

b) If $\liminf_{t \to 0} t^n/\varphi(t) = \infty$, then assertion ii) is equivalent to $f \in N_\Lambda$.

Concerning the proof of part a) we observe that by Lemma 2 assertion ii) here is equivalent to

iii) $K(t,f;p,D^n) = O[\varphi(t)], \ t \to 0,$

which by a result in [5] is equivalent to iii). Part b) follows from Lemma 3.

Note that in case $\varphi(t) = t^{\alpha}$, $0 < \alpha \leq n$, the spaces Lip $(\varphi,n;p)$ coincide with the familiar Lipschitz or Besov spaces Lip $(\alpha,n;p) = B_p^{\alpha,\infty}[a,b]$ (cf. [3]).

From the corollary above it follows in particular that for a sequence $\{\Delta_k\}_{k=1}^{\infty}$ of (strongly) mixed partitions there accurs saturation for the corresponding sequence of best approximations, i.e. $E_{\Delta_k}^{(\Lambda,m)}(f;p) = O(\overline{\Delta}_k^{\,n})$, $k \to \infty$, implies that $f \in N_{\Lambda}$ or that $E_{\Delta_k}^{(\Lambda,m)}(f;p)$ vanishes identically for all k. The (saturation) class of functions for which the rate of convergence is optimal, namely $O(\overline{\Delta}_k^{\,n})$, $k \to \infty$, is the class $\mathrm{Lip}(n,n;p) = B_p^{n,\infty}[a,b]$.

For the sequences of partitions in part b) of Theorem 5 there cannot occur saturation. Indeed, by a familiar theorem of Bernstein (cf.[1o, p. 4o]), since $Sp(\Lambda,\Delta_k,m) \subset Sp(\Lambda,\Delta_{k+1},m)$, we can find for each positive non-decreasing function φ on $(0,1)$ with $\lim_{x \to 0} \varphi(x) = 0$ a function $f \in L_p(a,b)$ such that $E_{\Delta_k}^{(\Lambda,m)}(f;p) = \varphi(1/k)$ for $k = 1,2,\ldots$. In particular, the rate of convergence can be arbitrarily fast. The question arises now how these functions can be characterized. The following theorem gives a negative result in this direction and can also be regarded as a weak form of a saturation theorem (see [1 ,p. 174] in the case of polynomial splines).

<u>Theorem 6</u>: If $f \in W_p^n(a,b)$, $p > 1$, and if $\{\Delta_k\}_{k=1}^{\infty}$ is a sequence of nested partitions with $\overline{\Delta}_k/\underline{\Delta}_k \leq M < \infty$, then $E_{\Delta_k}^{(\Lambda,m)}(f;p) = o(\overline{\Delta}_k^{\,n})$, $k \to \infty$, for some m, $0 \leq m \leq n-1$, implies $f \in N_{\Lambda}$.

Proof: Passing over to a suitable subsequence if necessary we may assume that in addition $\Delta_k/\Delta_{k+1} \le M' < \infty$ holds. Then we proceed as in the proof of Theorem 3 now writing $s_k = \sum_{l=k}^{r-1}(s_l - s_{l+1}) + s_r$. Then it follows that

$$K(\Delta_k, f; p, \Lambda) \le C \sum_{j=m}^{n-1} \Delta_k^j \sum_{l=k}^{r} \Delta_l^{-j} E_{\Delta_l}^{(\Lambda,m)}(f;p)$$
$$+ C \sum_{j=m}^{n-1} \Delta_k^{j+1/p} \{ \sum_{i=1}^{N-1} |[s_r]_i^j|^p \}^{1/p} .$$

We show that second term tends to zero for $r \to \infty$. To this end we introduce the operator $\tilde{\Lambda} = D\Lambda$ and $\tilde{s}_r \in Sp(\tilde{\Lambda}, \Delta_r, n)$ with $||f - \tilde{s}_r||_p = E_{\Delta_r}^{(\tilde{\Lambda},n)}(f;p)$. Then $[\tilde{s}_r]_i^j$ vanishes for $m \le j \le n-1$ and at the same time $\tilde{s}_r$ is a function which is piecewise in $N_{\tilde{\Lambda}}$. With a similar argument as in Theorem 1 (application of Lemma 4 to $\tilde{\Lambda}$) we then have

$$\{ \sum_{i=1}^{N-1} |[s_r]_i^j|^p \}^{1/p} = \{ \sum_{i=1}^{N-1} |[s_r - \tilde{s}_r]_i^j|^p \}^{1/p}$$
$$\le C\, \Delta_r^{-j-1/p} \{ E_{\Delta_r}^{(\Lambda,m)}(f;p) + E_{\Delta_r}^{(\tilde{\Lambda},n)}(f;p) \}$$

Theorem 5 in connection with its corollary then shows that for $f \in W_p^n(a,b)$, $p > 1$, the right-hand side tends to zero for $r \to \infty$. Hence we arrive at

$$K(\Delta_k, f; p, \Lambda) \le C \sum_{j=m}^{n-1} \Delta_k^j \sum_{l=k}^{\infty} \Delta_l^{-j} E_{\Delta_l}^{(\Lambda,m)}(f;p).$$

Now, assuming $E_{\Delta_l}^{(\Lambda,m)}(f;p) = o(\overline{\Delta}_l^n)$, $l \to \infty$, it follows that

$K(\Delta_k, f; p, \Lambda) = o(\Delta_k^n)$, $k \to \infty$, and hence by Lemma 3 the assertion.

Combining Theorem 3 and 6 we conclude that the class of non-trivial functions $f (\notin N_\Lambda)$ with $E_{\Delta_l}^{(\Lambda,m)}(f;p) = o(\overline{\Delta}_l^n)$, $l \to \infty$ for some m, $0 \le m \le n-1$ is contained in $W_p^m(a,b)$ but does not contain $W_p^n(a,b)$ if $1 < p \le \infty$. This gives a rough description of this class of functions. A characterization in terms of necessary and sufficient conditions remains an open question.

References

] J.H.AHLBERG-E.NILSON-J.L.WALSH: The Theory of Splines and their Applications, Academic Press, New York, 1967.

] A.FRIEDMAN: Partial Differential Equations, Holt, Rinehart, and Winston, New York 1969.

] G.W.HEDSTROM-R.S.VARGA: Application of Besov spaces to spline approximation, J. Approximation Theory 4 (1971), 295-327.

] J.W.JEROME: On uniform approximation by certain generalized spline functions. J. Approximation Theory 7 (1973), 143-154.

] H.JOHNEN: Inequalities connected with the moduli of smoothness, Mat. Vesnik, 9(24) (1972), 289-303.

] F.RICHARDS-R.de VORE: Saturation and inverse theorems for spline approximation. Spline Functions Approx. Theory, Proc. Sympos. Univ. Alberta, Edmonton 1972, ISNM 21 (1973), 73-82.

] K.SCHERER: Characterization of generalized Lipschitz classes by best approximation with splines, SIAM J.Numer. Anal. 11 (1974), 283-304.

] K.SCHERER: Best approximation by Chebychevian Splines and generalized Lipschitz spaces, Proceedings of the Conference on "Constructive Theory of Functions", Cluj (1973).

] B.K.SWARTZ-R.S.VARGA: Error bounds for spline and L-spline interpolation, J. Approximation Theory 6 (1972), 6-49.

0] A.F.TIMAN: Theory of Approximation of Functions of a Real Variable. Pergamon Press, New York 1963.

University of Bielefeld
Fakultät für Mathematik

COMPUTATION OF PERIODIC M-SPLINES WITH EQUI-SPACED NODES

Heinz-Walter Kösters

This note presents a method for the construction and computation of periodic M-Splines related to bilinear forms induced by certain differential operators of even order with constant coefficients. In case of equi-spaced nodes an improved algorithm is described.

Introduction

In [3] we discussed the construction of periodic M-Splines (cf. Lucas [6] and Schaback [8]) for arbitrary nodes related to inner products determined by a class of differential operators of even order with constant coefficients, which had been considered in [2] with respect to Lidstone-boundary conditions.

Here we describe an improved procedure for the computation of the a.c. periodic M-Splines in case of equi-spaced nodes.

For the convenience of the reader we list some of the results concerning the differential operators under consideration, for details we refer to [3].

Let $L_2(I)$ be the (complex) space of all (equivalence classes of) square integrable functions over the open interval $I =]0,1[$ equipped with the usual inner product

$$(u|v)_0 = \int_I u(x)\overline{v(x)}dx .$$

Moreover, let $W^k(I)$ be the (complex) Sobolev space with the inner product

$$(u|v)_k = \sum_{j=0}^{k} (D^j u | D^j v)_0 \ , \ D^j = \frac{d^j}{dx^j} . \qquad (1)$$

By

$$W^k_\pi(I) = \{u \in W^k(I) : (D^j u|1)_0 = 0,\ j = 1(1)k\}$$

we denote the subspace of all functions of $W^k(I)$ being 1-periodic with their derivatives up to order k-1 inclusive.

Composing the differential operators

$$L_i = -D^2 + q_i D^0,\ (q_i \in \mathbb{R},\ q_i > 0),$$

$$\mathrm{dom}(L_i) = W^2_\pi(I), \tag{2}$$

we define the formal operator

$$A_k = L_k \circ \cdots \circ L_1,\ A_0 = D^0. \tag{3}$$

By Vieta's theorem we obtain

$$A_k = \sum_{j=0}^{k} (-1)^j w_j D^{2j},$$

$$w_k = 1,\ w_{k-r} = \sum_{1 \le i_1 < \ldots < i_r \le k} q_{i_1} \ldots q_{i_r},\ (r = 1(1)k), \tag{4}$$

hence

$$\mathrm{dom}(A_k) = W^{2k}_\pi(I)$$

and we have

<u>THEOREM 1. The unbounded linear operator A_k in $L_2(I)$ is a positive definite, selfadjoint operator. It's spectrum is given by the point spectrum</u>

$$\sigma_p(A_k) = (r_m)_{m \in \mathbb{Z}},\ r_m = \prod_{i=1}^{k} (q_i + (2\pi m)^2), \tag{5}$$

<u>corresponding to the total orthonormal system of the eigen-</u>

functions

$$(e_m)_{m\in\mathbb{Z}},$$

$$I \ni x \leadsto e_m(x) = \exp(2\pi imx),\ i^2 = -1, \tag{6}$$

in $L_2(I)$.

The energetic space of A_k and it's reproducing kernel are determined in

THEOREM 2. The (complex) Hilbert space $W_{\pi}^{k}(I)$ *equipped with the inner product*

$$(u|v)_{\pi} = \sum_{j=0}^{k} w_j (D^j u|D^j v)_0 \tag{7}$$

is the energetic space of A_k. $W_{\pi}^{k}(I)$ *admits a reproducing kernel, which is the Green function* G_k *of* A_k, *and may be computed recursively by*

$$G_k(x,y) = (G_{k-1}(x,.)|g_k(.,y))_0,\ G_1(x,y) = g_1(x,y). \tag{8}$$

Here g_k *denotes the Green function of the operator* L_k *and is given by*

$$g_k(x,y) = \frac{\cosh(p_k(\frac{1}{2} - |y-x|))}{2p_k\sinh(\frac{1}{2}p_k)} \tag{9}$$

with

$$p_k = q_k^{1/2}.$$

If the coefficients q_k *are mutually distinct*, G_k *takes the form*

$$G_k(x,y) = \sum_{i=1}^{k} \frac{g_i(x,y)}{\prod_{\substack{j=1\\ j\neq i}}^{k}(q_j-q_i)}\ . \tag{10}$$

Let the nodes

$$0 < x_1 < \dots < x_N = 1, \quad (N \in \mathbb{N}),$$

be given.

Then an application of Theorem 3 from Delvos [1] yields as an analogon to Theorem 5 in [2] the following

THEOREM 3. For any $u \in W^k(I)$, $k \in \mathbb{N}$, the function $S_{k,N}(u)$ on I determined by

$$I \ni x \rightsquigarrow S_{k,N}(u)(x) = \sum_{i=1}^{N} a_i G_k(x,x_i), \tag{11}$$

$$\sum_{i=1}^{N} G_k(x_j,x_i)a_i = u(x_j), \quad (j=1(1)N), \tag{12}$$

is the unique one under all $v \in W^k(I)$ with $v(x_i) = u(x_i)$, $(i=1(1)N)$, which minimizes the functional

$$v \rightsquigarrow (v|v)_\pi .$$

In the terminology of Lucas [6], $S_{k,N}(u)$ is a periodic M-Spline corresponding to the bilinear form (7). Investigations concerning integral relations and error bounds for these splines may be found in [3].

Computation in case of equi-spaced nodes

It is the purpose of this section to present an improved algorithm for the computation of the M-Splines constructed above by means of an explicit form for the inverse of the coefficient matrix $G = [\alpha_{jr}]_{j,r=1(1)N}$ of the linear system (12) in case of equi-spaced nodes, see Golomb [5] for a similar procedure. Put

$$q_j^* = \frac{q_j}{(2\pi)^2}, \quad (j=1(1)k), \tag{13}$$

$$b_k = \frac{1}{(2\pi)^{2k}}, \tag{14}$$

$$c_k = \frac{1}{\prod_{j=1}^{k} q_j^*}. \tag{15}$$

Since

$$G_k(x,y) = \sum_{m \in \mathbb{Z}} \frac{e_m(x)\overline{e_m(y)}}{r_m}$$

$$= \frac{1}{\prod_{j=1}^{k} q_j} + 2\cdot\sum_{m=1}^{\infty} \frac{\cos(2\pi m(x-y))}{r_m}$$

we obtain for the equi-spaced nodes

$$x_j = \frac{j}{N}, \quad (j=1(1)N)$$

the relation

$$\alpha_{jr} = G_k(x_j, x_r)$$

$$= b_k\Big(c_k + 2\cdot\sum_{m=1}^{\infty} \frac{\cos(\frac{2\pi m}{N}(j-r))}{\prod_{j=1}^{k}(q_j^*+m^2)}\Big), \quad (j,r=1(1)N).$$

The substitution

$$m = s + tN, \quad (s=1(1)N;\ t=0,1,\ldots)$$

together with the cosine-additions-theorems and the agreements

$$\gamma(s) = 2b_k\cdot\sum_{t=0}^{\infty} \frac{1}{\prod_{j=1}^{k}(q_j^*+(s+tN)^2)}, \quad (s=1(1)N),$$

$$d_k = b_k c_k \tag{16}$$

lead to

$$\alpha_{jr} = d_k + \sum_{s=1}^{N} \gamma(s)\cos(\frac{2\pi s}{N}(j-r)), \ (j,r=1(1)N).$$

Since G_k is a kernel of difference-type, with

$$\alpha_j = \alpha_{1j} = \alpha_{j1} \ , \ (j=1(1)N)$$

one easily establishes

$$\alpha_{jr} = \alpha_{|j-r|+1} \ , \ (j,r=1(1)N),$$

$$\alpha_j = \alpha_{N+2-j} \ , \ (j=1(1)N).$$

Hence the matrix G is symmetric and circulant. By means of the polynomial

$$g(\lambda) = \sum_{j=1}^{N} \alpha_j \lambda^{j-1}$$

and the N×N permutation matrix

$$P = \begin{bmatrix} 0 & 1 & 0 & \dots & 0 \\ 0 & 0 & 1 & \dots & 0 \\ \cdot & \cdot & \cdot & \cdot & \cdot \\ \cdot & \cdot & \cdot & & \cdot \\ \cdot & \cdot & \cdot & & 1 \\ 1 & 0 & 0 & \dots & 0 \end{bmatrix}$$

we get

$$G = g(P).$$

The spectrum of P is given by

$$\sigma(P) = \{\omega^j : \omega = \exp(\frac{2\pi i}{N}), \ j=1(1)N\} ,$$

and the corresponding orthonormal eigenvectors are

$$e_j = \frac{1}{\sqrt{N}}\left[1,\omega^j,\omega^{2j},\ldots,\omega^{(N-1)j}\right]^T , \quad (j=1(1)N).$$

Hence the spectrum of G is

$$\sigma(G) = \{g(\omega^j):j=1(1)N\} ,$$

and with the unitary matrix

$$T = \left[e_1,\ldots,e_N\right] = \left[\tau_{jr}\right]_{j,r=1(1)N}$$

$$= \left[\frac{\omega^{(j-1)r}}{\sqrt{N}}\right]_{j,r=1(1)N}$$

we conclude

$$T^{-1}GT = \operatorname{diag}\left[g(\omega^1),\ldots,g(\omega^N)\right] .$$

In view of

$$g(\lambda) = \sum_{j=1}^{N} \alpha_j \lambda^{j-1}$$

$$= \alpha_1 + \tfrac{1}{2}\Big\{\sum_{j=2}^{N} \alpha_j \lambda^{j-1} + \sum_{j=2}^{N} \alpha_{N+2-j} \lambda^{N-(j-1)}\Big\}$$

we have

$$g(\omega^j) = \alpha_1 + \tfrac{1}{2}\cdot\sum_{r=2}^{N} \alpha_r \{\omega^{j(r-1)} + \omega^{-j(r-1)}\}$$

$$= \sum_{r=1}^{N} \alpha_r \cos\left(\frac{2\pi j}{N}(r-1)\right)$$

$$= d_k\cdot\sum_{r=1}^{N} \cos\left(\frac{2\pi j}{N}(r-1)\right)$$

$$+ \sum_{s=1}^{N} \gamma(s) \sum_{r=1}^{N} \cos(\frac{2\pi s}{N}(r-1)) \cos(\frac{2\pi j}{N}(r-1)) \ ,$$

i.e.

$$g(\omega^j) = \begin{cases} \frac{N}{2} \cdot \{\gamma(j) + \gamma(N-j)\} \ , \ j=1(1)N-1 \\ N \cdot \{d_k + \gamma(N)\} \ , \ j=N \end{cases} \tag{17}$$

and

$$g(\omega^{N-j}) = g(\omega^j) \ , \ (j=1(1)N-1). \tag{18}$$

The inverse of the matrix G is also symmetric and circulant, we denote it by

$$G^{-1} = \begin{bmatrix} \beta_1 & \beta_2 & \dots & \beta_{N-1} & \beta_N \\ \beta_N & \beta_1 & \dots & \beta_{N-2} & \beta_{N-1} \\ \cdot & \cdot & \cdots & \cdot & \cdot \\ \cdot & \cdot & \cdots & \cdot & \cdot \\ \cdot & \cdot & \cdots & \cdot & \cdot \\ \beta_3 & \beta_4 & \dots & \beta_1 & \beta_2 \\ \beta_2 & \beta_3 & \dots & \beta_N & \beta_1 \end{bmatrix} .$$

Because of

$$G^{-1} = T \cdot \text{diag} \left[\frac{1}{g(\omega^1)} , \dots , \frac{1}{g(\omega^N)} \right] \cdot T^{-1}$$

we find

$$\beta_j = \frac{1}{N} \cdot \sum_{s=1}^{N} \frac{\omega^{(j-1)s}}{g(\omega^s)} .$$

Hence the identity

$$\overline{\omega^{(j-1)s}} = \omega^{(j-1)(N-s)} \ , \ (s=1(1)N-1)$$

leads to

$$\beta_j = \frac{1}{N} \cdot \frac{1}{g(\omega^N)} + \frac{1}{2N} \cdot \sum_{s=1}^{N-1} \left(\frac{\omega^{(j-1)s}}{g(\omega^s)} + \frac{\omega^{(j-1)(N-s)}}{g(\omega^{N-s})} \right)$$

$$= \frac{1}{N}\cdot\frac{1}{g(\omega^N)} + \frac{1}{N}\cdot\sum_{s=1}^{N-1}\frac{\frac{1}{2}(\omega^{(j-1)s} + \overline{\omega^{(j-1)s}})}{g(\omega^s)}$$

$$= \frac{1}{N}\cdot\frac{1}{g(\omega^N)} + \frac{1}{N}\cdot\sum_{s=1}^{N-1}\frac{\cos(\frac{2\pi s}{N}(j-1))}{g(\omega^s)} ,$$

i.e.

$$\beta_j = \frac{1}{N}\cdot\sum_{s=1}^{N}\frac{\cos(\frac{2\pi s}{N}(j-1))}{g(\omega^s)} , \quad (j=1(1)N), \tag{19}$$

$$\beta_j = \beta_{N+2-j} , \quad (j=1(1)N). \tag{20}$$

The coefficients a_i of the linear system (12) are then determined by

$$a_i = \sum_{j=1}^{i-1}\beta_{N-i+1+j}u(x_j) + \sum_{j=i}^{N}\beta_{j-i+1}u(x_j) , \quad (i=1(1)N). \tag{21}$$

Concerning the computation of the infinite series for $\gamma(s)$, let us denote by

$$\gamma_M(s) = 2b_k\cdot\sum_{t=0}^{M}\frac{1}{\prod_{j=1}^{k}(q_j^*+(s+tN)^2)} , \quad (s=1(1)N) \tag{22}$$

the M-th partial sum of $\gamma(s)$. Then by Cauchy's integral criterion (cf. Meschkowski [7]) we have

$$R_M(s) = \gamma(s) - \gamma_M(s)$$

$$= 2b_k\cdot\sum_{t=M+1}^{\infty}\frac{1}{\prod_{j=1}^{k}(q_j^*+(s+tN)^2)}$$

$$\leq 2b_k\cdot\sum_{t=M+1}^{\infty}\frac{1}{(tN)^{2k}}$$

$$= \frac{2b_k}{N^{2k}} \cdot \sum_{t=M+1}^{\infty} \frac{1}{t^{2k}}$$

$$\leq \frac{2b_k}{N^{2k}} \cdot \frac{1}{2k-1} \cdot \frac{1}{M^{2k-1}},$$

i.e.

$$R_M(s) \leq \frac{2}{2k-1} \cdot \frac{1}{(2\pi N)^{2k}} \cdot \frac{1}{M^{2k-1}}, \quad (s=1(1)N;\ M \in \mathbb{N}). \qquad (23)$$

Summarizing we get

ALGORITHM 4. For the computation of the M-Spline $S_{k,N}(u)$ perform the following steps:

(i) Compute $\gamma_M(s)$, $s=1(1)N$, by (22), (13), (14), where M is to choose according to (23) for a prescribed accuracy.

(ii) Determine $g(\omega^N)$ and $g(\omega^j)$, $j=1(1)\left[\frac{N}{2}\right]+1$, by (16) and and (17), and $g(\omega^j)$, $j=\left[\frac{N}{2}\right]+2(1)N-1$, by (18).

(iii) Get β_j, $j=1(1)\left[\frac{N}{2}\right]+1$, from (19) and β_j, $j=\left[\frac{N}{2}\right]+2(1)N$, from (20).

(iv) Compute a_i, $i=1(1)N$, by (21).

(v) Determine $S_{k,N}(u)(x)$ by (11).

REMARK 5. Similar to Runge's scheme in case of trigonometric interpolation with equi-spaced nodes the number N of nodes should be a multiplicity of 4.

Numerical example

As an illustration for the procedure described in Algorithm 4 let us consider the function

$$I \ni x \rightsquigarrow u(x) = \sin(2\pi x),$$

the nodes

$$x_j = \frac{j}{N}, \ (j=1(1)N;\ N=4,8,16,32),$$

and the coefficients

$$q_j = j^2, \ (j=1,2,\ldots).$$

If

$$e_{k,N} = \max\left\{\left| u(\tfrac{j}{64}) - S_{k,N}(\tfrac{j}{64})\right| : j=1(1)64\right\}$$

denotes the discrete error, $t_{k,N}$ [sec] the computation time (TR 440, Rechenzentrum, Ruhr-Universität Bochum) and $M_{k,N}$ the number of iterations for the infinite series $\gamma_M(s)$, we obtain the following results:

k	$M_{k,4}$	$e_{k,4}$	$t_{k,4}$	$M_{k,8}$	$e_{k,8}$	$t_{k,8}$
1	800	$2,11.10^{-1}$	1,40	400	$7,19.10^{-2}$	1,48
2	40	$2,18.10^{-2}$	0,67	20	$1,20.10^{-3}$	0,91
3	5	$2,84.10^{-3}$	0,73	4	$2,64.10^{-5}$	1,05
4	3	$4,22.10^{-4}$	0,82	3	$6,98.10^{-7}$	1,24
5	2	$7,13.10^{-5}$	0,92	2	$2,19.10^{-8}$	1,44

k	$M_{k,16}$	$e_{k,16}$	$t_{k,16}$	$M_{k,32}$	$e_{k,32}$	$t_{k,32}$
1	200	$1,18.10^{-2}$	2,10	100	$5,18.10^{-3}$	2,23

2	10	$2,21.10^{-4}$	1,95	8	$5,23.10^{-6}$	2,78
3	4	$4,43.10^{-7}$	2,47	3	$7,31.10^{-9}$	3,12
4	3	$5,57.10^{-9}$	3,11	2	$1,20.10^{-11}$	4,30
5	2	$1,31.10^{-10}$	3,89	1	$6,54.10^{-12}$	5,41

References

1. Delvos, F.J.: Optimale Interpolation mit der Methode von Ritz. Zeitschr. Angew. Math. Mech. 55(1975), T234-T235.
2. Delvos, F.J., Kösters, H.W.: Zur Konstruktion von M-Splines höheren Grades. Computing 14(1975), 173-182.
3. Delvos, F.J., Kösters, H.W.: On periodic M-Splines. Computing (In print).
4. Delvos, F.J., Schempp, W.: On optimal periodic spline interpolation. J.Math.Anal.Appl. (In print).
5. Golomb, M.: Approximation by periodic splines on uniform meshes. J. Approximation Theory 1(1968), 26-65.
6. Lucas, T.R.: M-Splines. J. Approximation Theory 5(1972), 1-14.
7. Meschkowski, H.: Unendliche Reihen. Mannheim, Bibliographisches Institut 1962.
8. Schaback, R.: Konstruktion und algebraische Eigenschaften von M-Spline-Interpolierenden. Numer. Math. 21(1973), 166-180.

Dr. Heinz-Walter Kösters
Rechenzentrum
Ruhr-Universität
Universitätsstraße 150 NA
D-4630 Bochum
Bundesrepublik Deutschland

DISCRETE POLYNOMIAL SPLINE APPROXIMATION METHODS

Tom Lyche

Discrete splines were introduced by Mangasarian and Schumaker [12] as solutions to certain minimization problems involving differences. They can be defined as piecewise polynomials where the ties between each polynomial piece involve continuity of differences instead of derivatives. We study discrete analogs of local spline approximations, least squares spline approximations, and even order spline interpolation at knots. Error bounds involving differences over a finite number of points are given in each case. These contain classical error bounds as a special case.

§ 1 Introduction

Most of the theory of polynomial splines deals with the case where the pieces are tied together by continuity of certain derivatives at the knots. In this paper the ties will involve differences instead of derivatives. We will talk about the continuous case when derivatives are involved, and the discrete case when differences are involved.

Discrete splines were introduced by Mangasarian and Schumaker [12] as solutions to certain minimization problems involving differences. They were studied constructively in [15]. Their connection with best summation formulae was investigated in [13]. In [11] discrete cubic splines was used to compute nonlinear splines iteratively. Discrete L-splines have also been introduced [1].

In this paper we study several discrete spline approximation methods for fitting functions and data. The notation has been chosen to reveal as much as possible the connection between the discrete and the continuous case. In fact many of the known results (and a few new ones) for the continuous case follow if we set a discretization parameter h equal to zero.

In § 2 discrete B-splines are introduced as kernels in representing a divided difference over a coarse mesh in terms of differences over a equidistant finer mesh containing the coarse one. The nonnegativity of the discrete B-splines then leads to a discrete mean value theorem for divided differences.

In § 3 we study discrete analogs of the local spline approximation methods of [2],[4] and [10]. The error bounds derived require no smoothness assumtions, and we maximize (or sum) a difference over a finite number of points. Moreover we can get the continuous bounds as a special case.

In § 4 we give discrete error bounds for polynomial interpolation.

In § 5 we extend a result of Douglas,Dupont and Wahlbin [7] to the discrete case. This is used to derive discrete error bounds for least squares approximation by discrete splines.

Finally we consider in § 6 even order discrete spline interpolation at knots. To give a typical error bound let g_h be a discrete spline of order $2k$ interpolating data f given on I_0 where

$$(1.1) \qquad I_j = \{a, a+h, \ldots, b-jh\} \qquad j = 0,1,2,\ldots$$

Moreover let forward differences be given by

$$(1.2) \qquad D_h^j\varphi(x) = h^{-j} \sum_{i=0}^{j} (-1)^{j-i}\binom{j}{i}\varphi(x+ih),$$

and let Δ be the largest distance between to neighbouring knots x_{i-1} and x_i. Then there is a constant K depending on the global mesh ratio $\max(x_{i+1}-x_i)/\min(x_{i+1}-x_i)$, but not on f such that

$$\max_{t\in I_r} |D_h^r(f-g_h)(t)| \leq K\,\Delta^{2k-r} \max_{t\in I_{2k}} |D_h^{2k}f(t)| \quad 0 \leq r < 2k$$

This paper is a summary of the authors Ph.D. dissertation at the University of Texas at Austin. Proofs of the theorems can be found in [8].

§ 2 Discrete B-splines.

We begin by introducing some notation that will be used throughout. For $h > 0$, $\theta \in \mathbb{R}$ define

$$\mathbb{R}_{h\theta} = \{\theta + ih\colon i \text{ is an integer}\}$$

and let $\mathbb{R}_{0\theta} = \mathbb{R}$. For any $a,b \in \mathbb{R}$ and $h \geq 0$ let

$$[a,b]_h = [a,b] \cap \mathbb{R}_{ha}$$

with $(a,b)_h$, $[a,b)_h$, and $(a,b]_h$ analogously defined.

We note that if $D_h^j\varphi(x)$ is given by (1.2) then

$$\lim_{h\to 0} D_h^k\varphi(x) = \varphi^{(k)}(x)$$

provided φ is k times differentiable at x.

D_h^k has the property

$$D_h^m D_h^n \varphi(x) = D_h^{m+n}\varphi(x) .$$

We next define for $h > 0$ and $a,b \in \mathbb{R}$

$$(2.1)\quad \int_a^b f(x)d_h x = \begin{cases} h \sum_{x\in[a,b)_h} f(x) & b-a > h \\ 0 & |b-a| \leq h \\ -\int_b^a f(x)d_h x & b-a < -h \end{cases}$$

Note that b is excluded in the sum.
This operator has the usual properties of an integral i.e. if $b \in \mathbb{R}_{ha}$

$$(2.2)\quad \int_a^b f(x)d_h x + \int_b^c f(x)d_h x = \int_a^c f(x)d_h x ,$$

$$(2.3)\quad \int_a^b D_h f(x)d_h x = f(b) - f(a) ,$$

$$(2.4)\quad \int_a^b f(x)D_h^m g(x)d_h x = w(f;g;b) - w(f;g;a) + (-1)^m \int_a^b g(x+mh)D_h^m f(x)d_h x$$

where

$$(2.5)\quad w(f;g;x) = \sum_{i=0}^{m-1} (-1)^i D_h^i f(x) D_h^{m-1-i} g(x+ih),$$

and if $x^{(n)_h}$ denotes the usual factorial function given by

$$x^{(n)_h} = \begin{cases} 1 & n = 0 \\ x(x-h)\ldots(x-(n-1)h) & n \geq 1 \end{cases}$$

then

(2.6) $$\int_a^b x^{(n-1)_h} d_h x = (b^{(n)_h} - a^{(n)_h})/n \ .$$

We also note that if $f \in C[a,b]$ then

$$\lim_{h\to 0} \int_a^b f(x) d_h x := \int_a^b f(x) d_o x = \int_a^b f(x) dx$$

<u>LEMMA 2.1</u> (Discrete Taylor formula)

<u>Given</u> $a \in \mathbb{R}$ <u>and</u> $h > 0$ <u>let</u> $x \in \mathbb{R}_{ha}$. <u>Then</u>

(2.7) $$f(x) = \sum_{i=o}^{n} \frac{(x-a)^{(i)_h}}{i!} D_h^i f(a) + R_n(x)$$

<u>where</u>

(2.8) $$R_n(x) = \int_a^x (x-t-h)^{(n)_h} D_h^{n+1} f(t) d_h t/n!$$

$$= \theta \frac{(x-a)^{(n+1)_h}}{(n+1)!} D_h^{n+1} f(z)$$

<u>Here</u> $0 \leq \theta \leq 1$ <u>and</u> $z \in [a, x-(n+1)h]_h$ <u>if</u> $x \geq a$ <u>and</u> $z \in [x,a)_h$ <u>if</u> $x < a$.

This was proved in [13] using a different notation. We note that it follows from the fact that

$$R_{n-1}(x) = -\left[(x-t)^{(n)_h} D_h^n f(t)\right]_a^x / n! + R_n(x)$$

Among the many possible ways of representing R_n we mention

$$(2.8b)\ R_n(x) = (-1)^{n+1}\int_b^a (t+nh-x)_+^{(n)_h} D_h^{n+1} f(t)\, d_h t/n! \quad b \leq x \leq a$$

were as usual $(\text{expression})_+ = \max\ \{\text{expression},0\}$.

Suppose $\langle x_i \rangle_{i=-\infty}^{\infty} \subset \mathbb{R}_{hx_0}$ is a strictly increasing sequence. Taking divided differences in (2.7), (2.8b) with $n = k-1$ we find

$$(2.9)\ [x_i,\dots,x_{i+k}]f = \int_{x_i}^{x_{i+k}} M_{ik}^h(t+(k-1)h) D_h^k f(t)\, d_h t/k!$$

where

$$(2.10) M_{ik}^h(x) = (-1)^k k[x_i,\dots,x_{i+k}](x-\cdot)_+^{(k-1)_h}$$

For $h = 0$ M_{ik}^h is the Curry/Schoenberg B-spline [6]. For $h > 0$ they are called discrete B-splines and were introduced by Schumaker [15]. They were defined essentially on $\mathbb{R}_{hx_i}$.

We let

$$(x-y)_+^{(n)_h} := (x-y)_+(x-y-h)\cdots(x-y-(n-1)h),$$

and define M_{ik}^h by (2.10) for all $x \in \mathbb{R}$. It was established in [15] that

$$M_{i1}^h(x) = \begin{cases} 1/(x_{i+1}-x_i) & x_i \leq x < x_{i+1} \\ 0 & \text{otherwise} \end{cases}$$

and that $M_{ik}^h(x) = 0$ for x off (x_i, x_{i+k}) if $k \geq 2$. Moreover

$$M_{ik}^h(x_i+jh) = 0 \quad j = 0,1,\dots,k-2 \quad k \geq 2,$$

and M_{ik}^h is positive on $(x_i + (k-2)h,\ x_{i+k})$. This follows from the recursion

$$(2.11)\quad \frac{k-1}{k} M_{ik}^{h}(x) = \frac{(x-x_i-(k-2)h)M_{i,k-1}^{h}(x)}{x_{i+k}-x_i} + \frac{(x_{i+k}-x+(k-2)h)\,M_{i+1,k-1}(x)}{x_{i+k}-x_i} \qquad k \geq 2$$

We note that (2.11) holds for all $x \in \mathbb{R}$. Finally it was shown that

$$(2.12)\quad \int_{x_i}^{x_{i+k}} M_{ik}^{h}(t)d_h t = 1 \qquad k \geq 1$$

Using (2.12) and the nonnegativety of M_{ik}^{h} on $[x_i,x_{i+k}]_h$ we obtain the following mean-value theorem.

THEOREM 2.2

Let $x_i,\ldots,x_{i+k}$ and $h > 0$ be such that $x_j \in \mathbb{R}_{hx_i}$ $j = i+1,\ldots,i+k$ and $x_i<\ldots<x_{i+k}$. Then there is a number θ, $0 \leq \theta \leq 1$ and $z \in [x_i,x_{i+k}-kh]_h$ such that for any f defined on $[x_i,x_{i+k}-kh]_h$

$$(2.13)\qquad [x_i,\ldots,x_{i+k}]f = \theta\, D_h^k f(z)/k!$$

We will use discrete analogs of the normalized B-splines [3]. If we define

$$N_{ik}^{h}(x) = (x_{i+k}-x_i)\, M_{ik}^{h}(x)/k$$

then ([8] p.21)

$$\sum_i N_{ik}^{h}(x) = 1 \qquad x \in \mathbb{R}.$$

We next introduce discrete splines. Let $\mathbb{P}_k$ denote the class of polynomials of degree less than k. Let $a < b$ and $\Delta = \langle y_i \rangle_{i=0}^n$ be given with

$$(2.14) \qquad a = y_0 < y_1 < \ldots < y_n = b$$

If $g: [a,b] \to \mathbb{R}$ is a piecewise polynomial of degree $k-1$ ($g \in \mathbb{P}_k(\Delta)$) on Δ then we can differentiate g at any point $x \in [a,b)$ by defining

$$D_h^j g(x) = D_h^j g_i(x) \qquad y_{i-1} \leq x < y_i \ , \ 1 \leq i \leq n$$

If $d = \langle d_i \rangle_{i=0}^n$ where $d_0 = d_n = 1$, is a sequence of integers (multiplicities) then we define

$$S_k^h(\Delta;d) = \{g \in \mathbb{P}_k(\Delta): D_h^j(g_{i+1}-g_i)(y_i) = 0, i = 1,\ldots,n-1 \\ j = 0,1,\ldots,k-1-d_i\}$$

For $h = 0$ $S_k^h(\Delta;d)$ is the space of <u>polynomial</u> splines of <u>order</u> k with <u>knots</u> Δ of <u>multiplicity</u> d. For $h > 0$ we call these functions discrete splines. Thus rather than being vectors as in [15] we take them to be functions defined for all x. If $d_i = 1$ $i = 1,\ldots,n-1$ then the j-th difference is continuous for $j = 0,1,\ldots,k-2$.

An example of a discrete spline is furnished by piecewise polynomial interpolation. That is let $d_i = 1$, $i = 1,\ldots,n-1$ and $y_i - y_{i-1} = h$, $i = 1,\ldots,n$. Moreover we extend $\Delta = \langle y_i \rangle_{i=0}^n$ to $\langle y_i \rangle_{i=0}^{n+k-2}$ such that $y_n < \ldots < y_{n+k-2}$. Then if $z_0,\ldots,z_{n+k-2}$ are data, the function g given by

$$g|_{[y_{i-1},y_i)} = g_i|_{[y_{i-1},y_i)} \quad i = 1,\ldots,n$$

$$g_i \in \mathbb{P}_k \quad \text{and} \quad g_i(y_j) = z_j \qquad j = i-1,\dots,i+k-2$$

is a discrete spline.

We remark that every discrete spline has a unique representation in the form ([8] p.25)

$$g(x) = \sum_{j=0}^{k-1} \frac{D_h^j g(y_o)}{j!} (x-y_o)^{(j)_h} + \sum_{i=1}^{n-1} \sum_{j=k-d_i}^{k-1} (\lambda_{ij} g)(x-y_i)_+^{(j)_h}$$

where

$$\lambda_{ij} g = D_h^j (g_{i+1} - g_i)(y_i)/j! \qquad j = 0,1,\dots,k-1$$

describe the jump in $D_h^j g$ at y_i.

To see the relationship between discrete splines and discrete B-splines let

$$N = \sum_{i=1}^{n-1} d_i + k$$

be the number of free parameters of the discrete spline. We extend Δ to a partition

(2.15) $\pi = \langle x_i \rangle_{i=k}^{N+1} = \langle y_i, y_i - h, \dots, y_i - (d_i - 1)h \rangle_{i=o}^{n}$

Thus multiplicity d_i when $h > 0$ means d_i points h appart. Finally with $x_1,\dots,x_{k-1}$ and $x_{N+2},\dots,x_{N+k}$ increasing we let

(2.16) $\pi_e = \langle x_i \rangle_{i=1}^{N+k}$. We assume

(2.17) $y_i - y_{i-1} \geq d_i h$

so that π_e is increasing. It can now be shown that if

$\pi_e \subset \mathbb{R}_{hy_0}$ then

$$N_{ik}^h \subset S_k^h(\Delta;d) ,$$

$$\langle N_{ik}^h \rangle_{i=m+1-k}^m \text{ is linearly independent on } (x_m, x_{m+1})$$

$$m = k,\dots,N$$

(See [8]).

Thus it follows that every $g \in S_k(\Delta;d)$ can be written

$$g(x) = \sum_{i=1}^N \alpha_i N_{ik}^h(x) \qquad x \in \mathbb{R}$$

A similar fact was proved in [15].

We end this section by defining a more general class of discrete splines. We recall that in the piecewise polynomial interpolation example, the function $g_i \in \mathbb{P}_k$ interpolated the data at $y_{i-1},\dots,y_{i+k-2}$, and we used the interval $[y_{i-1}, y_i)$ to define the spline g. In applications it might be better to use a piece of g_i more in the center of the interval spanned by the interpolation points.

Thus let e be an integer $0 \leq e \leq k-1$. If $g = \langle g_i \rangle_{i=1}^n$ is a piecewise polynomial function of order k we define

$$(2.18)\quad S_{ke}^h(\Delta) = \{g \in \mathbb{P}_k(\Delta) : g_i(y_i+jh) = g_{i+1}(y_i+jh)$$

$$j = -e,\dots,k-2-e,\ i=1,\dots,n-1\}$$

(we have assumed for simplicity that the knots Δ are simple).

Since

$$D_h^j f(x) = 0 \quad j = 0,1,\ldots,m \iff f(x+ih) = 0 \quad i = 0,1,\ldots,m$$

we note that $S_{ko}^h(\Delta) = S_k^h(\Delta)$ when $e = 0$. When $h > 0$ the forward differences $D_h^j g(x)$ will no longer be continuous for all j up to $k-2$. Sometimes, however we can define different difference operators which preserve continuity up to $j = k-2$.

EXAMPLE [9].

Let $k = 4$. Then with $e = 1$ we have $g = \langle g_i \rangle_{i=1}^n$ where $g \in \mathbb{P}_4$ and $g_i(y_i+jh) = g_{i+1}(y_i+jh)$ $j = -1,0,1$ $i = 1,\ldots,n-1$. Now we define for $j = 0,1,2$

$$D_h^{\{j\}} g(x) = D_h^{\{j\}} g_i(x) \quad y_{i-1} \leq x < y_i \quad i = 1,\ldots,n$$

$$D_h^{\{j\}} g(b) = D_h^{\{j\}} g_n(b)$$

where

$$D_h^{\{0\}} f(x) = f(x), \quad D_h^{\{1\}} f(x) = \frac{f(x+h)-f(x-h)}{2h}$$

and

$$D_h^{\{2\}} f(x) = [f(x+h)+f(x-h)-2f(x)]/h^2$$

Then $D_h^{\{j\}} g$ is continuous for $j = 0,1,2$.

For the general $S_{ke}^h(\Delta)$ given by (2.33) we can define B-splines by

$$N_{ik}^{he} = (-1)^k (x_{i+k}-x_i)[x_i,\ldots,x_{i+k}](x+eh-.)_+^{(k-1)_h}$$

We have drawn some cubic B-splines in figure 2.1.

On the left we have N_{i4}^{h1}, and on the right $N_{i4}^{h} = N_{i4}^{ho}$. For comparison we have also included the usual cubic B-spline $N_{i4} = N_{i4}^{oo}$ together with N_{i4}^{h1}.

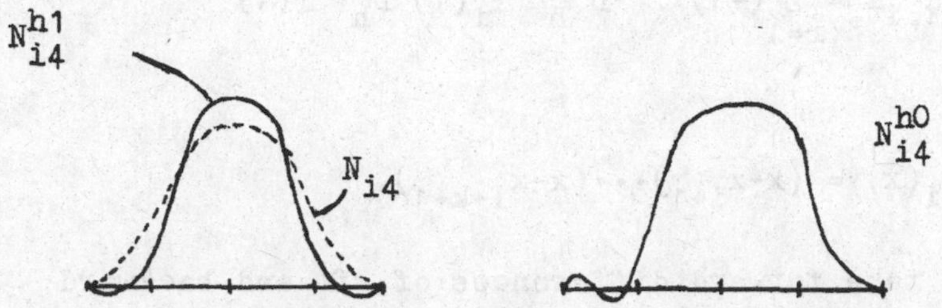

Figure 2.1

We see that N_{i4}^{h1} is more symmetrical than N_{i4}^{ho}. However these "centered" discrete B-splines are more difficult to handle analytically.

We also see that N_{i4}^{h1} "peaks" more than N_{i4}. This indicates among other things that the B-splines for $h > 0$ will be better numerically conditioned than for $h = 0$.

The B-splines N_{i4}^{he} have properties analogous to N_{i4}^{h} (see [8] p.33)

§ 3 Local discrete spline approximations.

Let $N_{1k}^{h},\ldots,N_{Nk}^{h}$ be the normalized discrete B-splines on $\langle x_i \rangle_{i=1}^{N+k}$ given by (2.14),(2.15),(2.16). Let $\lambda_1,\ldots,\lambda_N$ be N linear functionals with local support. As in [10] we consider on $[a,b]$ an approximation Qf to f in the form

$$(3.1) \qquad Qf = \sum_{i=1}^{N} \lambda_i f N_{ik}^{h}$$

Here λ_i is taken so that Q reproduces polynomials up to a certain degree $\mu-1$

$$(3.2) \qquad Qp = p \qquad p \in \mathbb{P}_\mu$$

EXAMPLE 3.1 (Discrete quasi-interpolant)

Let for $\tau \in \mathbb{R}$

$$\lambda_{\tau,i} f = \sum_{r=1}^{k} (-1)^{k-r} D_{-h}^{r-1} \Psi_i(\tau) D_h^{k-r} f(\tau)$$

where

$$\Psi_i(x) = (x-x_{i+1})\cdots(x-x_{i+k-1}).$$

Thus we take forward differences of f and backward differences of Ψ_i. Define

$$P_h f = \sum_{i=1}^{N} \lambda_{\tau_i,i} f N_{ik}^h \qquad \tau_i \in [x_i, x_{i+k})_h$$

For $h = 0$ P_h is the quasi-interpolant introduced in [4]. P_h not only reproduces $\mathbb{P}_k$, but is actually a linear projector onto $S_k^h(\Delta;d)$ for any $h \geq 0$. In particular taking f to be a polynomial we have

$$(3.3) \qquad \sum_{i=1}^{N} \xi_{ik}^{(r)} N_{ik}^h(x) = x^{(r-1)_h} \qquad r = 1,\ldots,k$$

where

$$(3.3a) \qquad \begin{aligned} &\xi_{ik}^{(1)} = 1, \ \xi_{ik}^{(2)} = (x_{i+1}+\ldots+x_{i+k-1})/(k-1)+(k-2)h/2, \\ &\xi_{ik}^{(r)} = (-1)^{r-1}(r-1)!\, D_{-h}^{k-r} \Psi_i(0)/(k-1)! \quad r \geq 1. \end{aligned}$$

We return to the general case (3.1) and consider λ_i's of the form

$$(3.4) \qquad \lambda_i = \sum_{j=1}^{\mu} \alpha_{ij} \lambda_{ij}$$

where $\lambda_{i1},\dots,\lambda_{i\mu}$ are given linear functionals and $\alpha_{i1},\dots,\alpha_{i\mu}$ are constants chosen so that (3.2) holds. Now since the N^h_{ik}'s are linearly independent it follows from (3.3) that (3.2) is equivalent to the linear system

$$\sum_{j=1}^{\mu} \alpha_{ij}\lambda_{ij}x^{(r-1)}h = \xi_{ik}^{(r)} \qquad r = 1,\dots,\mu$$

If $\lambda_{\tau,i}$ is the functional in example 3.1 it can be shown ([8] p.41) that the solution of this linear system can be written

$$\text{(3.4a)} \qquad \alpha_{ij} = \lambda_{\tau,i}p_{ij} \qquad \tau \in \mathbb{R} \text{ arbitrary}$$

where p_{ij} is the polynomial of degree less than μ such that $\lambda_{ir}p_{ij} = \delta_{rj}$. p_{ij} exists iff (3.3) is non-singular. It follows that any local spline approximation method of the form (3.1),(3.4) can be written

$$Qf = \sum_{i=1}^{N} \lambda_{\tau_i,i}(L_i f)N^h_{ik}\ , \quad L_i f = \sum_{j=1}^{\mu} (\lambda_{ij}f)p_{ij}$$

EXAMPLE 3.2

Let $\lambda_{ij} = [\tau_{i1},\dots,\tau_{ij}]$ be the divided difference functional. Then

$$p_{ij}(x) = \begin{cases} 1 & j = 1 \\ (x-\tau_{i1})\cdots(x-\tau_{i,j-1}) & j = 2,\dots,\mu \end{cases}$$

and $L_i f$ becomes the Newton form of the interpolation polynomial. If we take $\mu = 2$ and $\tau_{i1} = \xi_{ik}^{(2)}$ then $\alpha_{i2} = 0$ and we get

$$(3.5) \qquad Qf(x) = \sum_{i=1}^{N} f(\xi_{ik}^{(2)}) N_{ik}^{h}(x)$$

This is a discrete analog of the variation diminishing spline approximation of Schoenberg and Marsden. It reproduces straight lines ([14]).

EXAMPLE 3.3

We have an averaging analog of (3.5) by taking

$$Qf(x) = \sum_{i=1}^{N} \nu_i f N_{ik}^{h}(x)$$

where

$$\nu_i f = \int_{x_{i+1}}^{x_{i+k-1}} f(t - \tfrac{k-4}{2}h)\, M_{i+1,k-2}^{h}(t) d_h t \quad k \geq 3$$

It can be shown that Q reproduces straight lines ($\mu = 2$).

We shall give error bounds for

$$Qf(x) = \sum_{i=1}^{N} \lambda_i f N_{ik}^{h}$$

where

$$(3.6a) \qquad \lambda_i f = \sum_{j=1}^{\mu} \alpha_{ij} [\tau_{i1}, \ldots, \tau_{ij}] f$$

$$(3.6b) \qquad \tau_{ij} \in [x_i, x_{i+k})_h \cap [a,b]_h$$

$$(3.6c) \qquad \tau_{ir} \neq \tau_{is}, \quad r \neq s, \quad i = 1, \ldots, N$$

$$(3.6d) \qquad \alpha_{i1}, \ldots, \alpha_{i\mu} \text{ is such that (3.2) holds.}$$

Thus if $t \in J_m := [x_m, x_{m+1})_h$ then

$$(3.7) \qquad Qf(t) = \sum_{i=m+1-k}^{m} \lambda_i f N_{ik}^h(t)$$

depends only on values of f from

$$(3.8) \qquad U_m := [x_{m+1-k}, x_{m+k})_h .$$

Following [10] we choose $1 \leq s \leq \mu$ and bound the quantity

$$E_{rs}(x) = \begin{cases} D_h^r(f-Qf)(x) & 0 \leq r < s \\ D_h^r Qf(x) & s \leq r < k \end{cases}$$

Let g be the discrete Taylor polynomial of order s,

$$g(x) = \sum_{q=0}^{s-1} (x-t)^{(q)_h} D_h^q f(t)/q! \quad .$$

Since Q reproduces g, $D_h^r(f-g)(t) = 0$, $r \leq s-1$, and $D_h^r g(t) = 0$ $r = s,\ldots,k-1$ we have

$$|E_{rs}(t)| \leq |D_h^r Q(f-g)(t)| \quad 0 \leq r < k$$

Using (3.7)

$$|E_{rs}(t)| \leq \sum_{i=m+1-k}^{m} \sum_{j=1}^{\mu} |\alpha_{ij}| |[\tau_{i1},\ldots,\tau_{ij}](f-g)| |D_h^r N_{ik}^h(t)|$$

Thus the problem of bounding E_{rs} has been separated into three different problems. That of bounding α_{ij}, $[\tau_{i1},\ldots,\tau_{ij}](f-g)$, and $D_h^r N_{ik}^h(t)$. The details of this can be found in [8]. The estimates are given using the norms

$$\|f\|_{L_h^p[c,d]} = \begin{cases} \left(\int_c^d |f(x)|^p d_h x\right)^{1/p} & 1 \leq p < \infty \\ \max\limits_{x\in[a,b)_h} |f(x)| & p = \infty \end{cases}$$

and the usual modulus of continuity

$$\omega(f;\Delta;I) = \sup_{\substack{x,x+d\in I \\ 0<d\leq\Delta}} |f(x+d)-f(x)|$$

We give five theorems, two local and three global results. We first need some notation related to the partition. Let $k \leq m \leq N$. Define

$$\rho_{rm} = \max_{m+1-k\leq i\leq m} (x_{i+k}-x_i)/\min_{1\leq q\leq \mu-r} (\tau_{i,r+q}^{(\mu)} - \tau_{i,q}^{(\mu)})$$

where $\tau_{i1}^{(\mu)},\ldots,\tau_{i\mu}^{(\mu)}$ is an increasing rearrangement of $\tau_{i1},\ldots,\tau_{i\mu}$. Let

$$\pi_{m,k-r} = \max_{m+1-k+r\leq q\leq m} (x_{m+1}-x_m)/(x_{q+k-r}-x_q)$$

$$\bar{\Delta}_m = \max_{m+1-k\leq i\leq m+k-1} (x_{i+1}-x_i)$$

We also define global quantities

$$\rho_r = \max_{1 \leq i \leq N} (x_{i+k}-x_i) / \min_{1 \leq q \leq \mu-r} (\tau^{(\mu)}_{i,r+q}-\tau^{(\mu)}_{iq})$$

$$\pi_{k-r} = \max_{r < q \leq N} \bar{\Delta}/(x_{q+k-r}-x_q) \quad \text{where}$$

$$\bar{\Delta} = \max_{1 \leq i \leq N+k-1} (x_{i+1}-x_i).$$

Finally let

$$I_j = [a, b-(j-1)h]_h \qquad j = 0,1,2,\ldots$$

<u>THEOREM 3.4</u> (Local estimate without meshindependence)
<u>Suppose</u> $1 \leq s \leq \mu \leq k$ <u>and</u> $1 \leq q \leq \infty$. <u>Then</u> <u>for</u> $0 \leq r < k$ <u>and</u> $k \leq m \leq N$

$$(3.9) \quad \|E_{rs}\|_{L^q_h(J_m)} \leq K_m \bar{\Delta}_m^{\,s-r-1+1/q} \omega(D_h^{s-1}f; \bar{\Delta}_m; U_m)$$

<u>Moreover</u> <u>for</u> $1 \leq p \leq \infty$

$$(3.10) \quad \|E_{rs}\|_{L^q_h(J_m)} \leq K_m \bar{\Delta}_m^{\,s-r+1/q-1/p} \|D_h^s f\|_{L^p_h(U_m)}$$

<u>The</u> <u>constant</u> K_m <u>depends</u> <u>on</u> k, s, μ, r, <u>and</u> <u>the</u> <u>local</u> <u>mesh</u> <u>ratios</u> ρ_{rm} <u>and</u> $\pi_{m,k-r}$.

THEOREM 3.5 (global estimate without meshindependence)

Suppose $1 \leq s \leq \mu \leq k$ and $1 \leq q \leq \infty$. Then for $0 \leq r < k$

$$(3.11)\quad \|E_{rs}\|_{L_h^q(I_r)} \leq K\bar{\Delta}^{\,s-r-1}\,\omega(D_h^{s-1}f;\bar{\Delta};I_s)$$

Moreover if $1 \leq p \leq q \leq \infty$ then

$$(3.12)\quad \|E_{rs}\|_{L_h^q(I_r)} \leq (2k-1)K\bar{\Delta}^{\,s-r-1/p+1/q}\,\|D_h^s f\|_{L_h^p(I_s)}$$

The constant K depends on $k,s,\mu,r,$ and the mesh-ratios ρ_r and π_{k-r}.

Note that the constant K in this theorem depends on $\alpha_{k-r} = \min\limits_{r < q \leq N} (x_{q+k-r}-x_q)$. If simple knots (i.e. $d_i = 1 \quad i = 1,\ldots,n-1$) then by choosing the extra knots at the ends properly we have

$$\alpha_{k-r} = \min_{1 \leq i \leq n+r-k} (y_{i+k-r}-y_i)$$

The situation is less satisfactory for multiple knots. For example if $d_i = 2$ for some i then $\alpha_1 = h$. Hence if $r = k-1$ then $K = O(\bar{\Delta}/h)$. The problem arises because the spline g is not "smooth" enough to be differentiated $k-1$ times at y_i-h. I.e., if $h = 0$ and $d_i = 2$ then g is only C^{k-3} in a neighborhood of y_i and $D^{k-1}g(y_i)$ does not exist. For $h = 0$ this problem is circumvented by defining a modified norm (See [16]p.8)

$$\|g\|_{L^p[a,b]} = \left(\sum_{i=1}^{n} \|g\|^p_{L^p[y_{i-1},y_i]}\right)^{1/p}$$

This suggests that in the discrete case we remove the knots that are causing trouble. Thus for $0 \le r \le k$ let

$$Y_{ir} = \{y_i - jh:\ 1 \le j \le r+d_i-k\}$$

$$Y_r = \bigcup_{i=1}^{n-1} Y_{ir} .$$

In particular if the knots are simple $(d_i = 1)$ then $Y_r = \emptyset \quad r = 0,1,\ldots,k-1$. We now state

THEOREM 3.6 (global estimate without meshindependence)

Suppose in theorem 3.5 that expression $\|E_{rs}\|_{L_h^q(I_r)}$ in (3.11) and (3.12) is replaced by

$$\|E_{rs}\|_{L_h^q(I_r \setminus Y_r)}$$

then (3.11) and (3.12) holds, but with K depending on ρ_r and on

$$\max \bar{\Delta}/(x_{q+k-r} - x_q)$$

where the maximum is taken over q such that $r < q \le N$ and $[y_{i-1}, y_i-(d_i-1)h] \subset [x_q, x_{q+k-r}]$ for some i.

As in [10] we can obtain mesh independence results by restricting the location of the τ_{ij}'s. For $1 \le i \le N$ and $0 \le r < k$ define

$$J_{ir} = \begin{cases} [x_{i+r}, x_{k+1}]_h & 1 \le i \le r \\ [x_{i+r}, x_{i+k-r}]_h & r < i \le N-r \\ [x_N, x_{i+k-r}]_h & N-r < i \le N \end{cases}$$

THEOREM 3.7 (Local estimate with meshindependence)

Suppose in theorem 3.4 *that*

$$\tau_{ij} \in J_{ir} \qquad j = 1,\dots,\mu \qquad i = 1,\dots,N$$

In addition suppose $r \le s-1$ *and that* $2r \le s+1$ *if* $s < \mu$. *Then* (3.9) *and* (3.10) *hold with* K_m *depending only on* $k, s, \mu,$ *and* r.

We end this section with the following global result.

THEOREM 3.8 (Global estimate with meshindependence)

Suppose in theorem 3.5 *that*

$$\tau_{ij} \in J_{ir} \cap [a,b] \quad j = 1,\dots,\mu \quad i = 1,\dots,N$$

In addition suppose $r \le s-1$ *and that* $2r \le s+1$ *if* $s < \mu$. *Then* (3.11) *and* (3.12) *hold with* K *depending only on* $k, s, \mu,$ *and* r.

Explicit expressions for the constants K_m and K can be found in [8].

§ 4. Discrete error bounds for polynomial interpolation

Given distinct $z_1,\dots,z_k \in \mathbb{R}_{hz_1}$ let $g \in \mathbb{P}_k$ be the polynomial interpolating a given function f at $z_1,\dots,z_k$. Since

$$f(x) = g(x) + \prod_{i=1}^{k} (x-z_i)[z_1,\dots,z_k,x]f$$

it follows from the discrete mean value theorem 2.2 that for $x \in \mathbb{R}_{hz_1}$, $x \notin \{z_1,\dots,z_k\}$

$$f(x) - g(x) = \theta \prod_{i=1}^{k} (x-z_i) D_h^k f(z_x)/k!$$

where $0 \leq \theta \leq 1$, $z_x \in [y_o, y_k - kh]_h$, and $y_o,\dots,y_k$ is an increasing rearrangement of $z_1,\dots,z_k,x$. Let

$$E_{rs} = \begin{cases} D_h^r (f-g) & 0 \leq r < s \\ D_h^r g & s \leq r < k \end{cases}$$

We then have the following discrete bounds on $f-g$.

THEOREM 4.1

Suppose $1 \leq s \leq k$ *and* $1 \leq q \leq \infty$. *Given* $h > 0$ *let* $z_1,\dots,z_k$ *be such that* $z_1 < \dots < z_k$ *and* $z_i \in \mathbb{R}_{hz_1}$ $i = 2,\dots,k$. *If* $g \in \mathbb{P}_k$ *is the polynomial interpolating a given function* f *at* $z_1,\dots,z_k$ *then for* $0 \leq r < k$

$$\|E_{rs}\|_{L_h^q(I_r)} \leq K\bar{\Delta}^{\,s-1-r+1/q}\, \omega(D_h^{s-1} f; \bar{\Delta}; I_s)$$

Moreover for $1 \leq p \leq q \leq \infty$

$$\|E_{rs}\|_{L_h^q(I_r)} \leq K\bar{\Delta}^{\,s-r+1/q-1/p} \|D_h^s f\|_{L_h^p(I_s)}$$

Here

$$I_j = [z_1, z_k-(j-1)h]_h \qquad j = 0,1,2,\ldots$$

$$\bar{\Delta} = \max_{2 \leq i \leq k} (z_i - z_{i-1})$$

moreover K *depends on* r,s,k *and*

$$(4.1) \qquad \rho_r = \frac{z_k - z_1 + (k-1)_h}{\min\limits_{1 \leq j \leq k-s} (z_{j+s} - z_j)} .$$

Proof

Define a partition $\pi_e = \langle x_i \rangle_{i=1}^{2k}$ by

$$x_{k-i} = z_1 - ih, \quad x_{k+i+1} = z_k + ih \quad i = 0,1,\ldots,k-1.$$

On π_e consider the local spline approximation method $\tilde{Q}$ given by

$$\tilde{Q}f = \sum_{i=1}^{k} \tilde{\lambda}_i f N_{ik}^h$$

where

$$\tilde{\lambda}_i f = \sum_{j=1}^{k} \alpha_{ij} [z_1, \ldots, z_j] f$$

and α_{ij} is given by (3.4a) with $p_{ij}(x) = \prod_{r=1}^{j-1}(x-z_r)$

Thus from the results of section 3 we know that $\tilde{Q}p = p$ for all $p \in \mathbb{P}_k$. It follows that $\tilde{Q}f = \tilde{Q}g = g$. Hence we can use theorem 3.5 with $N = k$, $\bar{\Delta} = z_k - z_1$, $\pi_{k-r} = 1$ and ρ_r given by (4.1). Since $z_k - z_1 \leq k\bar{\Delta}$ we get the first inequality for $q = \infty$. Integrating gives the inequality for arbitrary q. ∎

An explicit expression for the constant K can be found in [8].

§ 5 Least squares discrete spline approximations

Given an integer μ, $h > 0$, $z = \langle z_i \rangle_{i=1}^{\mu} \subset [a,b]$, and $f: z \to \mathbb{R}$ let $L_k^h f \in S_k^h(\Delta;d)$ be such that

$$(5.1)\quad \|L_k^h f - f\|_2 \leq \|g - f\|_2 \qquad g \in S_k^h(\Delta;d)$$

where

$$\|g\|_2^2 = \sum_{i=1}^{\mu} w_i(z_i)[g(z_i)]^2$$

and $w: z \to \mathbb{R}$ is positive. This is a linear least squares problem which always has a solution. It can be shown([8]) that the solution is unique if and only if there is a subsequence $\hat{z} = \langle z_{r_j} \rangle_{j=1}^{N} \subset z$ such that

$$N_{jk}^h(z_{r_j}) \neq 0 \qquad 1 \leq j \leq N.$$

Computational considerations for this problem is discussed in [8]. We restrict ourselves to state some error bounds for the case where the datapoints z are equally spaced with spacing h and that $w = 1$. More specifically we assume that the inner product is given by

$$(5.2) \qquad \|g\|_2^2 = \int_a^b [g(t)]^2 d_h t$$

We first bound the operator L_k^h. This generalizes a result of [7] to the discrete setting.

LEMMA 5.1

Suppose $\Delta = \langle y_i \rangle_{i=o}^n$ is a partition such that $a = y_o < \ldots < y_n = b$ and $\Delta \subset \mathbb{R}_{ha}$ where $h > 0$. Let $d = \langle d_i \rangle_{i=o}^n$ be a sequence of multiplicities such that for $k \geq 1$, $1 \leq d_i < k$, $d_o = d_n = 1$, $i = 1,\ldots,n$. Furthermore suppose $y_i - y_{i-1} \geq d_i h$, $i = 1,\ldots,n$. Then if L_k^h is the L^2- projection onto $S_k(\Delta, d)$ we have with $I^h = [a,b]_h$

$$\|L_k^h\|_q := \sup_{f: I^h \to \mathbb{R}} \|L_k^h f\|_{L_h^q(I^h)} / \|f\|_{L_h^q(I^h)} \leq K,$$

$$2 \leq q \leq \infty$$

where K only depends on k and the global meshratio

$$\sigma = \max_{1 \leq i,j \leq n} |y_i - y_{i-1}| / |y_j - y_{j-1}| .$$

The proof of this lemma and an expression for K can be found in [8]p.84.

As a corollary we get error bounds. Let $1 \leq s \leq k$

and define

$$E_{rs}(t) = \begin{cases} D_h^r\,(f-L_k^h f)(t) & 0 \le r < s \\ D_h^r L_k^h f(t) & s \le r < k \end{cases}$$

Recall that $I_j - [a,b-(j-1)h]_h$ and $\bar{\Delta} = \max\limits_{1\le i\le n} (y_i - y_{i-1})$.
Then we have

THEOREM 5.2

Under the assumptions of Lemma 5.1 we have for $2 \le q \le \infty$, $1 \le s \le k$, and $0 \le r < k$

$$\|E_{rs}\|_{L_h^q(I_r)} \le K\bar{\Delta}^{\,s-1-r}\,\omega(D_h^{s-1}f;\bar{\Delta};I_s)$$

Moreover if $1 \le p \le q$ then

$$\|E_{rs}\|_{L_h^q(I_r)} \le (2k-1)K\bar{\Delta}^{s-r+1/q-1/p}\,\|D_h^s f\|_{L_h^p(I_s)}\,.$$

The constant K depends on r,s,k and the global mesh ratio σ.

We sketch a proof. The idea is to subtract and add a suitable local discrete spline approximation Qf which is reprodused by L_k^h. Thus

$$\|D_h^r(f-L_k^h f)\|_{L_h^q(I_r)} \le \|D_h^r(f-Qf)\|_{L_h^q(I_r)} + \|D^r L_k^h(f-Qf)\|_{L_h^q(I_r)}$$

The first term can be estimated by the results of section 3. For the second term we use a discrete version of Markhoff's inequality to get rid of D_h^r. Then lemma 5.1 and the results of section 3 bound the second term.

§ 6. Discrete polynomial spline interpolation

Let $k \geq 1$ and $h > 0$.

In this section we assume that the partition Δ and the multiplicities d are such that

$$\text{(6.1a)}\quad a = y_o < \ldots < y_n = b-(k-1)h, \quad y_i \in \mathbb{R}_{ha} \quad i = 1,\ldots,n$$

$$\text{(6.1b)}\quad d_o = d_n = k \quad 1 \leq d_i \leq k \quad i = 0,1,\ldots,n$$

$$\text{(6.1c)}\quad y_i - y_{i-1} \geq d_i h \qquad i = 1,\ldots,n\,.$$

Let π and π_e be increasing sequences such that

$$\text{(6.1d)}\quad \pi = \langle x_i \rangle_{i=k+1}^{N+k} = \langle y_i-(d_i-1)h,\ldots,y_i \rangle_{i=o}^{n} \quad \left(N = \sum_{i=o}^{n} d_i\right)$$

$$\text{(6.1e)}\quad \pi_e = \langle x_i \rangle_{i=1}^{N+2k} \quad \text{where} \quad \pi = \langle x_i \rangle_{i=k+1}^{N+k}$$

Let $g \in S_{2k}^h(\Delta;d)$ be the discrete spline of order $2k$ interpolating a function f defined on π

$$\text{(6.2)}\quad D_{-h}^{r-1}(g-f)(y_i+(k-1)h) = 0 \quad i = 0,\ldots,n, \quad r = 1,\ldots,d_i$$

Note that $D_{-h}^{r-1}\varphi(x) = 0$ $r=1,2,\ldots,m$ implies $\varphi(x-jh) = 0$

$j = 0,1,\ldots,m-1$. Thus if we define

$$z_i = x_{i+k} + (k-1)h \qquad i = 1,2,\ldots,N \tag{6.3}$$

then

$$z_{i+1} = a + ih, \quad z_{N-i} = b-ih \quad i = 0,1,\ldots,k-1$$

and (6.2) can be stated as

$$g(z_i) = f(z_i) \quad i = 1,2,\ldots,N$$

Since $N^h_{i,2k}(z_i) \neq 0 \quad i = 1,2,\ldots,N$ it follows that there exists a unique $g \in S_{2k}(\Delta;d)$ satisfying (6.2) ([8]p.92).

We next state discrete analogs of the first and second integral relation, the best approximation property, and minimum norm property.

Thus if Δ and Δ satisfies (6.1), $f:[a,b]_h \to \mathbb{R}$ is given and $g \in S^h_{2k}(\Delta;d)$ satisfies (6.2) then the first integral relation is

$$\int_{y_0}^{y_n} [D^k_h f(t)]^2 d_h t = \int_{y_0}^{y_n} [D^k_h g(t)]^2 d_h t + \int_{y_0}^{y_n} [D^k_h (f-g)(t)]^2 d_h t.$$

The second integral relation takes the form

$$\int_{y_0}^{y_n} [D^k_h (f-g)]^2 d_h t = (-1)^k \int_{y_0}^{y_n - hk} (f-g)(t) D^{2k}_h f(t) d_h t.$$

Moreover for any $w \in S^h_{2k}(\Delta;d)$ we have the best approximation property

$$\int_{y_o}^{y_n} [D_h^k(f-g)(t)]^2 d_h t \leq \int_{y_o}^{y_n} [D_h^k(f-w)(t)]^2 d_h t$$

with equality only if $w-g \in \mathbb{P}_k$. Finally the minimum norm property is

$$\int_{y_o}^{y_n} [D_h^k g(t)]^2 d_h t < \int_{y_o}^{y_n} [D_h^k w(t)]^2 d_h t$$

for any $w: [a,b]_h \to \mathbb{R}$ such that w interpolates f according to (6.2) and $w \neq g$.

All these properties follow using the integration by parts formula (2.4) (See [8]p.94-97).

To state some error bounds we introduce the usual E_{rs}, i.e. for $1 \leq s \leq 2k$

$$E_{rs}(t) = \begin{cases} D_h^r(f-g)(t) & 0 \leq r < s \\ D_h^r g(t) & s \leq r \leq 2k \end{cases}$$

Recall that $I_j = [a, b-(j-1)h]_h$ and $\bar{\Delta} = \max_{1 \leq i \leq n} (y_i - y_{i-1})$

<u>THEOREM 6.1</u>

<u>Suppose</u> $1 \leq k < s \leq 2k$ <u>and</u> $2 \leq q \leq \infty$. <u>Let</u> Δ <u>and</u> d <u>satisfy</u> (6.1). <u>Given</u> $f: [a,b]_h \to \mathbb{R}$ <u>let</u> $g \in S_{2k}^h(\Delta;d)$ <u>be given by</u> (6.2). <u>Then for</u> $0 \leq r < k$

$$\|E_{rs}\|_{L_h^q(I_r)} \leq K\bar{\Delta}^{s-1-r}\, \omega(D_h^{s-1} f; \bar{\Delta}; I_s)$$

<u>Moreover if</u> $1 \leq p \leq q$ <u>then for</u> $1 \leq k \leq s \leq 2k$ <u>and</u>

$0 \leq r < 2k$

$$\|E_{rs}\|_{L_h^q(I_r)} \leq K\bar{\Delta}^{-s-r+1/q-1/p} \|D_h^s f\|_{L_h^p(I_s)}.$$

K <u>is a constant that only depends on</u> k, r, <u>and</u> s <u>and the global meshratio</u> $\max_{i,j} (y_i - y_{i-1})/(y_j - y_{j-1})$

This theorem follows from theorem 4.1, and theorem 5.2 ([8]p.99).

We also have a theorem involving differences of order less than k on the right hand side, and interpolation to approximate data as in [16].

Thus let $g \in S_{2k}^h(\Delta;d)$ be given by

$$(6.4)\quad D_{-h}^j g(y_i+(k-1)h) = \alpha_{ij} \quad 0 \leq j < d_i, \quad 0 \leq i \leq n$$

For fixed s we assume that functions $F_i(f;\Delta)$ exists such that for $i = 0,1,\ldots,n$

$$(6.5)\quad K\bar{\Delta}^{s-r} F_i(f;\Delta) \geq \begin{cases} D_{-h}^r |f(y_{i-1}+(k-1)h)-\alpha_{ir}| & 0 \leq r < \min(d_i,s) \\ |\alpha_{ir}| & \min(d_i,s) \leq r < d_i \end{cases}$$

Here K depends only on k, r, s, and the global meshratio, and f is the function we really want to interpolate. Let

$$\|F(f;\Delta)\|_p = \begin{cases} \left(\bar{\Delta} \sum_{i=0}^{n} [F_i(f;\Delta)]^p\right)^{1/p} & 1 \leq p < \infty \\ \max_{0 \leq i \leq n} F_i(f;\Delta) & p = \infty \end{cases}$$

Then we have

THEOREM 6.2

Suppose $2 \le q \le \infty$, $1 \le s \le 2k$, *and* $0 < r < 2k$. *Let* Δ *and* d *satisfy* (6.1) *and assume in addition that* $y_i - y_{i-1} \ge (2k+1)h$ $i = 1,\dots,n$. *Given* $f\colon [a,b]_h \to \mathbb{R}$ *let* $g \in S^h_{2k}(\Delta;d)$ *satisfy* (6.4) *where* α_{ij} *satisfy* (6.5). *Then*

$$(6.6)\quad \|E_{rs}\|_{L^q_h(I_r)} \le K\bar{\Delta}^{s-r-1}\{\omega(D^{s-1}_h f;\Delta;I_s) + \bar{\Delta}\|F(f;\Delta)\|_\infty\}$$

where K *only depends on* k,r,s, *and the global mesh-ratio. Moreover for* $1 \le p \le q$

$$\|E_{rs}\|_{L^q_h(I_r)} \le K\bar{\Delta}^{s-r+1/q-1/p}\{\|D^s_h f\|_{L^p_h(I_s)} + \|F(f;\Delta)\|_p\}$$

We give an example of application of this theorem

COROLLARY 6.3

Suppose we let g *be given by* (6.4) *where*

$$\alpha_{ij} = D^j_{-h} f(y_i + (k-1)h) \quad 0 \le j < d_i \quad 1 \le i \le n-1$$

$$\alpha_{oj} = D^j_{-h} Qf(a+(k-1)h),\ \alpha_{nj} = D^j_{-h} Qf(b) \quad 0 \le j \le k-1$$

where Q *is the local spline approximation method in theorem* 3.4. *Then if* $2 \le q \le \infty$, $1 \le s \le 2k$, $0 \le r < 2k$, Δ, d, *satisfy* (6.1), *and* $y_i - y_{i-1} \ge (2k+1)h$ *we have*

$$\|E_{rs}\|_{L^q_h(I_r)} \le K\bar{\Delta}^{s-r-1}\omega(D^{s-1}_h f;\bar{\Delta};I_s)$$

<u>where</u> K <u>depends</u> <u>only</u> <u>on</u> k,r,s, <u>and</u> <u>the</u> <u>global</u> <u>mesh-ratio</u>. <u>Moreover</u> <u>for</u> $1 \leq p \leq q$

$$\|E_{rs}\|_{L_h^q(I_r)} \leq K\bar{\Delta}^{s-r+1/q-1/p}\|D_h^s f\|_{L_h^p(I_s)}$$

By using theorem 4.1 a similar theorem holds if we smooth f by Lagrange interpolation at the ends.

It follows from corollary 6.3 that there exists a discrete interpolating spline g_h such that

$$\|D_h^r(f-g_h)\|_{L_h^\infty(I_r)} \leq K\bar{\Delta}^{s-r}\|D_h^s f\|_{L_h^\infty(I_s)} \qquad 0 \leq r < s \leq 2k$$

where K depends only on k,r,s, and the global mesh-ratio. We can let h go to zero here. Hence if $f \in C^s[a,b]$ there exists a spline g og order 2k interpolating f at the interior knots $y_1,\ldots,y_{n-1}$, such that

$$\|D^r(f-g)\|_{L^\infty[a,b]} \leq K\bar{\Delta}^{s-r}\|D^s f\|_{L^\infty[a,b]}, \qquad 0 \leq r < s \leq 2k$$

References

1. Astor, P.H.and C.S.Duris, Discrete L-splines, Numer. Math. 22 (1974), 393-402.
2. de Boor, C., On uniform approximations by splines, J. Approximation Th. 1 (1968), 219-235.
3. de Boor, C., On calculating with B-splines, J. Approximation Th. 6 (1972), 50-62.
4. de Boor, C., and G.J.Fix, Spline Approximation by Quasi interpolants, J.Approximation Th. 8 (1973).19-45.
5. de Boor, C.,The quasi-interpolant as a tool in elementary polynomial spline theory, in Approximation Theory, G.G.Lorentz,ed,.Academic Press,New York,1973, 269-276.

6. Curry, H.B., and I.J.Schoenberg, On Pólya frequency functions IV: The Fundamental spline functions and their limits, J.d'Analyse 17 (1966), 71-107.

7. Douglas, J., Jr., T.Dupont, and L.Wahlbin, Optimal L_∞ error estimates for Galerkin approximations to solutions of two-point boundary value problems, Math. Comp. 29 (1975), 475-483.

8. Lyche, T., Discrete polynomial spline approximation methods, report RRI 2, Univ. of Oslo, 1975.

9. Lyche, T., Discrete cubic spline interpolation, report RRI 5, Univ. of Oslo, 1975.

10. Lyche, T., and L.L.Schumaker, Local spline approximation methods, to appear in J.Approximation Th., also MRC TSR 1417, Univ. of Wisconsin, Madison, 1974.

11. Malcolm, M.A., Nonlinear spline functions, report Stan-CS-73-372, Stanford University, 1973.

12. Mangasarian, O.L., and L.L.Schumaker, Discrete splines via Mathematical Programming, SIAM J.Control 9 (1971), 174-183.

13. Mangasarian, O.L., and L.L.Schumaker, Best summation formulae and discrete splines, SIAM J. Numer.Anal. 10 (1973), 448-459.

14. Schoenberg, I.J., On spline functions, in Inequalities, O.Shisha, ed., Academic Press, New York, 1967, 255-291.

15. Schumaker, L.L., Constructive aspects of discrete polynomial spline functions, in Approximation Theory, G.G.Lorentz, ed., Academic Press, New York, 1973, 469-476.

16. Swartz, B.K., and R.S.Varga, Error bounds for spline and L-spline interpolation, J.Approximation Th. 6 (1972), 6-49.

Aknowledgement

I am indepted to my supervisor prof. L.L.Schumaker.

Dr. T. Lyche,
Department of Mathematics
University of Oslo
Oslo 3, Norway.

Periodische Splinefunktionen

von

Günter Meinardus

Periodische Splinefunktionen sind für die Behandlung periodischer Funktionen von besonderem Interesse. Die aus der Periodizität entspringenden strukturellen Besonderheiten haben es ermöglicht, bei vielen Fragen zu übersichtlichen Resultaten zu gelangen. Dies gilt speziell für den Fall äquidistanter Knoten (vgl. etwa [1], [3], [6], [10], [11]). In der vorliegenden Arbeit soll neben einer kurzen Behandlung von Basisdarstellungen das Interpolationsproblem und, im kubischen Fall, die Abschätzung der Normen zugehöriger Operatoren bei beliebiger Knotenverteilung untersucht werden. Hierbei ergibt sich u.a. auch ein Beweis für eine auf M. MARSDEN [9] zurückgehende Vermutung (die kürzlich auf anderem Wege schon von DE BOOR [5] bewiesen wurde).

1. Definition und Basisdarstellung

Wir beschränken uns auf einfache Knoten. Es seien n und m natürliche Zahlen mit $n \geq 1$ und $m \geq 2$. Es sei ferner eine Knotenmenge K_n vorgegeben, bestehend aus reellen Zahlen $x_\nu, \nu \in \mathbb{Z}$, mit

$$x_\nu < x_{\nu+1} \text{ und } x_\mu = x_\nu + r(x_n - x_0) \text{ für } \mu = \nu + rn.$$

Eine reelle Funktion s der reellen Variablen x heißt eine

periodische Splinefunktion der Ordnung m zur Knotenmenge K_n, wenn sie die folgenden Eigenschaften besitzt:

1. $s \in C^{m-2}(-\infty,+\infty)$,
2. $s(x+(x_n-x_0)) = s(x)$ für alle x,
3. die Restriktion von s auf das Intervall $[x_{\nu-1},x_\nu]$ gehört zum Raum Π_{m-1} der Polynome vom Grad $\leq m-1$ $(\nu \in \mathbb{Z})$.

Der lineare Raum dieser Funktionen sei mit $\mathcal{S}_m(K_n)$ bezeichnet. Die Restriktion einer Funktion $s \in \mathcal{S}_m(K)$ auf das Intervall $[x_{\nu-1},x_\nu]$ heiße p_ν. Auf Grund der Eigenschaften 1. und 2. gibt es eindeutig bestimmte Zahlen c_ν, so daß

$$p_{\nu+1}(x) = p_\nu(x)+c_\nu(x-x_\nu)^{m-1} \tag{1}$$

und

$$p_{n+\nu}(x) = p_\nu(x+x_n-x_0) \tag{2}$$

für alle $\nu \in \mathbb{Z}$ gilt.

Zur Vereinfachung der Schreibweise nehmen wir hier ohne Beschränkung an, daß

$$x_0 = 0 \quad \text{und} \quad x_n = 1$$

sei.

Spezielle Funktionen aus $\mathcal{S}_m(K_n)$ sind

$$\varphi_0(x) \equiv 1$$

und $\quad \varphi_\nu(x) = \frac{1}{m}\{B_m^*(x-x_\nu)-B_m^*(x)\}, \; \nu = 1,2,\ldots,n-1.$

Dabei ist $B_m^*(x)$ die periodische Fortsetzung des m-ten Bernoulli'schen Polynoms:

$$B_m^*(x) = B_m(x-[x]),$$

wobei [x] die größte ganze Zahl $\leq x$ bedeutet. Für $\nu = 1,2,..,n-1$ ist

$$\varphi_\nu(x) = \begin{cases} \frac{1}{m}(B_m(x+1-x_\nu)-B_m(x)) & \text{für } 0\leq x\leq x_\nu \\ \frac{1}{m}(B_m(x-x_\nu)-B_m(x)) & \text{für } x_\nu\leq x\leq 1. \end{cases}$$

Insbesondere folgt (vgl. [8]):

$$\frac{1}{m}(B_m(x+1-x_\nu)-B_m(x))-\frac{1}{m}(B_m(x-x_\nu)-B_m(x)) = (x-x_\nu)^{m-1}.$$

Offensichtlich sind die Funktionen φ_ν ($\nu=0,1,...,n-1$) linear unabhängig.

Satz 1. Jede Funktion $s\in \mathcal{S}_m(K_n)$ besitzt eine eindeutige Darstellung der Form

$$s = a_0 + \sum_{\nu=1}^{n-1} a_\nu\varphi_\nu$$

(Für n=1 ist die Summe durch Null zu ersetzen).

Beweis: Für n=1 gilt

$$p_2(x) = p_1(x+1) = p_1(x)+c_1(x-1)^{m-1},$$

somit

$$c_1 = 0 \quad \text{und} \quad p_1(x+1) = p_1(x).$$

Hieraus folgt

$$p_1 = \text{const} = a_0 \equiv s.$$

Für n>1 ist mit der oben definierten Zahl c_{n-1} sicher

$$s+c_{n-1}\varphi_{n-1}$$

eine Splinefunktion aus $\mathfrak{S}_m(K_n)$, bei der die Knoten $x_{n-1+rn}, r\in\mathbb{Z}$, irrelevant sind. Durch vollständige Induktion nach n ergibt sich die Behauptung.

<u>Korollar</u>: Es gilt

$$\dim \mathfrak{S}_m(K_n) = n.$$

Bemerkung: Es ist wegen der bekannten Eigenschaften der Bernoulli'schen Polynome

$$a_0 = \int_0^1 s(x)dx$$

Die Basisdarstellung aus Satz 1 liefert auch für die anderen Fourierkoeffizienten von s einfache Beziehungen.

Eine zweite Art der Basisdarstellung in $\mathfrak{S}_m(K_n)$ läßt sich aus sog. B-Splines gewinnen (vgl. [4]). Wir verzichten auf eine explizite Angabe. Die algebraische Struktur der zwischen diesen Darstellungen bestehenden Transformation wäre für viele (z.B. auch für numerische) Zwecke von Interesse.

<u>2. Das Interpolationsproblem.</u>

Die Aufgabe (in der Literatur häufig behandelt) lautet: Zu gegebenen Zahlen $y_\nu, \nu\in\mathbb{Z}$, mit

$$y_\mu = y_\nu \quad \text{für } \mu \equiv \nu \bmod n$$

soll ein $s\in\mathcal{S}_m(K_n)$ gefunden werden mit der Eigenschaft:

$$s(x_\nu) = y_\nu \quad , \ \nu\in\mathbb{Z}.$$

Für gerade Ordnung m hat dieses Problem stets eine und nur eine Lösung. Aus einer weiter unten folgenden Betrachtung ergab sich die Vermutung, daß auch für ungerade Ordnung stets genau eine Lösung existiert, falls nur n auch ungerade ist. Im äquidistanten Fall ist dies lange bekannt (vgl. [1]). Ein Beweis für diese Vermutung wurde von Herrn DE BOOR in der in diesem Tagungsbericht enthaltenen Arbeit gegeben. In der Zwischenzeit machte mich Herr G. MERZ darauf aufmerksam, daß diese Aussage bereits von F. KRINZESZA [7] bewiesen wurde.

Der Einfachheit halber sei von nun an vorausgesetzt, daß die Interpolationsaufgabe stets genau eine Lösung besitzt. Für die lösende Splinefunktion s gilt dann:

$$p_{\nu+1}(x) = p_\nu(x)+(y_{\nu+1}-p_\nu(x_{\nu+1}))\left(\frac{x-x_\nu}{x_{\nu+1}-x_\nu}\right)^{m-1} \tag{3}$$

und

$$p_{n+1}(x) = p_1(x+1). \tag{4}$$

Die Transformation

$$x = x_{\nu-1}+t(x_\nu-x_{\nu-1})$$

führt das Intervall $x_{\nu-1}\leq x\leq x_\nu$ in das Intervall $0\leq t\leq 1$ über. Mit den Bezeichnungen ($\nu\in\mathbb{Z}$):

$$d_\nu = \frac{x_{\nu+1}-x_\nu}{x_\nu - x_{\nu-1}}$$

und

$$q_\nu(t) = p_\nu(x_{\nu-1}+t(x_\nu - x_{\nu-1}))$$

folgt aus (3) und (4) sofort

$$q_{\nu+1}(t) = q_\nu(1+d_\nu t)+(y_{\nu+1}-q_\nu(1+d_\nu))t^{m-1} \quad . \tag{5}$$

Man beachte noch die Gültigkeit der Beziehungen

$$q_\mu(t) = q_\nu(t)$$

und

$$d_\mu = d_\nu$$

für $\mu \equiv \nu$ mod n sowie

$$d_1 d_2 \ldots d_n = 1. \tag{6}$$

Wir führen die Vektoren

$$q(t) = \begin{pmatrix} q_1(t) \\ q_2(t) \\ \cdot \\ \cdot \\ \cdot \\ q_n(t) \end{pmatrix} \quad \text{und } y = \begin{pmatrix} y_1 \\ y_2 \\ \cdot \\ \cdot \\ \cdot \\ y_n \end{pmatrix}$$

ein. Wegen der vorausgesetzten eindeutigen Lösbarkeit der Interpolationsaufgabe existiert eine quadratische n-reihige Matrix

$$W(t) = W(t,D)$$

mit

$$q(t) = W(t,D)y.$$

Die Elemente von W sind Polynome aus Π_{m-1}. Weiter steht D für die Diagonalmatrix

$$D = \text{diag } (d_\nu) \quad , \nu = 1,2,\ldots,n.$$

Es gilt

$$D = I = \text{Einheitsmatrix}$$

genau für den Fall der Äquidistanz der Knoten:

$$x_{\nu+1}-x_\nu = h, \text{ unabhängig von } \nu.$$

Dieser Fall wurde häufig untersucht. Eine Konstruktion von W(t,I) für gerade Ordnung ist in [10] angegeben.

Für die weiteren Betrachtungen benötigen wir die Permutationsmatrix

$$T = ((t_{\nu\mu})) \text{ mit } t_{\nu\mu} = \begin{cases} 1 \text{ für } \nu-\mu \equiv 1 \bmod n, \\ 0 \text{ sonst,} \end{cases}$$

$(\nu,\mu = 1,2,\ldots,n)$. Es gilt

$$T^n = I.$$

Eine n-reihige quadratische Matrix R heißt zirkulant, wenn sie mit T vertauschbar ist, d.h. wenn

$$TR = RT$$

gilt. In diesem Fall gibt es Zahlen $r_0,r_1,\ldots,r_{n-1}$, so daß

$$R = r_0I+r_1T+\ldots+r_{n-1}T^{n-1}$$

gilt. Eine zirkulante Matrix kann somit durch eine, von dem speziellen R unabhängige, unitäre Matrix diagonalisiert werden.

Uns interessiert die Matrix W(t,D).

<u>Satz 2</u>. Die Matrix W(t,D) besitzt die Eigenschaften

1. $W(1,D) = I,$ (7)

2. $W^{(\mu)}(0,D) = TD^{\mu}W^{(\mu)}(1,D), \ \mu=0,1,\ldots,m-2$ (8)

Sie ist durch (7) und (8) eindeutig bestimmt.

Beweis: Die Eigenschaft (7) folgt unmittelbar aus der Interpolationsbedingung

$$q_\nu(1) = y_\nu \quad , \quad \nu\in\mathbb{Z}.$$

Die Zugehörigkeit von s zur Klasse C^{m-2} läßt sich vermöge (5) algebraisch als

$$q_{\nu+1}^{(\mu)}(0) = d_\nu^{\mu}\, q_\nu^{(\mu)}(1);\ \nu = 1,2,..,n;\ \mu = 0,1,\ldots,m-2$$

formulieren. Vektoriell heißt dies

$$q^{(\mu)}(0) = TD^{\mu}q^{(\mu)}(1),$$

woraus (8) folgt. - Die eindeutige Festlegung von W(t,D) durch (7) und (8) ist durch die Annahme der eindeutigen Lösbarkeit des Interpolationsproblems gegeben.

Aus Satz 2 ziehen wir einige Folgerungen.

Satz 3. Die Matrix $W(t,I)$ ist zirkulant.

Bemerkung: Diese Aussage begründet, weshalb der äquidistante Fall einer so einfachen Behandlung zugänglich ist wie in [10] (vgl. hierzu bereits [1]).

Beweis: Man rechnet leicht nach, daß die Matrix

$$TW(t,I)T'$$

die für $W(t,I)$ durch (7) und (8) gegebenen Bedingungen erfüllt. Also ist

$$TW(t,I)T' = W(t,I)$$

und somit $W(t,I)$ zirkulant.

Satz 4. Für jede Matrix D gilt

$$TW(t,T'DT)T' = W(t,D).$$

Beweis: Wieder sieht man, daß die Matrix $TW(t,T'DT)T'$ die gleichen Bedingungen (7) und (8) erfüllt wie die Matrix $W(t,D)$.

Bemerkung: Der Satz 4 ergibt für $D = I$ noch einmal den Satz 3. Seine allgemeine Aussage ist äquivalent zur Auswirkung einer Knotenverschiebung auf den interpolierenden Splines. Man betrachtet die neuen Knoten

$$\tilde{x}_\nu = x_{\nu+1}.$$

Dann wird

$$\tilde{d}_\nu = \frac{\tilde{x}_{\nu+1}-\tilde{x}_\nu}{\tilde{x}_\nu-\tilde{x}_{\nu-1}} = d_{\nu+1},$$

also

$$\tilde{D} = \begin{pmatrix} \tilde{d}_1 & & & 0 \\ & \tilde{d}_2 & & \\ & & \ddots & \\ 0 & & & \tilde{d}_n \end{pmatrix} = \begin{pmatrix} d_2 & & & & 0 \\ & d_3 & & & \\ & & \ddots & & \\ & & & d_n & \\ 0 & & & & d_1 \end{pmatrix} = T'DT.$$

Satz 5. Mit der Permutationsmatrix

$$S = ((s_{\nu\mu})), \; s_{\nu\mu} = \begin{cases} 1 \text{ für } \nu+\mu \equiv 1 \bmod n, \\ 0 \text{ sonst,} \end{cases}$$

$(\nu,\mu = 1,2,..,n)$, gilt für jede Matrix D die Relation

$$TSW(1-t, SD^{-1}S)S = W(t,D).$$

Beweis: Man beachte, daß S und TS symmetrische Permutationsmatrizen sind, woraus

$$S^2 = I \quad \text{und} \quad TSTS = I$$

folgt. Ansonsten verwende man den Satz 2.

Bemerkung: Die Aussage von Satz 5 basiert auf einer Spiegelung der Knoten am Punkt $\frac{1}{2}$:

$$\hat{x}_\nu = 1-x_{n-\nu}$$

und gibt die Wirkung dieser neuen Knoten auf die interpolierende Splinefunktion. Es ist (in naheliegender Bezeichnung):

$$\hat{d}_\nu = \frac{1}{d_{n-\nu}}$$

und

$$\hat{D} = SD^{-1}S.$$

Die in den Sätzen 4 und 5 ausgesprochenen Transformationseigenschaften sollen noch zu einigen Betrachtungen Anlaß geben. Wir definieren eine Matrix H(t,D) durch die Summe

$$H(t,D) = \frac{1}{2n}\sum_{\nu=0}^{n-1}\{W(t,T^{-\nu}DT^{\nu})+W(t,ST^{\nu}D^{-1}T^{-\nu}S)\}$$

Die Sätze 4 und 5 ergeben die Darstellung

$$H(t,D) = \frac{1}{2n}\sum_{\nu=0}^{n-1}\{T^{-\nu}W(t,D)T^{\nu}+ST^{\nu-1}W(1-t,D)T^{-\nu}S\}.$$

Es folgt

1. $H(t,I) = W(t,I)$,
2. $H(0,D) = T$, $H(1,D) = I$,
3. $TH(t,D) = H(t,D)T$, d.h. $H(t,D)$ ist zirkulant.
4. $ST^{-1}H(1-t,D)S = H(t,D)$.

Die Eigenschaften 2.,3. und 4. hat H(t,D) mit W(t,I) gemeinsam. Ferner gilt mit der Zeilensummennorm

$$\sup_{0\leq t\leq 1}\|H(t,D)\| \leq \sup_{0\leq t\leq 1}\|W(t,D)\|.$$

Die rechte Seite dieser Ungleichung ist gleich der (Supremum-)Norm des zugehörigen Spline-Interpolationsoperators. Die Vermutung

$$\sup_{0\leq t\leq 1}\|H(t,D)\| \geq \sup_{0\leq t\leq 1}\|W(t,I)\| \tag{9}$$

würde zeigen, daß die minimale Operatornorm für D = I, d.h. für die äquidistante Verteilung angenommen wird. Mir ist der Nachweis von (9) jedoch nicht gelungen.

3. Abschätzung der Operatornorm im kubischen Fall (m=4).

Durch die Beziehungen (7) und (8) kann man im Prinzip die Matrix W(t,D) berechnen. Für allgemeine Werte von n ist dies schwierig, auch für den kubischen Fall, auf den wir uns hier beschränken. Unser Ziel ist eine Abschätzung der Operatornorm

$$L(D) = \sup_{0 \leq t \leq 1} \|W(t,D)\|.$$

Für D = I ist diese Norm explizit bekannt (vgl. [3], [10]). Auch für allgemeines D gibt es eine Reihe von Aussagen (vgl. [2], [9]).

Unter Benutzung von

$$W(0,D) = T,$$

$$W(1,D) = I$$

und

$$W'(0,D) = TDW'(1,D)$$

ergibt sich das der Hermite'schen Interpolationsformel entsprechende

Lemma 1. Es gilt

$$W(t,D) = (1-t)^2(2t+1)T+t^2(3-2t)I+t(1-t)\{(1-t)TD-tI\}W'(1,D), \quad (1$$

Weiter folgt

Lemma 2. Es ist

$$L(D) \leq 1+\frac{1}{4}\|D\|\circ\|W'(1,D)\|. \qquad (11)$$

Beweis: Zunächst ist

$$\|(1-t)^2(2t+1)T+t^2(3-2t)I\| = (1-t)^2(2t+1)+t^2(3-2t) = 1.$$

Weiter folgt

$$\|(1-t)TD-tI\| = t+(1-t)\operatorname{Max}_{\nu} d_{\nu} \leq \operatorname{Max}_{\nu} d_{\nu},$$

da mindestens eine der Zahlen d_{ν} wegen (6) größer oder gleich 1 ist.

Bemerkung: Der Wert von $\|W'(1,I)\|$ ist relativ leicht berechenbar. Es ist stets

$$\|W'(1,I)\| \leq 3,$$

so daß aus (11) die Abschätzung

$$L(I) \leq \frac{7}{4}$$

folgt.

Die kleinste obere Schranke ist für alle äquidistanten Zerlegungen die Zahl

$$\frac{1}{4}(1+3\sqrt{3}) \approx 1,549...$$

(vgl. [2], [10])

Lemma 3. Die Matrix $W'(1,D)$ genügt der Beziehung

$$(I+2TD+2TD^2+TD^2TD)W'(1,D) = 3(I+TD^2)(I-T). \qquad (12)$$

Beweis: Man benutze (10) und die Relation

$$W''(0,D) = TD^2W''(1,D).$$

Im folgenden bezeichne ρ die kleinste reelle Zahl mit $\rho \geq 1$ mit der Eigenschaft, daß

$$\frac{1}{\rho} \leq d_\nu \leq \rho$$

für alle ν gilt.

<u>Lemma 4</u>. Es gibt eine und nur eine n-reihige Diagonalmatrix

$$B = \text{diag}\,(b_\nu) \quad , \quad b_\nu > 2 \text{ für alle } \nu,$$

so daß die folgende Zerlegung gilt:

$$(I+2TD+2TD^2+TD^2TD) = (TD^2+B^{-1})(TD+B). \tag{13}$$

Die Elemente b_ν genügen den Beziehungen

$$b_\nu = 2+\frac{2}{d_\nu}-\frac{1}{d_\nu b_{\nu+1}}, \quad \nu = 1,2,..,n, \tag{14}$$

mit $b_{n+1} = b_1$.

Ferner bestehen die Ungleichungen

$$\frac{\lambda(\rho)}{\rho} \leq b_\nu \leq \lambda(\rho) \quad \text{für alle } \nu \tag{15}$$

mit $\lambda(\rho) = 1+\rho+\sqrt{1+\rho+\rho^2}$.

Bemerkung: Im äquidistanten Fall ist $\rho = 1$ und somit nach (15):

$$B = (2+\sqrt{3})I.$$

Dies ist wegen $D = I$ aus der (13) entsprechenden Zerlegung

$$I+4T+T^2=(T+(2-\sqrt{3})I)(T+(2+\sqrt{3})I)$$

sofort ersichtlich.

Für allgemeines $\rho \geqq 1$ ist noch

$$2 < \frac{\lambda(\rho)}{\rho} \leq 2+\sqrt{3}$$

von Interesse.

Beweis von Lemma 4: Im $\mathbb{R}^n$ mit den Vektoren

$$b = \begin{pmatrix} b_1 \\ b_2 \\ \cdot \\ \cdot \\ b_n \end{pmatrix}$$

betrachten wir die Abbildung f des Quaders

$$\frac{\lambda(\rho)}{\rho} \leqq b_\nu \leqq \lambda(\rho) \quad , \quad \nu = 1,2,..,n,$$

komponentenweise definiert durch

$$f_\nu(b) = 2+\frac{2}{d_\nu} - \frac{1}{d_\nu b_{\nu+1}} \quad , \quad \nu = 1,2,...,n,$$

wobei $b_{n+1} = b_1$ gesetzt werden. Diese Abbildung ist im Quader stetig. Sie bildet diesen in sich ab:

Es ist

$$2+\frac{2}{d_\nu} - \frac{1}{d_\nu b_{\nu+1}} \leqq 2 + \frac{1}{d_\nu}\left(2 - \frac{1}{\lambda(\rho)}\right)$$

$$\leqq 2 + \rho\left(2 - \frac{1}{\lambda(\rho)}\right) = \lambda(\rho)$$

und

$$2 + \frac{2}{d_\nu} - \frac{1}{d_\nu b_{\nu+1}} \geq 2 + \frac{1}{d_\nu}\left(2 - \frac{\rho}{\lambda(\rho)}\right)$$

$$\geq 2 + \frac{1}{\rho}\left(2 - \frac{\rho}{\lambda(\rho)}\right) = \frac{\lambda(\rho)}{\rho} .$$

Nach dem Brouwerschen Fixpunktsatz gibt es daher im Quader einen Fixpunkt. Dieser ist sogar im Bereich

$$b_\nu > 2 \quad , \quad \nu = 1,2,..,n,$$

der einzige. Denn für zwei dort existierende Fixpunkte b und $\tilde{b}$ mit den Komponenten b_ν und $\tilde{b}_\nu$ muß gelten

$$b_1 - \tilde{b}_1 = \frac{b_2 - \tilde{b}_2}{d_1 \circ b_2 \tilde{b}_2}$$

$$= \frac{b_3 - \tilde{b}_3}{d_1 d_2 \cdot b_2 b_3 \tilde{b}_2 \tilde{b}_3} = \ldots$$

$$= \frac{b_1 - \tilde{b}_1}{(b_1 b_2 \ldots b_n)(\tilde{b}_1 \tilde{b}_2 \ldots \tilde{b}_n)}$$

wegen (6) und $b_{n+1} = b_1$, $\tilde{b}_{n+1} = \tilde{b}_1$. Hieraus folgt $b_1 = \tilde{b}_1$ und insgesamt $b = \tilde{b}$.

Die mit dem Fixpunkt b gebildete Matrix

$$B = \text{diag}\ (b_\nu)$$

genügt nach (14) der Beziehung

$$T'B^{-1}T + DB = 2(I+D)$$

oder

$$B^{-1}TD+TD^2B = 2TD+2TD^2,$$

was mit (13) gleichbedeutend ist.

Man erkennt aus der eben bewiesenen Identität, daß die Matrix

$$I+2TD+2TD^2+TD^2TD$$

nicht singulär ist, denn es ist unter Verwendung von (6):

$$\det(TD^2+B^{-1}) = \frac{1}{b_1b_2\ldots b_n} + (-1)^{n-1} \neq 0$$

und

$$\det\ (TD+B) = b_1b_2\ldots b_n+(-1)^{n-1} \neq 0.$$

Man kann dies aber auch ohne die Zerlegung (13) zeigen.

Zur Vereinfachung der Schreibweise seien die Zahlen b_ν für alle $\nu\in\mathbb{Z}$ durch

$$b_\mu = b_\nu \quad \text{für} \quad \mu \equiv \nu \bmod n$$

definiert.

<u>Lemma 5</u>. Für alle $\nu\in\mathbb{Z}$ und alle $r\in\mathbb{Z}$ mit $r\geq 0$ gelten die Ungleichungen

$$\frac{d_\nu}{b_\nu} \leq \frac{\rho^2}{\lambda(\rho)} \tag{16}$$

und

$$\frac{1}{d_\nu^2b_\nu d_{\nu+1}^2b_{\nu+1}\cdots d_{\nu+r}^2b_{\nu+r}} \leq \frac{4\rho-1}{3\rho}\left(\frac{\rho^2}{\lambda(\rho)}\right)^{r+1}. \tag{17}$$

Beweis: Zunächst gilt mit (15) sofort

$$\frac{d_\nu}{b_\nu} \leq \frac{d_\nu \rho}{\lambda(\rho)} \leq \frac{\rho^2}{\lambda(\rho)} .$$

Der Nachweis von (17) ist komplizierter. Zur Abkürzung setzen wir

$$\lambda_1 = \lambda(\rho) \quad \text{und} \quad \lambda_2 = \frac{\rho}{\lambda_1}$$

Wir zeigen induktiv das Bestehen der Ungleichung

$$\frac{1}{d_\nu^2 b_\nu \ldots d_{\nu+r}^2 b_{\nu+r}} \leq \frac{2\rho^{2r} \cdot \sqrt{1+\rho+\rho^2}}{d_{\nu+r}^2 \{(\lambda_1^{r+1} - \lambda_2^{r+1}) b_{\nu+r} - \rho(\lambda_1^r - \lambda_2^r)\}} . \tag{18}$$

Für r=0 gilt in (18) sogar das Gleichheitszeichen. Zur Durchführung der Induktion braucht man nur zu beachten, daß beim Einsetzen von

$$b_{\nu+r} = 2 + \frac{1}{d_{\nu+r}} \left(2 - \frac{1}{b_{\nu+r+1}}\right)$$

in die rechte Seite von (18) der Wert höchstens vergrößert wird, wenn $d_{\nu+r}$ durch $\frac{1}{\rho}$ ersetzt wird. So ergibt sich die Gültigkeit von (18) leicht.
Aus

$$b_\nu \geq \frac{\lambda(\rho)}{\rho}$$

folgt mit (14) sofort

$$b_{\nu+r} \geq 2 + \frac{1-\rho+\sqrt{1+\rho+\rho^2}}{d_{\nu+r}} .$$

Setzt man dies in (18) ein, so darf wieder $d_{\nu+r}$ durch $\frac{1}{\rho}$ ersetzt werden. Insgesamt folgt

$$\frac{1}{d_\nu^2 b_\nu \ldots d_{\nu+r}^2 b_{\nu+r}} \leq \frac{2\rho^{2r+2} \circ \sqrt{1+\rho+\rho^2}}{(\lambda_1^{r+1}-\lambda_2^{r+1})(2+\rho(1-\rho+\sqrt{1+\rho+\rho^2}))-\rho(\lambda_1^r-\lambda_2^r)}$$

$$\leq \left(\frac{\rho^2}{\lambda_1}\right)^{r+1} \frac{2\sqrt{1+\rho+\rho^2}}{(1+\rho)(1-\rho+\sqrt{1+\rho+\rho^2})}$$

$$\leq \left(\frac{4\rho-1}{3\rho}\right)\left(\frac{\rho^2}{\lambda(\rho)}\right)^{r+1} .$$

<u>Lemma 6</u>. Die folgenden Abschätzungen sind gültig:

$$\|(TD+B)^{-1}\| \leq \frac{\rho \circ v_n(\rho)}{\lambda(\rho)(1-2^{-n})}, \tag{19}$$

$$\|(TD^2+B^{-1})^{-1}\| \leq \frac{(4\rho-1)\circ\rho v_n(\rho)}{3\circ(1-2^{-n})} . \tag{20}$$

Hier ist

$$v_n(\rho) = \sum_{\nu=0}^{n-1} \left(\frac{\rho^2}{\lambda(\rho)}\right)^\nu . \tag{21}$$

Beweis: Es ist

$$\|(TD+B)^{-1}\| \leq \frac{1}{1+(-1)^{n-1}(b_1 b_2 \ldots b_n)^{-1}} \| B^{-1}\| \sum_{\nu=0}^{n-1} \|TDB^{-1}\|^\nu .$$

Aus

$$\|B^{-1}\| \leq \frac{\rho}{\lambda(\rho)} \quad \text{und} \quad b_\nu > 2$$

sowie aus (16) folgt sofort (19).

Weiter ist

$$\|(TD^2+B^{-1})^{-1}\| \leq \frac{1}{1+(-1)^{n-1}(b_1b_2\ldots b_n)^{-1}}\|D^2\| \; \|\sum_{\nu=0}^{n-1}(T'B^{-1}D^{-2})^{\nu}\|.$$

Mit

$$\|D^2\| \leq \rho^2$$

und (n>1)

$$\|\sum_{\nu=0}^{n-1}(T'B^{-1}D^{-2})^{\nu}\| = \operatorname*{Max}_{\nu}\{1+\frac{1}{d_\nu^2 b_\nu}+\frac{1}{d_\nu^2 b_\nu d_{\nu-1}^2 b_{\nu-1}}+\ldots+\frac{1}{d_\nu^2 b_\nu \ldots d_{\nu-n+2}^2 b_{\nu-n}}$$

$$\leq (\frac{4\rho-1}{3\rho})v_n(\rho),$$

nach (17), folgt die Abschätzung (20).

Endlich gelangen wir zu der angestrebten Abschätzung der Operatornorm.

Satz 6. Es sei ρ die kleinste Zahl mit $\rho \geq 1$ und

$$\frac{1}{\rho} \leq d_\nu \leq \rho \quad , \quad \nu = 1,2,..,n.$$

Dann gilt für die Norm L(D) des zugehörigen kubischen Spline-Interpolationsoperators die Abschätzung

$$L(D) \leq 1+\frac{(4\rho-1)\rho^4}{2\lambda(\rho)} \cdot \frac{v_n^2(\rho)}{(1-2^{-n})^2} \tag{22}$$

Beweis: Beachtet man, daß

$$\|(I+TD^2)(I-T)\| \leq 2\rho^2$$

ist, so folgt die Behauptung durch Zusammenfassung von (11), (12), (13), (19) und (20).

Man erkennt, daß

$$\frac{\rho^2}{\lambda(\rho)} \begin{cases} < 1 \text{ für } 1 \leq \rho < \frac{3+\sqrt{5}}{2} \\ = 1 \text{ für } \rho = \frac{3+\sqrt{5}}{2} \\ > 1 \text{ für } \rho > \frac{3+\sqrt{5}}{2} \end{cases}$$

ist.

Damit wird

$$v_n(\rho) = \begin{cases} \dfrac{1-\left(\frac{\rho^2}{\lambda(\rho)}\right)^n}{1-\frac{\rho^2}{\lambda(\rho)}} & \text{für } \rho \geq 1,\ \rho \neq \frac{3+\sqrt{5}}{2} \\ n & \text{für } \rho = \frac{3+\sqrt{5}}{2} \end{cases} .$$

Für $\rho < \frac{3+\sqrt{5}}{2}$ ist somit eine von n unabhängige Schranke gefunden. Dies beweist erneut die anfangs zitierte Vermutung von MARSDEN. Daß es sich um die optimale Konstante handelt, zeigt das ebenfalls von MARSDEN angegebene Beispiel mit $\rho = \frac{3+\sqrt{5}}{2}$ und unbeschränkter Operatorfolge.

Literatur

[1] Ahlberg, J.H., E.N. Nilson and J.L. Walsh: The theory of splines and their applications.
New York, Academic Press 1967.

[2] Cheney, E.W. and F. Schurer: A note on the operators arising in spline approximation
J A T 1, 94-102 (1968).

[3] Cheney, E.W. and F. Schurer: On interpolating cubic splines with equally spaced nodes.
Indag. Math. 30, 517-524 (1968).

[4] Curry, H.B. and I.J. Schoenberg: On Polya frequency functions IV.
J. Analyse Math. 17, 71-107 (1966).

[5] de Boor, C.: On cubic spline functions which vanish at all knots.
MRC Report No. 1424 (1974).

[6] Golomb, M.: Approximation by periodic splines on uniform meshes.
J A T 1, 26-65 (1968).

[7] Krinzeßa, F.: Zur periodischen Spline-Interpolation.
Dissertation, Bochum 1969.

[8] Nörlund, N.E.: Vorlesungen über Differenzenrechnung.
Chelsea Publ., New York, 1954.

[9] Marsden, M.: Cubic spline interpolation of continuous functions.
J A T 10, 103-111 (1974)

[10] Meinardus, G. und G. Merz: Zur periodischen Spline-Interpolation. Erschienen in: Spline-Funktionen, Hrsg. K. Böhmer, G. Meinardus und W. Schempp. BI-Verlag Mannheim 1974

[11] Richards, F.B.: Best bounds for the uniform periodic spline interpolation operator. J A T 7, 302-317 (1973).

BEMERKUNGEN ZUR NUMERISCHEN LÖSUNG VON ANFANGSWERTPROBLEMEN MIT HILFE NICHTLINEARER SPLINE-FUNKTIONEN

G. MICULA

In einigen vorhergehenden Veröffentlichungen (R.Schaback [1o], H.Werner [11], [12], R.Runge [9]) wurde eine Theorie nichtlinearer Klassen von Spline-Funktionen entwickelt und zur numerischen Lösung von Anfangswertproblemen angewendet. Wie schon Lambert und Show [1-4] bemerkt haben, lassen sich mit nichtpolynomialen Approximationsfunktionen, die ein ähnliches Verhalten wie die Lösungen zeigen, bessere Ergebnisse als mit linearen Methoden erzielen. Diese Idee wurde von R.Runge [9] und H.Werner [12] verfolgt, um neue Methoden für die numerische Lösung von Anfangswertproblemen bei Differentialgleichungen erster Ordnung angeben zu können mit Hilfe nichtlinearer Klassen von Spline-Funktionen.

Entsprechend zu den Methoden von H.Werner und R.Runge werden wir eine nichtlineare Spline-Funktionen konstruieren, welche die Lösung eines Anfangswertproblems für Differentialgleichungen zweiter Ordnung approximiert. Bei geeigneten Voraussetzungen sollen die Konvergenzeigenschaften untersucht werden und Fehlerabschätzungen gegeben werden.

Wir betrachten das Anfangswertproblem:

(1) $$y'' = f(x,y)$$

(2) $$y(x_0) = y_0,\ y'(x_0) = y_0'$$

wobei $x_0 \in I$ und $I \subset R$ ein nichttriviales kompaktes Intervall ist.

Die Funktion $f: I \times \mathbb{R} \to \mathbb{R}$ sei genügend glatt und Lipschitz-stetig.
Dann existiert für einen gegebenen Punkt (x_0, y_0, y_0') genau eine Lösung des Problems (1)-(2), die mit $y: [x_0, b] \to \mathbb{R}$ bezeichnet werde.

Es sei $f \in C^{k-2}(I \times \mathbb{R})$ wobei k eine ganze Zahl ist.
Gesucht wird als Näherungslösung der exakten Lösung y von (1)-(2) eine Spline-Funktion $s: [x_0, b] \to \mathbb{R}$ mit den Knoten

$$a = x_0 < x_1 < \dots < x_{m+1} = b.$$

Die Restriktionen

$$s_j := s_{|[x_j, x_{j+1}]} \quad (j = 0, \dots, m)$$

seien von k+2 Parametern abhängende Funktionen aus einer gewissen Klasse mindestens k-mal stetig differenzierbarer Funktionen.
Die Konstruktion der Näherungslösung wird analog zu R. Runge in [9] und H. Werner in [12] folgendermaßen durchgeführt:
Die Idee ist, daß die bekannten Ableitungen der Spline-Funktion am rechten Endpunkt eines Intervalls als Startwerte für das benachbarte Intervall benützt werden. Weil nur der Wert von $f(x,y)$ gebraucht wird, kann die Methode als Mehrschnittverfahren betrachtet werden.
In praktischen Anwendungen, wo Singularitäten auftreten, ist diese Methode gegenüber linearen Methoden überlegen.
Mit Hilfe der Anfangswerte und der Differentialgleichung beschaffe man sich die Werte:

$$s_0^{(r)} := y^{(r)}(x_0), \quad (r = 0, \dots, k)$$

und setze $j := 0$.

Liegen Werte $s_j^{(r)}$ für $r = 0,\ldots,k$, so bestimme man die k+2 Parameter der Näherungslösung $s_j(x)$ in $[x_j, x_{j+1}]$ aus den k+2 Gleichungen:

$$s_j^{(r)}(x_j) = s_j^{(r)} \;, \quad r = 0,\ldots,k$$

$$s_j''(x_{j+1}) = f(x_{j+1}, s(x_{j+1}))$$

Danach berechne man die Werte:

$$s_{j+1}^{(r)} = s_j^{(r)}(x_{j+1}), \; r = 0,\ldots,k$$

aus der dann bekannten Funktion $s_j(x)$ und wiederhole für $j \leq m$ den Iterationsschritt mit j+1 anstelle von j. Falls sich aus den Gleichungen (3) immer die k+2 Parameter berechnen lassen, erhält man durch dieses Verfahren eine k-mal stetig differenzierbare Spline-Funktion.
Im folgenden werden für eine gewisse Funktionenklasse T hinreichende Bedingungen für die Lösbarkeit von (3) angegeben. Ferner wird auch untersucht werden müssen, wann das beschriebene Verfahren konvergent ist.
Auf den ersten Blick sieht das Verfahren wie ein implizites Mehrschrittverfahren aus.
Numerisch scheint es sich auch so zu verhalten, denn es sind keine besonderen Startwerte nötig und eine Änderung der Schrittweite ist ohne zusätzlichen Rechenaufwand möglich.
Es seien

$$t_j = t_j(x,c,d)$$

von zwei Parametern c,d abhängige Funktionen, die auf dem Intervall $I_j := [x_j, x_{j+1}]$ definiert sind und k-mal stetig, differenzierbar bezüglich x sind. Die Klasse aller Funktionen $\{t_j\}$ wird mit T bezeichnet.

Die Klasse von nichtlinearen Spline-Funktionen wird definiert durch:

$$(5)\quad \mathcal{S} := (x_o,\ldots,x_m) = \{s|s \in C^{k-1}(I)\colon s|_{I_j} = p_j + t_j(x,c_j,d_j),$$

$$p_j \in \mathfrak{P}_{k-1},\ t_j \in T,\ j = 0,1,\ldots,m\}$$

wobei $\mathfrak{P}_{k-1}$ der lineare Raum polynomialer Funktionen deren Grad $\leq k-1$ ist.
Im folgenden soll eine "kubische" (k=3) nichtlineare Spline-Funktion $s \in \mathcal{S}$ effektiv konstruiert werden, um die Lösung y auf dem Intervall $[a,b]$ näherungsweise zu bestimmen.
Auf dem Teilintervall $I_j = [x_j,x_{j+1}]$ wird die nichtlineare Spline-Funktion $s \in \mathcal{S}$ definert durch:

$$s(x) = s_j(x) = \sum_{r=0}^{2} \frac{c_{j,r}}{r!}(x-x_j)^r + \frac{c_j}{x-d_j}\,,$$

$$x \in [x_j,x_{j+1}], j = 0,1,\ldots,m$$

Da $s \in C^2[a,b]$, haben wir

$$c_{j,0} = s_{j-1}(x_j),\ j = 0,1,\ldots,m,\ (s_{-1}(x_o) = y_o)$$

$$c_{j,1} = s'_{j-1}(x_j),\ j = 0,1,\ldots,m,\ (s'_{-1}(x_o) = y'_o)$$

$$c_{j,2} = s''_{j-1}(x_j),\ j = 0,1,\ldots,m,\ (s''_{-1}(x_o)=f(x_o,y_o))$$

Der unbekannte Koeffizient $c_j (j = 0,1,\ldots,m)$ wird so bestimmt, daß für jedes j, die Funktion s die Differentialgleichung (1) an der Stelle $x = x_{j+1}$ erfüllen soll, d.h.

$$(7) \qquad s_j''(x_{j+1}) = f(x_{j+1}, s_j(x_{j+1}))$$

Wenn die Funktionen $\{s_i : i = 0,1,\ldots,j-1\}$ bekannt sind, dann ist (7) eine algebraische nichtlineare Gleichung mit der Unbekannten c_j.

Bemerkung: Für den Fall $t_j \in T_3$, wobei

$$T_3 = \{t \mid t(x) = c(x+d)^3,\ c,d \in R\}$$

stimmt diese Methode mit der Methode von [6] überein. Nun soll gezeigt werden, daß für hinreichend kleines

$$h(h = \min_j h_j,\ h_j = x_{j+1} - x_j,\ j = 0,1,\ldots,m)$$

die Werte $c_j (j = 0,1,\ldots,m)$ eindeutig aus (7) bestimmt werden können.

Theorem 1. Ist die Funktion $f \in C^2(I \times R)$, dann gibt es ein $h_0 > 0$, so daß für jedes $h \leq h_0$ die Gleichung (7) genau eine Lösung $c_j (j = 0,\ldots,m)$ hat, d.h. die Spline-Funktion s, die durch (6) definiert ist, existiert und ist eindeutig.

Beweis: Die algebraische Gleichung (7) mit der Unbekannten c_j, läßt sich geschrieben werden

$$c_j = \Phi_j(c_j)$$

Durch eine übliche Beweisführung kann man zeigen, daß der Operator $\Phi_j : R \to R,\ a_j \to \Phi_j(a_j)$ für genügend klein h stark kontrahierend ist, und man erhält durch Iteration einen eindeutig bestimmten Fixpunkt c_j.

Das gegebene Verfahren kann man als implizites nichtlineares diskretes Zweischrittverfahren auffassen. In diesem Sinne erhalten wir folgende Konvergenzergebnisse des Verfahrens:

Theorem 2. Es sei für $f \in C^3(I \times R)$ mit Hilfe des obigen Verfahrens eine kubische nichtlineare Spline-Funktion s konstruiert. Dann gelten gleichmäßig für $x \in [a,b]$ die Relationen:

$$s^{(i)}(x) - y^{(i)}(x) = O(h^{3-i}), \quad 0 \leq i \leq 2$$

$$s'''(x) - y'''(x) = O(h)$$

falls die dritte Ableitung von $s(x)$ in den Knoten x_j als Mittelwert der entsprechenden rechtsseitigen bzw. linksseitigen Ableitungen definiert wird.

Der Beweis geht mit einigen kleinen Veränderungen genau so wie in [9]. Man beachte dabei, daß bei entsprechenden nichtlinearen Zweischrittverfahren die Konsistenzbedingungen und die Stabilitätsbedingungen erfüllt sind.

Man wird im allgemeinen keine höhere Konvergenzordnung erwarten können, denn für den Spezialfall der Polynomial-Spline-Funktionen ergibt sich dieselbe Konvergenzordnung wie man in (6) sehen kann.

Beispiel: Wir werden nun ein Beispiel zeigen, das die numerischen Eigenschaften des vorhergehenden Verfahrens zeigt.

Man betrachtet das Anfangswertproblem:

$$y'' = \frac{8y^2}{1-2x}, \quad y(0) = 1, \quad y'(0) = 2, \; x \in [0,2/5]$$

mit der exakten Lösung: $y(x) = (1-2x)^{-1}$

Wir wählen $h_j = h = 1/10$, d.h. die Knotenpunkte sind:

$$x_o = 0,\ x_1 = 1/10,\ x_2 = 2/10,\ x_3 = 3/10,\ x_4 = 4/10$$

Die Approximationsspline-Funktion hat die Gestalt:

$$s_k(x) = c_{k,o} + c_{k,1}(x - x_k) + c_{k,2}(x - x_k)^2 + \frac{c_k}{1-2x}$$

Für $x \in [0,1/10]$ haben wir:

$$s(x) = y(0) + y'(0)x + \frac{y''(0)}{2'}x^2 + \frac{c_o}{1-2x}$$

Aus der Gleichung

$$s''(1/10) = \frac{8\ s^2(1/10)}{1 - 2\cdot 1/10}$$

bestimmen wir $c_o = 0,00794$ und dann

$$s(x) = 1 + 2x + 4x^2 + \frac{0,00794}{1-2x},\ x \in [0,1/10]$$

Auf $[1/10,\ 2/10]$ ist die Spline-Funktion gegeben durch:

$$s(x) = s(1/10) + (x-1/10)s'(1/10) +$$

$$+ \frac{s''(1/10)}{2!}(x - 1/10) + \frac{c_1}{1-2x}$$

Aus der Gleichung

$$s''(2/10) = \frac{8\, s^2(2/10)}{1-2\cdot 2/10}$$

bestimmen wir $c_1 = 0{,}234712$.
Weitere Ergebnisse sind in der folgenden Tabelle enthalten:

x	y(x)	s(x)
3/10	2,50000	2,499873
4/10	5,00000	4,499998

Bei diesem Beispiel ist zu beobachten, daß die Näherungslösung mit Hilfe nichtlinearer Spline-Funktionen im Grunde bessere Werte liefern als die polynomialen Spline-Funktionen, die in [6], [7] verwendet werden. Wie R.Runge in [9] bemerkt hat, ergeben sich im Fall rationaler Spline-Funktionen in der Nähe des Pols von y die bei weitem besten Näherungswerte.

Anerkennung: Der Autor dankt Herrn Prof. H.Werner (Münster) und Herr Dr.R.Runge (Münster), die dem Verfasser über ihre neuen Ergebnisse benachrichtigt haben. Der Alexander von Humboldt-Stiftung spricht der Autor seinen Dank für die Unterstützung während der Abfassung dieser Arbeit aus.

Literatur

1.) LAMBERT J.D. and SHOW, B.: On numerical solution of $y' = f(x,y)$ by a class of formulae based on rational approximation.
Math.Comput. 19 (1965) pp. 456-462.

2.) LAMBERT J.D. and SHOW,B. : A method for the numerical solution of $y' = f(x,y)$ based on a self-adjusting non-polynomial interpolant,
Math.Comput. 2o (1966), pp. 11-2o.

3.) LAMBERT J.D. and SHOW, B.: A generalization of multistep methods for ordinary differential equation,
Numer. Math. 8 (1966) pp. 25o-263.

4.) LAMBERT, J.D.: Nonlinear methods for stiff systems of ordinary differential equations.
Proc. Dundee Conference on the Numerical Solution of Differential Equations, Springer Lecture Notes, 1973.

5.) LOSCALZO, F.R. and TALBOT, T.D.: Spline function approximation for solution of ordinary differential equations.
SIAM J. Numer. Anal. 4(1967) pp. 433-445.

6.) MICULA, G.: Approximate solution of differential equation $y'' = f(x,y)$ with spline functions.
Math.Comput. 27 (1973), pp. 8o7-816.

7.) MICULA, G.: Die numerische Lösung nichtlinearer Differentialgleichungen unter Verwendung von Spline-Funktionen. Proc.Conf."Numerische Behandlung nichtlinearer Integrodifferential- und Differentialgleichungen", Oberwolfach, 1973.
Lect. Notes in Mathematics 395, 57-83. Berlin-Heidelberg-New York: Springer, 1974.

8.) MICULA, G.: Über die numerische Lösung nichtlinearer Differentialgleichungen mit Splines von niedriger Ordnung,"Numerische Behandlung von Differentialgleichungen", ISNM 27 (1975), pp. 185-195,
Birkhäuser-Verlag, Basel-Stuttgart, 1975.

9.) RUNGE, R.: Lösung von Anfangswertproblemen mit Hilfe nichtlinearer Klassen von Spline-Funktionen, Dissertation. University of Münster, 1972.

1o.) SCHABACK,R.: Interpolation mit nichtlinearen Klassen von Spline-Funktionen J.Approximation Theory 8 (1973) pp. 173-188.

11.) WERNER, H.: Tschebyscheff-Approximation mit einer Klasse rationaler Spline-Funktionen, J. Approximation Theory, 1o (1974), pp. 74-92.

12.) WERNER, H.: Interpolation and integration of initial value problems of ordinary differential equations by regular splines, SIAM J.Numer.Anal. 12 (2975) pp 255-271

Dr. Gheorghe MICULA
Faculty of Mathematics
University of Cluj

34oo Cluj-Napoca (Romania)

z.Zt.

am Institut für Angewandte Mathematik
der Albert-Ludwigs-Universität
78 Freiburg i.Br. (BRD)
Hebelstr. 4o

ON THE RELATIONS BETWEEN FINITE DIFFERENCES AND DERIVATIVES OF CARDINAL SPLINE FUNCTIONS

Hennie ter Morsche

Let m be a natural number and let S_m denote the class of cardinal spline functions of degree m. The object of this note is to establish a linear relationship between the 2m + 2 quantities $s(i+x), s(i+1+x), \ldots, s(i+m+x)$, $s^{(k)}(i+y), s^{(k)}(i+1+y), \ldots, s^{(k)}(i+m+y)$, where $x,y \in [0,1]$, $i = 0,\pm1,\pm2,\ldots$, $s \in S_m$ and where $s^{(k)}$ denotes the k-th derivative of s ($k = 0,1,2,\ldots,m-1$). Using the shift operator E, we represent this relation in a simple form, involving the exponential Euler polynomials. The results are applied to cardinal spline interpolation.

1. Introduction and summary

A function s defined on the real line is said to be a cardinal spline function of degree m if it satisfies the following conditions:

a) In each interval $[i,i+1]$ ($i = 0,\pm1,\pm2,\ldots$) the function s coincides with a polynomial of degree at most m, i.e. $s(x) \in \pi_m$;

b) $s \in C^{(m-1)}(\mathbb{R})$.

The symbol S_m denotes the set of all cardinal spline functions of degree m. Let there be given a function $s \in S_m$. It is well known that the 2m quantities $s(i), s(i+1), \ldots, s(i+m-1), s^{(k)}(i), \ldots, s^{(k)}(i+m-1)$ are linearly dependent. This linear dependence was shown by D.J. Fyfe [3]. In this note we generaliz this linear relation to the 2m + 2 quantities $s(i+x), s(i+1+x), \ldots, s(i+m+$ $s^{(k)}(i+y), \ldots, s^{(k)}(i+m+y)$, where $x,y \in [0,1]$. In order to represent this in a simple form, we shall need the so-called exponential Euler polynomials (cf. [7], p. 20) and the shift operator E, defined by

$$(1.1) \qquad Ef(x) = f(x+1) ,$$

for all functions f defined on $\mathbb{R}$ and for all real x.

As a consequence of the definition of E, the linear combination $\sum_{j=0}^{n} a_j f(x+j$ can be written in the form

$$(1.2) \qquad \sum_{j=0}^{n} a_j f(x+j) = P(E)f(x) ,$$

where $P(E) = \sum_{j=0}^{n} a_j E^j$ and $E^0 := I$, the identity operator.

In section 3, we define the exponential Euler polynomials, denoted by $P_m(z,\lambda)$, and list some of their properties. Now the linear dependence of the 2m + 2 quantities $s(i+x), s(i+1+x), \ldots, s(i+m+x), s^{(k)}(i+y), \ldots, s^{(k)}(i+m+y)$, can be written as follows:

$$(1.3)\qquad \frac{1}{m!} P_m(E,y) s^{(k)}(i+x) = \frac{1}{(m-k)!} P_{m-k}(E,x)(E-I)^k s(i+y) ,$$

$x,y \in [0,1]$, $k = 0,1,\ldots,m-1$ and $i = 0,\pm1,\pm2,\ldots$.

This linear relation can be regarded as a relation between finite differences and derivatives if we set $\Delta := E - I$. We shall prove (1.3) in section 4. In section 5 we apply (1.3) to the problem of cardinal spline interpolation at particular inner points of the intervals $[i,i+1]$ $(i = 0,\pm1,\ldots)$, specifically at the points $x_i = i + \lambda$, where $\lambda \in (0,1]$. The same interpolation problem for periodic cubic splines is discussed in [4] and for periodic splines of arbitrary degree in [1] (p. 197-214). As a basis for the linear space S_m we use the B-splines. These will be the subject of the next section.

The cardinal B-spline

The function B_m, defined by

$$(2.1)\qquad B_m(x) := \frac{1}{m!} \Delta^{m+1}(x-m-1)_+^m \quad (x \in \mathbb{R}) ,$$

is called a cardinal B-spline of degree m and with knots $0,1,\ldots,m+1$. Here $x_+ := \max(0,x)$.

The function B_m has the following basic property, with respect to the linear space S_m (cf. [7], p. 11):
If $s \in S_m$, then there exists a unique sequence $(a_n)_{-\infty}^{\infty}$ of real numbers such that

$$(2.2)\qquad s(x) = \sum_{n=-\infty}^{\infty} a_n B_m(x-n) .$$

We now collect some further well-known properties of the spline function B_m

$$(2.3)\qquad B_m^{(k)}(x) = \Delta^k B_{m-k}(x-k), \quad (k = 0,1,\ldots,m-1) ,$$

$$(2.4)\qquad B_m(x) = B_m(m+1-x) ,$$

$$(2.5)\qquad B_m(x) > 0 \text{ if } 0 < x < m+1 \text{ and } B_m(x) = 0 \text{ elsewhere} .$$

3. The exponential Euler polynomials

We define a class of functions $P_m(z,\lambda)$ of two variables by way of their gene rating function

$$(3.1)\qquad \frac{e^{\lambda t}}{z - e^t} = \sum_{m=0}^{\infty} \frac{P_m(z,\lambda)}{(z-1)^{m+1}} \frac{t^m}{m!} .$$

They are related to the exponential Euler polynomials $A_m(\lambda, z)$ (cf. [7], p.2 as follows:

$$(3.2)\qquad P_m(z,\lambda) = (z-1)^m A_m(\lambda,z) .$$

In this note we only use the functions $P_m(z,\lambda)$ and call them exponential Euler polynomials as well. Now we list some properties of the functions $P_m(z$

$$(3.3)\qquad P_m(z,\lambda) = m! \sum_{r=0}^{m} B_m(r+1-\lambda) z^r, \quad (0 \le \lambda \le 1) ;$$

(3.4) $P_m(z,\lambda)$ is a polynomial of the two variables z and λ. With respect to the variable z, it is a polynomial of degree m, if $0 < \lambda \le 1$. $P_m(z,0)$ is a polynomial of degree $m-1$;

$$(3.5)\qquad P_m(z,0) = \Pi_m(z) ,$$

where $\Pi_m(z)$ is the so-called Euler-Frobenius polynomial of degree $m-1$ (cf. [7], p. 22);

$$(3.6)\qquad zP_m(z,\lambda) - P_m(z,\lambda+1) = (z-1)^{m+1}\lambda^m ;$$

$$(3.7)\qquad P_m(-1,\lambda) = (-2)^m E_m(\lambda) ;$$

where E_m is the classical Euler polynomial of degree m (cf. [6], p. 23);

$$(3.8)\qquad \frac{\partial}{\partial\lambda} P_m(z,\lambda) = m(z-1)P_{m-1}(z,\lambda) ;$$

$$(3.9)\qquad P_m(z,\lambda) = \sum_{k=0}^{m} \binom{m}{k} (z-1)^k \Pi_{m-k}(z)\lambda^k ;$$

$$(3.10)\qquad \frac{P_m(z,\lambda)}{(1-z)^{m+1}} = \sum_{n=0}^{\infty} (n+1-\lambda)^m z^n, \quad (|z| < 1) ;$$

$$(3.11)\qquad P_m(z,\lambda) = z^m P_m(\tfrac{1}{z}, 1-\lambda) ;$$

(3.12) If $0 < \lambda < 1$ the polynomial $P_m(z,\lambda)$ has, as a function of the variable z, m distinct negative zeros.
If $\lambda = 0$, then $P_m(z,0) = \Pi_m(z)$ has $m-1$ distinct negative zeros.
If $\lambda = 1$ then $P_m(z,1) = z\Pi_m(z)$ has $m-1$ distinct negative zeros, while in addition $z = 0$ is a zero;

(3.13) If $z < 0$ the polynomial $P_m(z,\lambda)$ either has, as a function of λ, exactly one zero in the interval $(0,1)$ or it has the two zeros $\lambda = 0$ and $\lambda = 1$. For $z = 0$ we have $P_m(0,\lambda) = (1-\lambda)^m$; then $\lambda = 1$ is the only zero;

(3.14) Let $\lambda \in (0,1]$ and let $\omega_1(\lambda) < \omega_2(\lambda) < \ldots < \omega_m(\lambda)$ be the zeros of $P_m(z,\lambda)$ with respect to the variable z. Then we have the following assertion about the location of these zeros:
If $0 < \lambda_1 < \lambda_2 \leq 1$, then

$$\omega_{i-1}(1) < \omega_i(\lambda_1) < \omega_i(\lambda_2) \leq \omega_i(1) \quad (i = 1,\ldots,m;\ \omega_0(1) = -\infty)\ .$$

We remark that the polynomials $P_m(z,\lambda)$, as defined in this note are closely related to the polynomials P_m as defined in [1], p. 203. They differ by a multiplicative constant m! and the fact that λ is replaced by $(1-\lambda)$. So the properties (3.3),...,(3.12) are direct consequences of theorem 3.1 stated in [1], p. 205 and therefore the proofs are omitted. Property (3.13) can easily be shown by induction with respect to m, using property (3.8) and the identity $P_m(z,1) = zP_m(z,0)$. Property (3.14) follows from the observation that $\omega_i(\lambda)$ is continuous with respect to the variable λ (cf. [8], p. 148) and the fact that $\omega_i(\lambda_1) \neq \omega_i(\lambda_2)$; if $\omega_i(\lambda_1) = \omega_i(\lambda_2)$ then $P_m(\omega_i(\lambda_1),\lambda_1) = P_m(\omega_i(\lambda_1),\lambda_2) = 0$ which contradicts property (3.13).

A relation between finite differences and derivatives

In this section we prove relation (1.3). This will be done by proving this relation for $k = 0$ first.

LEMMA 4.1. *Let* $s \in S_m$ *and let* $x,y \in [0,1]$ *be real numbers, then the following identity holds*:

$$P_m(E,y)s(i+x) = P_m(E,x)s(i+y)\ , \quad (i = 0,\pm 1,\pm 2,\ldots)\ .$$

<u>Proof</u>. As a consequence of (3.3) we have

$$P_m(E,y)s(i+x) = m! \sum_{r=0}^{m} B_m(r+1-y)s(i+r+x) =$$

$$= m! \sum_{r=-\infty}^{\infty} B_m(r+1-y)s(i+r+x) \ ,$$

since $B_m(r+1-y) = 0$ for $r > m$ and $r < 0$. From (2.2) it follows that

$$s(i+r+x) = \sum_{n=-\infty}^{\infty} a_n B_m(i+r+x-n) \ .$$

So we have

$$P_m(E,y)s(i+x) = m! \sum_{r=-\infty}^{\infty} \sum_{n=-\infty}^{\infty} a_n B_m(r+1-y)B_m(i+r+x-n) =$$

$$= m! \sum_{n=-\infty}^{\infty} \sum_{r=-\infty}^{\infty} a_n B_m(m-r+y)B_m(m+1-i-r-x+n) =$$

$$= m! \sum_{n=-\infty}^{\infty} \sum_{r=-\infty}^{\infty} a_n B_m(i+r+y-n)B_m(r+1-x) =$$

$$= m! P_m(E,x)s(i+y) \ ,$$

where in the course of writing down these identities we have applied property (2.4) and we have replaced r by $m+n-r-i$. Changing the order of summation provides no problems, since all but a finite number of elements are zero. □

Using property (3.8), the next theorem is an immediate consequence of lemma 4.1.

<u>THEOREM 4.2</u>. *Let* $s \in S_m$ *and let* $x,y \in [0,1]$ *be real numbers, then the following identity holds:*

$$\frac{1}{m!} P_m(E,y)s^{(k)}(i+x) = \frac{1}{(m-k)!} P_{m-k}(E,x)(E-I)^k s(i+y) \ ,$$

$(i = 0,\pm1,\pm2,\ldots;\ k = 0,1,2,\ldots,m-1)$.

Cardinal spline interpolation

Let a sequence of real numbers $(y_n)_{-\infty}^{\infty}$ be given and let $\lambda \in (0,1]$. We consider the following problem of cardinal spline interpolation. Find a function $s \in S_m$, such that

(5.1) $$s(i + \lambda) = y_i, \quad (i = 0,\pm 1,\pm 2,\dots) .$$

We wish to apply lemma (4.1) to this problem, therefore we need the following lemma.

LEMMA 5.2. *Let* p *be a polynomial of degree* n, *which does not vanish on the unit circle in the complex plane. Consider the difference equation:*

(5.3) $$p(E)x_i = a_i \quad (i = 0,\pm 1,\pm 2,\dots) ,$$

where the numbers a_i *are given in such a way, that* $|a_i| = O(|i|^{\nu})$ $(|i| \to \infty)$, *for a given value of* $\nu \geq 0$.
Then there exists exactly one solution $(x_i)_{-\infty}^{\infty}$ *of* (5.1) *with the property* $|x_i| = O(|i|^{\nu})$ $(|i| \to \infty)$, *and this solution is given by the expression:*

(5.4) $$x_i = \sum_{j=-\infty}^{\infty} \alpha_j a_{i+j} .$$

Here the numbers α_j *are the coefficients in the Laurent expansion*

$$\frac{1}{p(z)} = \sum_{j=-\infty}^{\infty} \alpha_j z^j$$

converging on the unit circle in the complex plane.

Proof. Because of the convergence of the series $\sum_{j=-\infty}^{\infty} |\alpha_j||j|^{\nu}$, it is easy to verify that (5.4) is a solution of (5.1) with the asserted property.
From the general solution of the homogeneous equation $p(E)x_i = 0$, we conclude that only the trivial solution $x_i = 0$ satisfies the condition $|x_i| = O(|i|^{\nu})$ $(|i| \to \infty)$. This completes the proof of lemma 5.1. □

With respect to the variable z the polynomial $P_m(z,\lambda)$ has only nonpositive zeros. Hence $z = -1$ is the only possible zero on the unit circle. Now we know that $P_m(-1,\lambda) = (-2)^m E_m(\lambda)$, so $P_m(-1,\lambda) = 0$ in the following two cases (cf. [6], p. 26):

(5.5) i) $\lambda = \frac{1}{2}$ and m is an odd number.

ii) $\lambda = 1$ and m is an even number.

Now we are ready to prove the following existence theorem.

THEOREM 5.6. *Let a sequence* $(y_n)_{-\infty}^{\infty}$ *of real numbers be given with the property* $|y_i| = O(|i|^{\nu})$ $(|i| \to \infty)$ *for a given value* $\nu \geq 0$. *Let* m *be a natural number and* $\lambda \in (0,1]$ *a real number, such that* $\lambda \neq \frac{1}{2}$ *if* m *is an odd number and* $\lambda \neq 1$ *if* m *is an even number.*

Then there exists a unique spline function $s \in S_m$ *with the properties:*

i) $s(i+\lambda) = y_i$, $(i = 0,\pm 1,\pm 2,\ldots)$,

ii) $|s(i+x)| = O(|i|^{\nu})$ $(|i| \to \infty)$ for all $x \in [0,1]$.

Proof. For m = 1 an explicit solution is easily derived. Therefore from now on we shall disregard this case. From lemma 4.1 it follows that

$$P_m(E,\lambda)s(i+x) = P_m(E,x)y_i, \quad (i = 0,\pm 1,\pm 2,\ldots) .$$

Since $P_m(-1,\lambda) \neq 0$ and $|P_m(E,x)y_i| = O(|i|^{\nu})$ $(|i| \to \infty)$, we can apply lemma 5.2 to this difference equation. The Laurent expansion

$$(5.7) \qquad \frac{P_m(z,x)}{P_m(z,\lambda)} = \sum_{n=-\infty}^{\infty} A_n(x,\lambda)z^n ,$$

converging on the unit circle, defines a function s given by the expression:

$$(5.8) \qquad s(i+x) = \sum_{n=-\infty}^{\infty} A_n(x,\lambda)y_{i+n}, \; x \in [0,1] \text{ and } i = 0,\pm 1,\ldots .$$

We have to verify that the function s solves the interpolation problem. We know that $A_n(x,\lambda)$ is given by the complex integral

$$(5.9) \qquad A_n(x,\lambda) = \frac{1}{2\pi i} \oint_{|z|=1} \frac{P_m(z,x)}{z^{n+1}P_m(z,\lambda)}\,dz .$$

Since $\lim_{|z|\to\infty} \dfrac{P_m(z,x)}{zP_m(z,\lambda)} = 0$, we have

$$(5.10) \qquad A_n(x,\lambda) = -\sum_{j=1}^{p} \frac{P_m(\omega_j(\lambda),x)}{P_m'(\omega_j(\lambda),\lambda)} \cdot \frac{1}{\omega_j^{n+1}(\lambda)} \qquad (n \geq 1) ,$$

where $\omega_1(\lambda),\ldots,\omega_p(\lambda)$ are those zeros of $P_m(z,\lambda)$ that are smaller than -1.

For $n \le -1$ we have

$$(5.11)\qquad A_n(x,\lambda) = \sum_{j=p+1}^{m} \frac{P_m(\omega_j(\lambda),x)}{P'_m(\omega_j(\lambda),\lambda)} \cdot \frac{1}{\omega_j^{n+1}(\lambda)} \qquad (n \le -1),$$

where $-1 < \omega_{p+1}(\lambda) < \ldots < \omega_m(\lambda) \le 0$.
Since

$$\lim_{|z|\to\infty} \frac{P_m(z,x)}{P_m(z,\lambda)} = \frac{B_m(m+1-x)}{B_m(m+1-\lambda)} = \frac{B_m(x)}{B_m(\lambda)} = (\frac{x}{\lambda})^m,$$

we obtain for $n = 0$

$$(5.12)\qquad A_0(x,\lambda) = (\frac{x}{\lambda})^m - \sum_{j=1}^{p} \frac{P_m(\omega_j(\lambda),x)}{P'_m(\omega_j(\lambda),\lambda)} \cdot \frac{1}{\omega_j(\lambda)}.$$

Substituting the expressions (5.10), (5.11) and (5.12) in (5.8) and using the properties of the polynomials $P_m(\omega_j(\lambda),x)$, we find that the function s solves the interpolation problem. □

If we apply (5.8) to the special sequence $(y_n)_{-\infty}^{\infty}$ with $y_n = \delta_{0,n}$, then we obtain the fundamental spline function L(x) for our interpolation problem. We note that

$$(5.13)\qquad L(i+x) = A_{-i}(x,\lambda),\ i = 0,\pm 1,\pm 2,\ldots \text{ and } x \in [0,1].$$

Using the relations (5.9), (5.10) and (5.11), we obtain similar expressions for the fundamental spline function as given by E.N. Nilson in [5].
The following theorem gives information about the sign structure of the fundamental spline function.

THEOREM 5.7. *Let* $x \in (\lambda+i,\lambda+i+1)$, *and let* $m \ge 2$, *then*

$$\operatorname{sgn} L(x) = \begin{cases} (-1)^i, & i = 0,1,2,\ldots \\ (-1)^{i+1}, & i = -1,-2,\ldots . \end{cases}$$

Proof. To obtain the sign structure of the functions $A_n(x,\lambda)$ we make use of property (3.14). From this property we conclude that

$$(5.14)\qquad \operatorname{sgn} \frac{P_m(\omega_i(\lambda_1),\lambda_2)}{P'_m(\omega_i(\lambda_1),\lambda_1)} = \begin{cases} -1 & \text{if } \lambda_2 > \lambda_1 \\ 0 & \text{if } \lambda_1 = \lambda_2 \\ 1 & \text{if } \lambda_2 < \lambda_1 \end{cases} \qquad (i = 1,2,\ldots,m).$$

From (5.10) and (5.14) it follows that for $n \geq 1$

$$\text{(5.15)} \qquad \operatorname{sgn} A_n(x,\lambda) = \begin{cases} (-1)^n & \text{if } 0 \leq x < \lambda \\ 0 & \text{if } x = \lambda \\ (-1)^{n+1} & \text{if } \lambda < x \leq 1 \,. \end{cases}$$

This proves the asserted sign structure of $L(x)$ for $x \leq 0$.
From (5.11) and (5.14) it follows that for $n \leq -1$

$$\text{(5.16)} \qquad \operatorname{sgn} A_n(x,\lambda) = \begin{cases} (-1)^{n+1} & \text{if } 0 \leq x < \lambda \\ 0 & \text{if } x = \lambda \\ (-1)^n & \text{if } \lambda < x \leq 1 \,. \end{cases}$$

This proves the sign structure of $L(x)$ for $x \geq 1$.
Now we have to prove that $L(x) > 0$ on $[0,1]$. Because of (5.12) and (5.14) we see that $A_0(x,\lambda) > 0$ for $x \in [0,\lambda]$. So $L(x) > 0$ for $x \in [0,\lambda]$. We note that, using property (3.11) and substituting $z := \frac{1}{z}$,

$$A_0(1-x, 1-\lambda) = \frac{1}{2\pi i} \oint_{|z|=1} \frac{P_m(\frac{1}{z}, x)}{zP_m(\frac{1}{z}, \lambda)} dz = \frac{1}{2\pi i} \oint_{|z|=1} \frac{P_m(z,x)}{zP_m(z,\lambda)} dz =$$
$$= A_0(x,\lambda) \,.$$

If $\lambda \leq x \leq 1$, then $0 \leq 1 - x \leq 1 - \lambda$ and hence $A_0(1-x, 1-\lambda) > 0$ and so $A_0(x,\lambda) = L(x) > 0$. This proves the theorem. □

<u>Remark</u>

Let m be an odd natural number and let $\lambda = 1$. Then by theorem 4.2 and lemma 5.2, the spline function s in theorem 5.6 has the following property

$$\text{(5.17)} \qquad s^{(k)}(0) = \sum_{n=-\infty}^{\infty} \beta_n s(n) \,.$$

Here the coefficients β_n are the coefficients in the Laurent expansion

$$\text{(5.18)} \qquad \frac{m!}{(m-k)!} (z-1)^k \frac{\Pi_{m-k}(z)}{\Pi_m(z)} = \sum_{n=-\infty}^{\infty} \beta_n z^n \,,$$

converging on the unit circle.
These coefficients appear in a differentiation formula used by C. de Boor and I.J. Schoenberg [2] in a proof of Kolmogorov's theorem.

References

[1] BÖHMER, K., MEINARDUS, G. and SCHEMPP, W.: Spline-Funktionen. Mannheim-Wien-Zürich, B.I.-Wissenshaftsverlag 1974.

[2] BOOR de, C. and SCHOENBERG, I.J.: Cardinal Interpolation and Spline functions VIII. The Budan-Fourier theorem for splines and applications. MRC T.S.R. 1546, May 1975.

[3] FYFE, D.J.: Linear Dependence Relations Connecting Equal Interval N-the Degree Splines and Their Derivatives. J. Inst. Maths. Applics. 7 (1971), 398-406.

[4] MEIR, A. and SHARMA, A.: Convergence of a Class of Interpolatory Splines. J. Approximation Theory 1 (1968), 243-250.

[5] NILSON, E.N.: Polynomial Splines and a Fundamental Eigenvalue Problem for Polynomials. J. Approximation Theory 6 (1972), 439-465.

[6] NÖRLUND, N.E.: Vorlesungen über Differenzenrechnung. Berlin, Springer, 1924.

[7] SCHOENBERG, I.J.: Cardinal Spline Interpolation, CBMS Vol. 12, Philadelphia, SIAM 1973.

[8] WEBER, H.: Lehrbuch der Algebra, Erster band. Braunschweig, Friedr. Vieweg & Sohn, 1912.

H.G. ter Morsche
Department of Mathematics
Technological University Eindhoven
Eindhoven, The Netherlands.

ON OPTIMAL APPROXIMATION

Arthur Sard

This note presents a new instance of spline approximation in which the observation of a function is its value on an interior contour or hypersurface and the coobservation is its gradient. There follow three comments relevant to the application of the theory of optimal approximation.

1. Introduction

In earlier papers I have described a theory of approximation of a given operator G in terms of an observation operator F and , in the background, a coobservation operator U [1, 2, 3].

An instance is the approximation of a function based on its values on an interior contour or hypersurface, as follows. Let D be a bounded, open domain in R^m to which Gauss's theorem applies. Let X be a space of functions on D to R, for example, $C_2(\bar{D})$, the space of functions on D which have C_2 extensions on an open domain containing $\bar{D}$. Let $c \subset D$ be a rectifiable curve or, more generally, an n-measurable hypersurface (self intersections are allowed.) Let Y be the space $L^2(c)$ of functions measurable on c, absolute square integrable, two such functions being equivalent if equal almost everywhere, with the inner product

$$(y,z) = \int_c y.z \quad , \quad y, z \in L^2(c) .$$

Let the observation of $x \in X$ be $Fx := x\upharpoonright c$, the restriction of x to c. Let Z be the space $L^2(D)$. Let the coobservation of x be the gradient $Ux := \nabla x$. Let G be any admissible * operator on X; for example the identity.

We envisage the problem of approximating Gx , $x \in X$, in terms of Fx alone. The approximation which is optimal relative to F, U is $G\xi$, where $\xi = A_o x = E_o \overline{F}x$ is the spline approximation of x. In the present instance ξ is the function which is harmonic on D - c , which agrees with x on c, and which has zero normal derivative on the boundary of D.

2. <u>The completeness condition</u>

The theory starts with a linear space X, perhaps <u>devoid of</u> topology, and inner product spaces Y and Z.

* The strict hypothesis on G is that it have an extension which is linear and continuous on the space $N \subset \overline{X}$. In the present note I use the notation of [3], which differs from that of [2] in minor respects. Note that A_o and E_o of [3] are respectively Π and e of [2].

It then constructs the Hilbert space $\overline{\mathcal{X}}$ as follows: $\mathcal{X}$ is the space X with the inner product

$$(x,y) \;=\; (Fx,Fy) + (Ux,Uy) \;, \quad x,\, y \in X,$$

induced by the inner products on Y, Z; and $\overline{\mathcal{X}}$ is the completion of $\mathcal{X}$. To carry out the completion may be complicated and tedious. It is therefore useful to be able to avoid the completion, where possible. Theorem 6 of [3] implies the following. Suppose that

X is a Banach space,

F, U are continuous on X, and

(1) $b < \infty$ exists such that

$$||x||_X^2 \leq b^2(||Fx||^2 + ||Ux||^2), \text{ all } x \in X.$$

Then the sets X and $\overline{\mathcal{X}}$ are the same, and the spaces X and $\overline{\mathcal{X}}$ have equivalent norms. Completions are unnecessary. Condition (1) implies that the topology of X is relevant to spline approximation.

Note that (1) holds if X is $\overline{\mathcal{X}}$ to start with and if F, U are continuous.

Since the constructed space $\overline{\mathcal{X}}$ always exists and contains X, one may wish to assume from the beginning that (1) holds. This is done, for example, in [4,5,6].

Thus if we start with the general problem of [2, 3], we may enlarge X and introduce a norm on X for which (1) holds.

3. The inner product on Y

The construction of the space $\overline{X}$ and hence of the splines depends on the given inner products in Y and Z.

Let us say that two inner products $(\, , \,)_1$ and $(\, , \,)_2$ on Y are equivalent if they induce equivalent norms, that is, if constants a, b > 0 exist such that

$$||y||_1^2 \leq a||y||_2^2 \leq b||y||_1^2 \, , \quad \text{all } y \in Y.$$

We may replace the inner product on Y by an equivalent inner product without affecting the closure $\overline{X}$ or $\overline{Y}$ – and without affecting the spline approximation $\xi = A_0 x$, $x \in \overline{X}$, of x. This follows from Theorem 2 of [3] (Theorem 2 of [2]), which characterizes ξ solely in terms of $\overline{F}x$ and $||\overline{U}y||$, $y \in \overline{X}$, both of which are unchanged by the change of inner product in Y.

On the other hand, changing the inner product in Z may change the splines, because $||\overline{U}y||$ is minimized.

4. Surjectivity

Suppose that $Y = FX = F\chi$, instead of that $Y \supset F\chi$ as heretofore. Then

$$\overline{Y} = \overline{F\chi} \supset \overline{F}\ \overline{\chi} \supset F\chi .$$

Now $F\chi$ is dense in $\overline{F\chi}$, hence in $\overline{F}\ \overline{\chi}$. Hence $\overline{Y} = \overline{F}\ \overline{\chi}$ if and only if $\overline{F}\ \overline{X}$ is closed. Hence surjectivity of $\overline{F}$, when F is surjective, is equivalent to continuity of E_o [3, Lemma 2].

Surjektivity of $\overline{U}$ seems not to be needed.

References

1. Sard, A.: Optimal approximation. J. Functional Analysis 1(1967), 222-244 and 2(1968), 368-369.
2. Sard, A.: Approximation based on nonscalar observations. J. Approximation Theory 8(1973), 315-334.
3. Sard, A.: Instances of generalized splines. In K. Böhmer, G. Meinardus, W. Schempp, Editors: Spline-Funktionen. Bibliographisches Institut, Mannheim, Wien, Zürich. 215-241, 1974.
4. Delvos, F.-J. and W. Schempp: Sard's method and the theory of spline systems. J. Approximation Theory 14(1975),230-243.

5. Delvos, F.-J.: On surface interpolation. To appear in J. Approximation Theory.

6. Delvos, F.-J. and Posdorf, H.: On optimal tensor product approximation. To appear in J. Approximation Theory.

Department of Mathematics
University of California
La Jolla, California 92037

Splineapproximation in intermediären Räumen

WERNER SCHÄFER und WALTER SCHEMPP

0. EINLEITUNG

Die vorliegende Arbeit versucht mit Hilfe des Ansatzes von Sard [13] und der von Delvos - Schempp [8] durchgeführten operatortheoretischen Erweiterung dieses Ansatzes, Fehlerabschätzungen abstrakter Spline Projektoren in intermediären Räumen zu gewinnen. Dabei benutzen wir als wesentliches Hilfsmittel den bei Berezanskij [1] bewiesenen Interpolationssatz für Hilbertraumskalen (vgl. Schäfer [14]). Am Beispiel der Lidstone - Spline Interpolation auf dem Intervall $I = [o,1]$ werden die abstrakten Ergebnisse realisiert.

Als eine weitere Anwendung der Theorie gelangen wir zu neuen L_2 - Fehlerabschätzungen der diskreten Greenschen Funktion zum Operator $A_k = (-1)^k D^{2k}$ bzgl. eines geeignet zu wählenden Spline - Unterraumes $Sp_o(A_k,\pi,z)$ des Sobolevraumes $W^k(I)$ (vgl. Ciarlet - Varga [3], Schäfer [15]) und zu Abschätzungen für den punktweisen Fehler bei der Lidstone - Spline Interpolation.

1. KONSTRUKTION EINER HILBERTRAUMSKALA ZUM ERWEITERTEN ANSATZ VON SARD

Es seien $(X,(\cdot|\cdot)_X)$, $(Y,(\cdot|\cdot)_Y)$ und $(Z,(\cdot|\cdot)_Z)$ separable Hilberträume über dem Körper $\mathbb{C}$ der komplexen Zahlen. Mit $||\cdot||_X$, $||\cdot||_Y$ und $||\cdot||_Z$ seien die induzierten Normen bezeichnet.

$$U : X \to Y \quad \text{und} \quad F : X \to Z \tag{1.1}$$

seien stetige lineare Abbildungen. Durch

$$((x|y))_X = (Ux|Uy)_Y + (Fx|Fy)_Z \tag{1.2}$$

werde ein zweites Skalarprodukt auf X definiert. Die hiervon induzierte Norm $|||.|||_X$ möge äquivalent zur ursprünglichen Norm $||.||_X$ sein (vgl. Sard [13]) .

Man definiert nun in $(X, ((.|.))_X)$ durch

$$Im(P) = Ker(F)^{\perp} \tag{1.3}$$

den sogenannten Spline - Projektor P, für den die folgende Eigenschaft charakteristisch ist.

<u>Satz 1:</u> (Sard [13]) Sei $x_o \in X$ gegeben. Unter allen $y \in X$ mit $Fx_o = Fy$ ist Px_o das eindeutig bestimmte Element, welches das verallgemeinerte Dirichletfunktional $y \rightsquigarrow ||Uy||_Y$ minimiert.

Es gelte nun weiter

$$Ker(F) \text{ liegt dicht in } Y \tag{1.4}$$

$$||x||_Y \leq B \; ||x||_X \qquad (x \in Ker(F),\; B > o) \tag{1.5}$$

In diesem Fall nennen wir das Tupel

$$(X,Y,Z;\; U,F)$$

ein <u>erweitertes Sard-System</u> (vgl. Delvos - Schempp [7] [8]).

<u>Satz 2:</u> (Delvos - Schempp [8]) Das Tupel (X,Y,Z; U,F) bilde ein erweitertes Sard - System, und es sei $U_o : = U|_{Ker(F)}$. Dann ist

$$A = U_o^* U_o \tag{1.6}$$

der eindeutig bestimmte positiv definite selbstadjungierte Operator in Y mit dem Energieraum

$$H_A = \mathrm{Ker}(F) = \mathrm{Dom}(\sqrt{A}) \tag{1.7}$$

und

$$(Ax|y)_Y = (Ux|Uy)_Y = ((x|y))_X \tag{1.8}$$

$$(x \in \mathrm{Dom}(A),\ y \in H_A).$$

Daraus ergibt sich der

Satz 3: Es ist $\sqrt{A}$ der eindeutig bestimmte positiv definite selbstadjungierte Operator in Y mit

$$\mathrm{Dom}(\sqrt{A}) = \mathrm{Ker}(F) \tag{1.9}$$

$$\mathrm{Im}(\sqrt{A}) = Y \tag{1.10}$$

$$(\sqrt{A}x|\sqrt{A}y)_Y = (Ux|Uy)_Y = ((x|y))_X \tag{1.11}$$

$$(x,y \in H_A).$$

Beweis: Es sei B ein weiterer positiv definiter selbstadjungierter Operator in Y mit den Eigenschaften (1.9) - (1.11). Dann folgt aus Satz 2

$$B^2 = A.$$

Wegen der Eindeutigkeit der positiven Wurzel folgt daraus

$$B = \sqrt{A}\ .-$$

Mit Hilfe des Operators $\sqrt{A}$ wollen wir nun eine Hilbertraumskala $\{X_\alpha\}_{\alpha \geq o}$ definieren. Dazu sei $\{E_\lambda\}_{\lambda \geq o}$ die Spektralzerlegung des Operators $\sqrt{A}$ und $\alpha \in \mathbb{R}$ $(\alpha \geq o)$. Wir wissen aus der Spektraltheorie (vgl. Riesz - Nagy [12]), daß der Operator

$$\sqrt{A}^{\alpha} = \int_{\gamma}^{\infty} \lambda^{\alpha}\, dE_{\lambda} \tag{1.12}$$

ebenfalls positiv definit und selbstadjungiert ist und einen in Y dichten Definitionsbereich $\mathrm{Dom}(\sqrt{A}^{\alpha})$ besitzt. Die Zahl $\gamma > 0$ sei die untere Grenze des Operators $\sqrt{A}$.
Wir schreiben

$$X_{\alpha} = \mathrm{Dom}(\sqrt{A}^{\alpha}) \ .$$

Definieren wir durch

$$(x|y)_{\alpha} = (\sqrt{A}^{\alpha}x|\sqrt{A}^{\alpha}y)_{Y}$$

eine Bilinearform auf X_{α}, so gilt der

<u>Satz 4:</u> Unter der Bilinearform $(.|.)_{\alpha}$ wird X_{α} zu einem separablen Hilbertraum.

<u>Beweis:</u> Es gilt für $x \in X_{\alpha}$

$$||\sqrt{A}^{\alpha}x||_{Y}^{2} = \int_{\gamma}^{\infty} \lambda^{2\alpha}\, d\,(E_{\lambda}x|x)_{Y}$$

$$\geq \gamma^{2\alpha}\, ||x||_{Y}^{2} \ . \tag{1.13}$$

Die Bilinearform $(.|.)_{\alpha}$ definiert demnach ein Skalarprodukt auf X_{α}. $||.||_{\alpha}$ sei die hiervon induzierte Norm auf X_{α}. Die Vollständigkeit von $(X_{\alpha}, ||.||_{\alpha})$ folgt aus der Ungleichung (1.13) und der Abgeschlossenheit des Operators $\sqrt{A}^{\alpha}$ in Y. Die Separabilität von X_{α} ergibt sich aus derjenigen von Y und aus der Surjektivität des Operators $\sqrt{A}^{\alpha}$.-

2. FEHLERABSCHÄTZUNGEN

Es sei nun $$G : H_{A} \to W \tag{2.1}$$

eine stetige lineare Abbildung von H_A in einen komplexen separablen Hilbertraum $(W,(\cdot|\cdot)_W)$.

Durch

$$\text{Im}(Q) = \text{Ker}(G)^{\perp} \tag{2.2}$$

definieren wir einen Orthogonalprojektor Q in H_A. Für den Projektor Q gilt analog zu Satz 1 der

Satz 5: (Delvos [5]). Es sei $x_o \in H_A$. Unter allen $y \in H_A$ mit $Gx_o = Gy$ minimiert Qx_o eindeutig das Funktional $y \rightsquigarrow ||Uy||_Y$.

Beweis: Wegen

$$|||x|||_X^2 = ||Ux||_Y^2 \leq ||Ux||_Y^2 + ||Gx||_W^2 \quad (x \in H_A)$$

wird durch

$$(x|y)_A = (U_o x|U_o y)_Y + (Gx|Gy)_W \qquad (x,y \in H_A)$$

ein zu $((\cdot|\cdot))_X$ äquivalentes Skalarprodukt definiert.
Es sei Q_A der analog zu (1.3) in $(H_A,(\cdot|\cdot)_A)$ durch

$$\text{Im}(Q_A) = \text{Ker}(G)^{\perp}$$

definierte Spline - Projektor. Man sieht nun die Beziehung

$$\text{Im}(Q) = \text{Im}(Q_A) \tag{2.3}$$

ein. Außerdem gilt wegen (2.1) für $x,y \in H_A$

$$\begin{aligned}(Qx|y)_A &= (UQx|y)_Y + (GQx|Gy)_W \\ &= (Ux|UQy)_Y + (GQx|Gy)_W.\end{aligned}$$

Es gilt weiter

$$\begin{aligned}(GQx|Gy)_W &= (GQx|Gy)_W + (G(x-Qx)|Gy)_W \\ &\quad + (Gx|G(Qy-y))_W\end{aligned}$$

$$= (Gx|Gy)_W + (Gx|G(Qy-y)_W$$

$$= (Gx|GQy)_W.$$

D.h. Q ist symmetrisch auf H_A auch bzgl. $(\cdot|\cdot)_A$.

Wegen (2.3) gilt deshalb

$$Q = Q_A.$$

Dann folgt die Behauptung des Satzes aus Satz 1.

Aufgrund von Satz 5 nennen wir Q den <u>Projektor der optimalen Approximation</u> in H_A.

Wir bezeichnen den zu Q supplementären Projektor mit R und mit C_Q die Norm der Abbildung

$$R : H_A \to Y$$

d.h.
$$C_Q = \sup_{\substack{x \in H_A \\ x \neq 0}} \frac{||x-Qx||_Y}{||Ux||_Y} \tag{2.4}$$

$$= \sup_{\substack{x \in Ker(G) \\ x \neq 0}} \frac{||x||_Y}{||Ux||_Y}$$

(vgl. Sard [13]).

Damit lassen sich die folgenden Abschätzungen für den Resprojektor R aufschreiben.

<u>Satz 6:</u> a) $||Rx||_Y \leq C_Q \cdot ||Ux||_Y \quad (x \in H_A)$

b) $||U \cdot Rx||_Y \leq ||Ux||_Y \quad (x \in H_A)$

c) $||U \cdot Rx||_Y \leq C_Q \cdot ||Ax||_Y \quad (x \in Dom(A))$

<u>Beweis:</u> Die Abschätzungen a) und b) ergeben sich unmittelbar aus der Definition von R bzw. C_Q. Die Abschätzung c)

findet sich in dem Beweis zu Theorem 3.3 bei Delvos - Schempp [8]. -

Wir betrachten nun in der Hilbertraumskala $\{X_\alpha\}_{\alpha \geq o}$ die Räume $X_o = Y$, $X_1 = H_A$ und $X_2 = \mathrm{Dom}(A)$ (vgl. Satz 3 und Satz 4). Dann folgt mit Satz 6 aus dem bei Berezanskij [1] bewiesenen <u>Interpolationssatz für Hilbertraumskalen</u> der

<u>Satz 7:</u> Es seien $o \leqq \alpha \leqq 1$ und $o \leqq \beta \leqq 1$ gegeben. Dann gilt für alle $x \in \mathrm{Dom}(A^{\frac{1+\beta}{2}})$ die Abschätzung

$$||A^{\frac{\alpha}{2}}(x-Qx)||_Y \leq C_Q^{1+\beta-\alpha} \cdot ||A^{\frac{1+\beta}{2}} x||_Y.$$

<u>Beweis:</u> Aus a) und b) in Satz 6 folgt die Stetigkeit der linearen Abbildungen

$$R : X_1 \to X_o$$

$$R : X_1 \to X_1.$$

Es sei $o \leqq \alpha \leqq 1$. Dann ergibt sich aus dem Interpolationssatz die Stetigkeit der linearen Abbildung

$$R : X_1 \to X_\alpha$$

und die Abschätzung

$$||Rx||_\alpha \leq C_Q^{1-\alpha} ||x||_1. \tag{2.5}$$

Aus den Abschätzungen b) und c) in Satz 6 folgt die Stetigkeit der linearen Abbildungen

$$R : X_1 \to X_1$$

$$R : X_2 \to X_1.$$

Es sei $o \leq \beta \leq 1$. Dann resultiert aus dem Interpolationssatz die Stetigkeit der linearen Abbildung

$$R : X_{1+\beta} \to X_1$$

und die Abschätzung

$$||Rx||_1 \leq C_Q^{\;\beta} \cdot ||x||_{1+\beta}. \tag{2.6}$$

Ersetzt man nun in (2.5) x durch Rx, so ergibt sich die Behauptung des Satzes durch Kombination von (2.5) und (2.6). -

Man vergleiche hierzu die in [15] dargestellte Fehlerabschätzung.

3. BEISPIEL EINES ERWEITERTEN SARD - SYSTEMS

Es sei $k \in \mathbb{N}$, $\mu = [\frac{k-1}{2}] + 1$ und $I = [o,1]$ das kompakte Einheitsintervall der reellen Zahlengeraden. $L_2(I)$ sei der Hilbertraum (der Äquivalenzklassen) der quadratintegrierbaren komplexwertigen Funktionen auf I, versehen mit dem üblichen Skalarprodukt

$$(f,g) \rightsquigarrow (f|g)_o = \int_I f(x) \cdot \overline{g(x)}\, dx. \tag{3.1}$$

$||\cdot||_o$ sei die hiervon auf $L_2(I)$ induzierte Norm. $W^k(I)$ sei der mit dem Skalarprodukt

$$(f,g) \rightsquigarrow (f|g)_k = \sum_{j=o}^{k} (D^j f|D^j g)_o \tag{3.2}$$

versehene Sobolevraum. Dabei ist D^j der Differentialoperator j-ter Ordnung auf $W^k(I)$. Für einen Punkt $s \in I$ bezeichne $\varepsilon_s^{(j)}$ die j-te distributionelle Ableitung des Diracmaßes ε_s $(o \leq j \leq k-1)$.

Definieren wir die Abbildung

$$F_\mu : W^k(I) \to \mathbb{C}^{2\mu} \tag{3.3}$$

durch

$$F_\mu(f) := (\varepsilon_o f, \varepsilon_o^{(2)} f, \ldots, \varepsilon_o^{(2\mu-2)} f; \varepsilon_1 f, \varepsilon_1^{(2)} f, \ldots, \varepsilon_1^{(2\mu-2)} f),$$

so können wir den folgenden Satz formulieren:

<u>Satz 8:</u> Das Tupel $(W^k(I), L_2(I), \mathbb{C}^{2\mu}; D^k, F_\mu)$ bildet ein <u>erweitertes</u> Sard - System.

<u>Beweis:</u> Die Stetigkeit der linearen Abbildungen

$$D^k : W^k \to L_2(I) \text{ und } F_\mu : W^k(I) \to \mathbb{C}^{2\mu}$$

ist offensichtlich. Wir müssen nun zeigen, daß die durch

$$|||f|||_k := (||D^k f||_o^2 + \sum_{j=o}^{\mu-1} (|\varepsilon_o^{(2j)} f|^2 + |\varepsilon_1^{(2j)} f|^2)^{1/2}$$

auf $W^k(I)$ definierte Norm äquivalent zu der durch (3.2) induzierten Norm $||\cdot||_k$ ist.

Dazu genügt es nach Smirnow [17] zu zeigen, daß für jedes $p \in \mathrm{Ker}(D^k)$ mit $F_\mu p = o$ folgt $p \equiv o$. Dies wird jedoch von Pólya [11] gezeigt.

Die Dichtheit von $\mathrm{Ker}(F_\mu)$ in $L_2(I)$ ergibt sich aus der Inklusion

$$\overset{o}{W}{}^k(I) \subset \mathrm{Ker}(F_\mu)$$

und der Dichtheit von $\overset{o}{W}{}^k(I)$ in $L_2(I)$. -

Mit $((\cdot|\cdot))_k$ sei das analog zu (1.2) auf $W^k(I)$ gebildete Skalarprodukt bezeichnet und P_μ sei der analog zu (1.3) definierte Spline Projektor.

<u>Satz 9:</u> Es sei $f \in W^k(I)$ und $x \in I$. Dann gilt die folgende Darstellungsformel für $P_\mu f$

$$P_\mu f(x) = \sum_{j=0}^{\mu-1} \frac{2^{2j+1}}{(2j+1)!} (\varepsilon_1{}^{(2j)} f\, B_{2j+1}(\tfrac{x+1}{2}) - \varepsilon_0{}^{(2j)} f\, B_{2j+1}(\tfrac{x}{2})).$$

Dabei bezeichnet $B_{2j+1}(x)$ des Bernoullipolynom (2j+1)-ten Grades.

<u>Beweis:</u> vgl. Delvos - Kösters [6]. -

<u>Satz 10:</u> Der zu dem erweiterten Sard - System $(W^k(I), L_2(I), ¢^{2\mu}; D^k, F_\mu)$ gemäß Satz 2 existierende Operator ist der Differentialoperator

$$A_k = (-1)^k D^{2k} \tag{3.4}$$

mit dem Definitionsbereich

$$Dom(A_k) = \{f \in W^{2k}(I);\ \varepsilon_0{}^{(2j)} f = \varepsilon_1{}^{(2j)} f = o;\ o \leq j \leq k-1\}$$

<u>Beweis:</u> Delvos - Kösters [6] zeigen, daß A_k ein positiv definiter und selbstadjungierter Operator in $L_2(I)$ ist. Ferner gilt $Dom(A_k) \subset Ker(F_\mu)$. Mittels partieller Integration folgt für alle $f \in Dom(A_k)$ und für alle $g \in Ker(F_\mu)$ analog zu (1.8)

$$(A_k f|g)_o = (D^k f|D^k g)_o.$$

Aus der Eindeutigkeit des Operators A ergibt sich die Behauptung. -

Den <u>Energieraum</u> $(Ker(F_\mu), ((\cdot|\cdot))_k)$ zum Operator A_k bezeichnen wir im weiteren mit H_k. Es gilt also

$$H_k = \{f \in W^k(I) : \varepsilon_0{}^{(2j)} f = \varepsilon_1{}^{(2j)} f = o;\ o \leq j \leq \mu-1\}. \tag{3.5}$$

4. DIE ZUGEHÖRIGE HILBERTRAUMSKALA

Wir wollen nun die durch den Operator A_k erzeugte Hilbertraumskala betrachten. Es gilt der

Satz 11: Für $1 \leqq \nu \leqq 2k-1$ ($\nu \in \mathbb{N}$) gilt

$$\mathrm{Dom}(A_k^{\frac{\nu}{2k}}) = \{f \in W^\nu(I) : \varepsilon_o^{(2j)}f = \varepsilon_1^{(2j)}f = o;\ o \leqq j \leqq [\tfrac{\nu-1}{2}]\}$$

Beweis: Wir gehen aus von dem positiv definiten und selbstadjungierten Differentialoperator $-D^2$ mit dem Definitionsbereich

$$\mathrm{Dom}(-D^2) = \{f \in W^2(I) : \varepsilon_o f = \varepsilon_1 f = o\} \subset L_2(I).$$

Es sei nun ν eine gerade Zahl ($\nu = 2\ell$, $1 \leqq \ell \leqq k-1$):
Dann folgt aus der Spektraltheorie (vgl. Riesz - Nagy [12])

$$A_k^{\frac{2\ell}{2k}} = (-D^2)^{\frac{k\cdot 2\ell}{2k}} = (-D^2)^\ell = (-1)^\ell D^{2\ell} .$$

Da $\frac{\nu-1}{2} = \frac{2\ell-1}{2} = \ell-1$ gilt, folgt

$$\mathrm{Dom}((-1)^\ell D^{2\ell}) = \{f \in W^\nu(I) : \varepsilon_o^{(2j)}f = \varepsilon_1^{(2j)}f = o;\ o \leqq j \leqq [\tfrac{\nu-1}{2}]\}$$
$$= \mathrm{Dom}(A_k^{\frac{\nu}{2k}}).$$

Es sei ν ungerade ($\nu = 2\ell-1$, $1 \leqq \ell \leqq k$) :

Dann sei B der ebenfalls positiv definite und selbstadjungierte Operator $(-D^2)^{1/2}$.

Es ergibt sich damit

$$A_k^{\frac{2\ell-1}{2k}} = (B^{2k})^{\frac{2\ell-1}{2k}} = B^{2\ell-1} .$$

Weiter gilt $\mathrm{Dom}(B^{2\ell-1}) = H_{(-D^2)^{2\ell-1}}$. Dabei bezeichnet $H_{(-D^2)^{2\ell-1}}$ den Energieraum zu dem Operator $(-D^2)^{2\ell-1}$.

Ersetzen wir nun in (3.5) k durch $2\ell - 1$ und definieren $\mu = [\frac{2\ell-1-1}{2}] + 1 = \ell$, so ergibt sich sofort die Beziehung

$$H_{(-D^2)^{2\ell-1}} = \{f \in W^{2\ell-1}(I) : \varepsilon_o^{(2j)} f = \varepsilon_1^{(2j)} f = o;\ o \leqq j \leqq \ell-1\}$$

$$= \{f \in W^{\nu}(I) : \varepsilon_o^{(2j)} f = \varepsilon_1^{(2j)} f = o;,\ o \leqq j \leqq [\frac{\nu-1}{2}]\}\ .$$

Eine weitere Eigenschaft der Operatoren $A_k^{\frac{\nu}{2k}}$ $(o \leqq \nu \leqq 2k,\ \nu \in \mathbb{N})$ beweisen wir in dem nun folgenden

<u>Satz 12:</u> Es sei $o \leqq \nu \leqq 2k$ $(\nu \in \mathbb{N})$ und $f \in \mathrm{Dom}(A_k^{\frac{\nu}{2k}})$.

Dann gilt

$$||A_k^{\frac{\nu}{2k}} f||_o = ||D^{\nu} f||_o\ .$$

<u>Beweis:</u> Für $\nu = o$ und $\nu = 2k$ ist die Behauptung offensichtlich. Es sei nun $1 \leqq \nu \leqq 2k-1$. Dann folgt aus den Sätzen 8 und 10, daß

$$A_{\nu} := (-D^2)^{\nu} = A_k^{\frac{2\nu}{2k}}$$

der gemäß Satz 2 zu dem erweiterten Sard - System

$$(W^{\nu}(I),\ L_2(I),\ \mathbb{C}^{2\rho};\ D^{\nu},\ F_{\rho})$$

mit $\rho = [\frac{\nu-1}{2}] + 1$ gehörige Operator ist.

Aus Satz 3 ergibt sich damit die Isometrie von

$$(A_k^{\frac{2\nu}{2k}})^{\frac{1}{2}} = A_k^{\frac{\nu}{2k}} \text{ und } D^{\nu} \text{ . -}$$

5. LIDSTONE - SPLINES

Es sei m eine beliebige aber fest vorgegebene natürlich Zahl und

$$\pi : o = s_o < s_1 < \dots < s_m < s_{m+1} = 1 \tag{5.1}$$

eine Zerlegung des Intervalls I,

$$h = \max_{o \leq i \leq m} |s_{i+1} - s_i| \tag{5.2}$$

ihre Feinheit. Dann definieren wir analog zu (2.1) die Abbildung

$$G_m : H_k \to \mathbb{C}^m$$

durch

$$G_m f = (\varepsilon_{s_1} f, \dots, \varepsilon_{s_m} f) \tag{5.3}$$

und analog zu (2.2) den Spline Projektor Q_m in $(H_k, ((\cdot|\cdot))_k)$ durch

$$\operatorname{Im}(Q_m) = \operatorname{Ker}(G_m)^{\perp}. \tag{5.4}$$

Aufgrund der durch die Abbildung F_μ (vgl. 3.3) gegebenen Randbedingungen nennen wir Q_m einen Lidstone - Spline Projektor (vgl. Davis [4]). In der Notation von Schultz - Varga [16] stellt für $f \in H_k$ die Funktion $Q_m f$ eine g - Spline vom Typ II dar. Nach Schultz - Varga [16] gilt nun für $f \in H_k$ die folgende Fehlerabschätzung

$$||f - Q_m f||_o \leq M \cdot h^k \, ||D^k f||_o \tag{5.5}$$

mit einer positiven, von f und der Zerlegung π unabhängigen Konstanten M.

Als Anwendung von Satz 7 und mit Hilfe von Satz 12 finden wir damit unter Berücksichtigung der Ungleichung (5.5) den

Satz 13: Es sei $o \leq j \leq k$ ($j \in \mathbb{N}$) und $f \in \mathrm{Dom}(A_k^{\frac{k+j}{2k}})$

Dann gilt für $o \leq i \leq k$ ($i \in \mathbb{N}$) die Abschätzung

$$||D^i(f - Q_m f)||_o \leq M\, h^{k+j-i}\, ||D^{k+j} f||_o .$$

Beweis: Man setze im Beweis zu Satz 7 $\alpha = \frac{i}{k}$ $(o \leq i \leq k)$ und $\beta = \frac{j}{k}$ $(o \leq j \leq k)$ und berücksichtige die Aussage von Satz 12. -

Man vergleiche hierzu die bei Hedstrom - Varga [9] mittels der K-Methode (vgl. Butzer - Berens [2]) gewonnenen Ergebnisse und das in [15] dargestellte Resultat.

6. DIE DISKRETE GREENSCHE FUNKTION ZUM OPERATOR A_k

Es sei $G_k(x,y)$ $(x,y \in I)$ die aufgrund der positiven Definitheit des Operators A_k zu dem Randwertproblem

$$\begin{aligned} A_k u &= f \qquad (f \in L_2(I)) \\ \varepsilon_o^{(2j)} u &= \varepsilon_1^{(2j)} u = o \qquad (o \leq j \leq k-1) \end{aligned} \tag{6.1}$$

existierende Greensche Funktion.

Delvos - Kösters [6] leiten die folgende Darstellungsformel für $G_k(x,y)$ her:

$$G_k(x,y) = C_k(B_{2k}{}^*(\tfrac{x-y}{2}) - B_{2k}(\tfrac{x+y}{2})) \qquad (6.2)$$

Dabei sind $B_{2k}(x)$ die Bernoullipolynome 2k-ten Grades, $B_{2k}{}^*(x) = B_{2k}(x - [x])$ die 1-periodischen Bernoullifunktionen, und es ist $C_k = \frac{(-1)^{k-1}2^{2k-1}}{(2k)!}$.

Mit $G_{k,x}$ wollen wir die durch

$$G_{k,x}(y) = G_k(y,x) \qquad (y \in I) \qquad (6.3)$$

auf I definierte Funktion bezeichnen.

Gemäß Ciarlet - Varga [3] und Varga [18] ist dann

$$Q_m G_k(y,x) = Q_m G_{k,x}(y) \qquad (y \in I) \qquad (6.4)$$

die diskrete Greensche Funktion zum Operator A_k bzgl. des Spline - Unterraumes $Sp_o(A_k,\pi,z) = Im(Q_m)$ von H_k. Dabei ist der Inzidenzvektor $z = (1,\ldots,1) \in \mathbb{C}^m$ gewählt und π die in (5.1) definierte Zerlegung des Intervalls I.

Darüberhinaus ist $G_k(x,y)$ der reproduzierende Kern von H_k und $Q_m G_k(x,y)$ derjenige von $Im(Q_m)$ (vgl. Ciarlet - Varga [3]). Wir wollen nun L_2-Fehlerabschätzungen für die diskrete Greensche Funktion $Q_m G_{k,x}$ herleiten. Dazu dient als Vorbereitung der

Satz 14: Für $f \in H_k$, $x \in I$ und $o \leq j \leq k-1$ $(j \in \mathbb{N})$ gilt

(a) $(-1)^j \varepsilon_x{}^{(j)}(f - Q_m f) = ((f|D^j(G_{k,x} - Q_m G_{k,x})))_k$

(b) $D^j Q_m G_{k,x} = Q_m D^j G_{k,x}$

<u>Beweis:</u> ad (a): Für jedes $x \in I$ und für $o \leq j \leq k-1$ gehören die Abbildungen

$$(-1)^j \varepsilon_x^{(j)} : H_k \to \mathbb{C}$$

zum stetigen Dual des Hilbertraumes H_k. Es gilt nun

$$D^j G_{k,x} = \mathrm{repr}[(-1)^j \varepsilon_x^{(j)}],$$

und ebenso folgt

$$D^j Q_m G_{k,x} = \mathrm{repr}[(-1)^j \varepsilon_x^{(j)} |_{\mathrm{Im}(Q_m)}]$$

(vgl. Davis [4]). Daraus ergibt sich für $f \in H_k$

$$(-1)^j \varepsilon_x^{(j)} f = ((f | D^j G_{k,x}))_k \tag{6.5}$$

$$(-1)^j \varepsilon_x^{(j)} Q_m f = ((Q_m f | D^j Q_m G_{k,x}))_k$$

$$= ((f | D^j Q_m G_{k,x}))_k \;. \tag{6.6}$$

Mit (6.5) und (6.6) finden wir sofort die Behauptung (a).

ad (b): Es gilt nun aber auch für $f \in H_k$

$$(-1)^j \varepsilon_x^{(j)} Q_m f = ((Q_m f | D^j G_{k,x}))_k$$

$$= ((f | Q_m D^j G_{k,x}))_k . \tag{6.7}$$

Da (6.6) und (6.7) für alle $f \in H_k$ gültig ist, folgt daraus die Eigenschaft (b). -

Wir formulieren nun die angekündigten und im Vergleich zu Ciarlet - Varga [3] und Varga [18] verbesserten und erweiterten Fehlerabschätzungen für die diskrete Greensche Funktion $Q_m G_{k,x}$.

<u>Satz 15:</u> Es sei $x \in I$. Dann gilt für $o \leq j \leq 2k-1$ $(j \in \mathbb{N})$

$$||D^j(G_{k,x} - Q_m G_{k,x})||_o \leq M\, h^{2k-j-\frac{1}{2}}$$

mit einer positiven, von j,m und x unabhängigen Konstanten M. "h" bedeutet hierbei die in (5.2) definierte Feinheit der Zerlegung π.

<u>Beweis:</u> Aufgrund von Satz 14(a) gilt für $f \in H_k$

$$|\varepsilon_x^{(j)}(f - Q_m f)| \leq |||D^j(G_{k,x} - Q_m G_{k,x})|||_k \, |||f|||_k \tag{6.8}$$

Nach Schultz - Varga [16] existiert eine positive, von f, j, m und x unabhängige Konstante M mit

$$|\varepsilon_x^{(j)}(f-Q_m f)| \leq M\, h^{k-j-\frac{1}{2}}\, ||D^k f||_o \quad (o \leq j \leq k-1). \tag{6.9}$$

Aus dem Satz von Riesz ergibt sich damit

$$||D^{k+j}(G_{k,x} - Q_m G_{k,x})||_o \leq M\, h^{k-j-\frac{1}{2}}. \tag{6.10}$$

Andererseits folgt aus Satz 14(a) mit Satz 2 für $f \in Dom(A_k)$ und $o \leq j \leq k-1$ die Beziehung

$$|\varepsilon_x^{(j)}(f - Q_m f)| \leq ||D^j(G_{k,x} - Q_m G_{k,x})||_o \, ||A_k f||_o. \tag{6.11}$$

Ersetzen wir nun in (6.8) f durch $f - Q_m f$, so ergibt sich aus Satz 6(c) mit (5.5) für $f \in Dom(A_k)$ die Abschätzung

$$|\varepsilon_x^{(j)}(f - Q_m f)| \leq M\, h^{2k-j-\frac{1}{2}}\, ||A_k f||_o. \tag{6.12}$$

Aus (6.11), (6.12) und der Surjektivität des Operators A_k ergibt sich wie oben mit dem Satz von Riesz die Beziehung

$$||D^j(G_{k,x} - Q_m G_{k,x})||_o \leq M\, h^{2k-j-\frac{1}{2}}. \tag{6.13}$$

Die Aussagen (6.10) und (6.13) ergeben zusammen die Behauptung des Satzes. -

Das Ergebnis dieses Satzes legt eine genauere Untersuchung des punktweisen Fehlers bei der Lidstone - Spline Interpolation nahe.

7. ZUM PUNKTWEISEN FEHLER

Eine Auskunft über den punktweisen Fehler bei der Lidstone - Spline Interpolation gibt der folgende

<u>Satz 16:</u> Es sei $o \leq j \leq k (j \in \mathbb{N})$ und $f \in \mathrm{Dom}(A_k^{\frac{k+j}{2k}})$.

Dann gilt für $o \leq i \leq k-1 (i \in \mathbb{N})$ und für $x \in I$ die Beziehung

$$|\varepsilon_x^{(i)}(f - Q_m f)| \leq M\, h^{k+j-i-\frac{1}{2}}\, ||D^{k+j} f||_o .$$

<u>Beweis:</u> Es sei $o \leq i \leq k-1 (i \in \mathbb{N})$ und $x \in I$. Dann folgt aus (6.8) und (6.11) die Stetigkeit der linearen Abbildungen

$$H_k \ni f \rightsquigarrow (-1)^i \varepsilon_x (f - Q_m f) \in \mathbb{C}$$

$$\mathrm{Dom}(A_k) \ni f \rightsquigarrow (-1)^i \varepsilon_x (f - Q_m f) \in \mathbb{C}$$

Mit dem Interpolationssatz für Hilbertraumskalen ergibt sich daraus die Stetigkeit der linearen Abbildung

$$\mathrm{Dom}(A_k^{\frac{j+k}{2k}}) \ni f \rightsquigarrow (-1)^i \varepsilon_x^{(i)} (f - Q_m f) \in \mathbb{C} \quad (o \leq j \leq k)$$

und die Abschätzung

$$|\varepsilon_x^{(i)}(f - Q_m f)| \leq ||D^{k+i}(G_{k,x} - Q_m G_{k,x})||_o^{\frac{k-j}{k}}\; ||D^i(G_{k,x} - Q_m G_{k,x})||_o^{\frac{j}{k}}\; ||A_k^{\frac{k+j}{2k}} f||_o$$

Aus den Sätzen 12 und 15 folgt dann die Behauptung. -

Korollar: Es sei u die Lösung des Randwertproblems (6.1). Dann gilt für $o \leq i \leq k-1$ ($i \in \mathbb{N}$) und $x \in I$ die Abschätzung

$$|\varepsilon_x^{(i)}(u - Q_m u)| \leq M\, h^{2k-i-\frac{1}{2}}\, ||f||_o .$$

Beweis: Der Beweis folgt aus Satz 16 mit j = k und der Beziehung $A_k u = f$ (vgl. 6.1). -

Abschließend sei noch folgendes bemerkt. Während wir zur Herleitung von Satz 13 allein die Ungleichung (5.5) benötigen, gehen bei der Behandlung des punktweisen Fehlers ebenso wesentlich die Abschätzungen (6.9) und (6.11) ein. Dies liegt daran, daß wir mit Hilfe des Interpolationssatzes für Hilbertraumskale keine Abschätzung etwa der Form

$$|\varepsilon_x^{(i)}(f - Q_m f)| \leq M\, h^{k-i-\frac{1}{2}}\, ||D^k f||_o$$

"simulieren" können.

Literaturverzeichnis

1. BEREZANSKIJ, J.M.: Expansions in Eigenfunctions of Selfadjoint Operators. Translations of Mathematical Monographs, Vol. 17. Rhode Island. Providence 1968.

2. BUTZER, P.L., BERENS, H.: Semi Groups of Operators and Approximation. Berlin - Heidelberg - New York. Springer Verlag 1967.

3. CIARLET, P.G., VARGA, R.S.: Discrete Variational Green's Function. II. One Dimensional Problem. Num. Math. 16, 115 - 128 (1970).

4. DAVIS, P.J.: Interpolation and Approximation. New York. Blaisdell 1965.

5. DELVOS, F.J.: Optimale Approximation mit der Methode von Ritz. ZAMM 55, T234 - T235 (1975).

6. DELVOS, F.J., KÖSTERS, H.W.: Zur Konstruktion von Splines höheren Grades. Erscheint demnächst.

7. DELVOS, F.J., Schempp, W.: Sard's method and the theory of spline systems. Erscheint im J. of Approximation Theory.

8. DELVOS, F.J., SCHEMPP, W.: An extension of Sard's method. Erscheint in diesem Band.

9. HEDSTROM, G.W., VARGA, R.S.: Application of Besov Spaces to Spline Approximation. J. Approx. Theory 4, 295 - 327 (1971).

10. MESCHKOWSKI, H.: Hilbertsche Räume mit Kernfunktion. Berlin - Göttingen - Heidelberg. Springer - Verlag 1962.

11. POLYA, G.: Bemerkungen zur Interpolation und zur Näherungstheorie der Balkenbiegung. ZAMM 11, 445 - 449 (1931).

12. RIESZ, F., NAGY, B. SZ.: Vorlesungen über Funktionalanalysis. Berlin. VEB Deutscher Verlag der Wissenschaften 1956.

13. SARD, A.: Optimal approximation. J. Functional Analysis 1, 222 - 244 (1967); 2, 368 - 369 (1968).

14. SCHÄFER, W.: Hilbertraumskalen und Konvergenz von Spline Approximationen. Diplomarbeit. 69 pp. Ruhr-Universität Bochum 1973.

15. SCHÄFER, W.: Splineapproximation in intermediären Räumen. Dissertation. 92 pp. Ruhr-Universität Bochum 1975.

16. SCHULTZ, M.H., VARGA, R.S.: L-Splines. Numer. Math. 10, 345 - 369 (1967).

17. SMIRNOW, W.I.: Lehrgang der höheren Mathematik,. Band V. Berlin. VEB Deutscher Verlag der Wissenschaften 1968.

18. VARGA, R.S.: Error Bounds für Spline Interpolation. In Schoenberg (ed.): Approximations with Special Emphasis on Spline Functions, 367 - 388. London - New York. Academic Press 1969.

Werner Schäfer
Walter Schempp
Lehrstuhl für Mathematik I
Universität Siegen
D-59 Siegen 21
Hölderlinstr. 3
Germany

Mehrdimensionale Spline-Interpolation mit Hilfe der Methode von Sard

Karl-Heinz Schloßer

In this paper we introduce a space of functions defined on sets $\Omega \subseteq \mathbb{R}^n$ $(n \in \mathbb{N})$ fullfilling a condition called n-parametric stellar. By means of Sard's method we are able to proof all classical minimal properties and Schoenberg's approximation theorem. Morever that function space considered with a graph scalarproduct is a Hilbertspace with reproducing kernel and therefore we are able to construct the spline approximation explicitly.

1. Einleitung.

Ziel dieser Arbeit ist die Einführung eines Funktionenraumes, der es ermöglicht für gewisse, auf Teilmengen des $\mathbb{R}^n$ definierte Funktionen Splineapproximierende zu konstruieren, die die üblichen Minimaleigenschaften erfüllen und einem zum Approximationssatz von Schoenberg analogen Satz genügen. Wesentliches Hilfsmittel ist dabei die von Sard [10] [11] [12] entwickelte funktionalanalytische Spline Theorie, die sich gerade auf unseren Funktionenraum anwenden läßt. Zusätzlich erweist sich der betrachtete Hilbertraum als Raum mit einem reproduzierenden Kern, der es ermöglicht, den Spline Projektor explizit zu bestimmen.

2. Sard Systeme

Wir definieren nun das folgende System, das wir nach seinem Schöpfer A. Sard [10] [11] [12] Sard System nennen wollen. Eine interessante Verallgemeinerung findet sich bei Delvos - Schempp [4] .

Definition 2.1:

Gegeben seien ein Banachraum X und die Hilberträume Y und Z. Dann nennen wir das System

$$(X,Y,Z,U,F)$$

ein Sard System, wenn $U \in L(X,Y)$ eine stetige und surjektive Abbildung ist und $F \in L(X,Z)$ stetig und surjektiv mit abgeschlessenem Bild.

Gilt zusätzlich die Ungleichung

$$\|x\|_X = B(\|U(x)\|_Y^2 + \|F(x)\|_Z^2) \qquad B > 0,$$

so erfüllt das Sard System die "completness - condition".
Unter diesen Bedingungen kann man das folgende Graphenskalarprodukt

$$(x,y) = (Ux,Uy)_Y + (Fx,Fy)_Z$$

konstruieren, dessen induzierte Norm äquivalent zur ursprünglichen Banach-Norm ist (Sard [10]). Dann existiert dazu ein eindeutig bestimmte Projektor - der Spline Projektor - der durch

$$Im(SP) = Kern(F)^{\perp}$$

definiert ist. Es fällt nun recht leicht, die folgenden Sätze zu beweisen (Sard [10] [11] [12], Schempp [14]).

<u>Theorem 1:</u>(Eindeutige Spline Interpolation)

Zu jedem Element $x_o \in X$ existiert eine eindeutig bestimmte Spline Approximation $s_{x_o} := SP(x_o)$ mit: $F(x_o) = F(s_{x_o})$.

<u>Theorem 2:</u>(Optimale Interpolation)

Zu jedem $x_o \in X$ ist die Spline Approximation s_{x_o} das eindeutig bestimmte Element unter allen $x \in X$ mit $F(x) = F(x_o)$, welches das Funktional

$$x \longmapsto \|Ux\|_Y$$

minimiert.

Wir betrachten nun ein Funktional G X' und die zugehörige Approximationsklasse

$$\mathrm{Appr}(G) := \left\{ A \in X' : A = E \circ F \text{ und } \mathrm{Kern}U \subseteq \mathrm{Kern}(G-A) \right\}$$

wobei $E \in Z'$. Dann folgt aus dem Quotiententheorem (Sard [13]) die Existenz eines eindeutig bestimmten $Q \in Y'$ mit:

$$R := G - A = Q \circ U.$$

Wendet man nun den Satz von Frèchet - Riesz an, so erhält man für jedes $w' \in Y'$ ein eindeutig bestimmtes $w \in W$ mit:

$$w'(f) = (f,w)_Y \quad \text{für alle} \quad f \in Y$$

und

$$\|w\|_Y = \|w'\| .$$

Somit existiert zu jedem $A \in \mathrm{Appr}(G)$ ein verallgemeinerter Peano-Kern $K_{G-A} \in Y$ mit:

$$\| K_{G-A}\|_Y \leq \| Q \| .$$

Eine genaue Darstellung dieses Zusammenhanges findet sich z.B. bei Schempp [15]. Wir sind jetzt in der Lage, das folgende Theorem zu formulieren, das eine direkte Konsequenz des optimalen Quotienten-Theorems (Sard [13]) ist und dessen Beweis sich z.B. bei Schempp [15] findet.

Theorem 3:

Sei $G \in X'$ gegeben. Dann gilt:

(i) $\langle x,G\rangle = \langle x,A\rangle + (Ux,K_{G-A})_Y$ für $A \in \mathrm{Appr}(G)$.

(ii) $A_o := G \circ SP$ liegt in Appr(G).

(iii) A_o ist optimale Approximation an G im Sinne von Sard, d.h.

$$\| K_{G-A_o} \|_Y \leq \| K_{G-A} \|_Y$$

für alle $A \in \mathrm{Appr}(G)$.

3. Der Raum $K^m(\Omega)$.

Sei $m = (m_1,\dots,m_n)$ ein Multiindex und J eine nichtleere Teilmenge der Menge $\{1,\dots,n\}$ mit der komplementären Menge J^c. Mit proj_J bezeichnen wir die Projektion

$$\mathrm{proj}_J : \mathbb{R}^n \longrightarrow \prod_{i \in J} R_i \qquad R_i = \mathbb{R}$$

für alle i.

Definition 3.1:

Sei $\Omega \subset \mathbb{R}^n$ eine kompakte Teilmenge mit $|\Omega|>1$. nennen wir n-parametrig sternförmig, wenn ein Punkt $Z = (z_1,\dots,z_n)$ existiert, so daß für jeden

Punkt $P = (p_1,\dots,p_n) \in \Omega$, $P \neq Z$ die Menge

$$\left\{ X \in R^n : x_i = \lambda_i z_i + (1-\lambda_i)p_i \ , \ 1 \le i \le n \ , \ 0 \le \lambda_i \le 1 \right\}$$

in Ω enthalten ist.

Bemerkung:

(i) Offensichtlich sind alle n-dimensionalen Kreise und achsenparallele Rechtecke sind Elemente der der Klasse aller n-parametrig sternförmigen Mengen.

(ii) Für alle $J \subset \{1,\dots,n\}$ ist $\mathrm{proj}_J(\Omega) =: \Omega_J$ J - parametrig sternförmig bezüglich $\mathrm{proj}_J(Z)$.

(iii) Jede n-parametrig sternförmige Menge ist sternförmig.

Jetzt sind wir in der Lage, den Funktionenraum $K^m(\Omega)$ zu definieren. Eine reellwertige Funktion f auf Ω gehört zu $K^m(\Omega)$, wenn für alle $X \in \Omega$ die folgende Taylorentwickelung gilt:

$$f(X) = \sum_{p=0}^{m-1} c_p \prod_{i=1}^{n} (x_i - z_i)^{p_i}(1/p_i!)$$

$$+ \sum_{\substack{p_J=0 \\ 1 \le |J| \le n-1}}^{m_J - 1_J} \underbrace{\int_{z_i}^{x_i} \cdots \int_{z_i}^{x_i}}_{i \in J^c} k_{p_J}(x, s_{J^c}) f_{p_j}(s_{J^c})\, ds_{J^c}$$

$$+ \int_{z_1}^{x_1} \cdots \int_{z_n}^{x_n} \prod_{i=1}^{n} \frac{(x_i - s_i)^{m_i - 1}}{(m_i - 1)!} f_m(s)\, ds$$

dabei ist $c_p \in \mathbb{R}$,

$$k_{p_J}(x, s_{J^c}) := \prod_{i \in J} \frac{(x_i - z_i)^{p_i}}{p_i!} \prod_{i \in J^c} \frac{(x_i - s_i)^{m_i - 1}}{(m_i - 1)!} ,$$

$f_{p_J} \in L_2(\bigtimes_{i \in J^c} \Omega_i)$ und $f_m \in L_2(\Omega)$.

Verwendet man die Heavyside-Funktion H, so ergibt sich sofort:

$$f(X) = \sum_{p=0}^{m-1} c_p \, e_p(X)$$

$$+ \sum_{\substack{p_J = 0 \\ 1 \le |J| \le n-1}}^{m_J - 1_J} \int \cdots \int_{\Omega_{J^c}} g_{p_J}(X, s_{J^c}) \, f_{p_J}(s_{J^c}) \, ds_{J^c}$$

$$+ \int \cdots \int_{\Omega} g_m(X, s) \, f_m(s) \, ds$$

mit $$e_p(X) = \prod_{i=1}^{n} \frac{(x_i - z_i)^{p_i}}{p_i!}$$

$$g_{p_J}(X, s_{J^c}) = \prod_{i \in J^c} \frac{(x_i - z_i)^{p_i}}{p_i!} \, G_{p_J}(X_{J^c}, s_{J^c})$$

$$g_m(X, s) = \prod_{1 \le i \le n} \frac{1}{(m_i - 1)!} \Big\{ (x_i - s_i)^{m_i - 1} H(x_i - s_i) H(s_i - z_i)$$

$$+ (-1)^{m_i} H(s_i - x_i) H(z_i - s_i) \Big\}$$

$$G_{p_J}(X_{J^c}, s_{J^c}) =$$

$$\prod_{i\in J^c} \frac{1}{(m_i-1)!}\left\{(x_i-s_i)^{m_i-1} H(x_i-s_i)H(s_i-z_i) + \right.$$

$$\left. + (-1)^{m_i}(s_i-x_i)^{m_i-1} H(s_i-x_i)H(z_i-s_i)\right\}.$$

Nun verifiziert man leicht, daß

$$f(X) = \sum_{p=0}^{m-1} c_p e_p(X) +$$

$$+ \sum_{\substack{p_J=0 \\ 1\le |J| \le n-1}}^{m_J-1} (g_{p_J}(X,.), f_{p_J})_{L_2(\times_{i\in J^c} \Omega_i)}$$

$$+ (g_m(X,.), f_m)_{L_2(\Omega)} .$$

Der Beweis des folgenden algebraischen Einbettungssatzes findet sich bei Scheffold-Schloßer [14] ,wo man auch eine ausführliche Diskussion des Raumes $K^m(\Omega)$ findet.

<u>Lemma 1:</u>

Es gelten die folgenden Inklusionen:

$$C^m(\Omega) \hookrightarrow K^m(\Omega) \hookrightarrow C^{m-1}(\Omega)$$

Dabei ist $C^m(\Omega)$ der Banachraum aller Funktionen F,

deren patielle Ableitungen

$$D^{i_1,\dots,i_n} F = \frac{\partial^{i_1+\dots+i_n}}{\partial x_1^{i_1}\dots\partial x_n^{i_n}} F \quad \text{für}$$

$0 \leq i_i \leq m_i$, $1 \leq i \leq n$ existieren und auf Ω stetig sind. Dabei ist $C^m(\Omega)$ mit der Norm

$$\|F\|_{s,m} := \sup_{1\leq i\leq n} \sup_{0\leq i_i\leq m_i} \sup_{x} \left| D^{i_1,\dots,i_n} F(x)\right|$$

versehen. Den Raum $K^m(\Omega)$ versehen wir nun mit der folgenden Norm

$$\|F\|_B = \max \begin{cases} \|F^\mu\|_{s,\mu} & : 0\leq \mu \leq m-1 \\ \|f_{p_J}\|_{L_2(I_{J^c})} & : 1\leq |J| \leq n-1 \ , \ 0\leq p_J \leq m_J-1 \\ \|f_m\|_{L_2(\Omega)} \end{cases}$$

wobei $I_{J^c} := \bigtimes_{i\in J^c} \Omega_i$

Es ist leicht zu verifizieren, daß $\|\cdot\|_B$ eine Norm ist, da die psitive Definitheit sofort aus der Taylorformel folgt. Sei nun (F_n) eine Cauchy-Folge, so bilden die zugehörigen "Koordinatenfolgen" Cauchyfolgen in den zugehörigen Banch-Koordinaten-Räumen. Dann sieht man natürlich recht leicht, daß die aus den Grenzwerten der Koordinatenfolgen konstruierte Koordinatengrenzfunktion gerade die gesuchte Grenzfunktion der Cauchyfolge (F_n) im Raum $K^m(\Omega)$ ist. Somit gilt das folgende

Lemma 2:

Der Vektorraum $K^m(\Omega)$ versehen mit der Norm $\|\cdot\|_B$ ist ein reeller Banachraum.

Wir wollen nun im folgenden Satz eine Charakterisierung des starken topologischen Duals des $K^m(\Omega)$ geben.

Theorem 4:

Jedse Element $L \in K^m(\Omega)'$ läßt sich folgendermaßen darstellen:

$$\langle F, L\rangle = \sum_{p=0}^{m-1} \int \cdots \int_{\Omega} F^{(p)}(X)\, d\mu_p(X)$$

$$+ \sum_{\substack{p_J=1 \\ 1\leq |J| \leq n-1}}^{m_J-1} \int \cdots \int_{\Omega_{J^c}} f_{p_J}(X_{J^c}) \tilde{g}_{p_J}(X_{J^c})\, d(X_{J^c})$$

$$+ \int \cdots \int_{\Omega} f_m(X) \tilde{g}_m(X)\, dX \quad ,$$

wobei die μ_p Radon-Maße auf Ω sind und $\tilde{g}_{p_J} \in L_2(\Omega_{J^c})$, $\tilde{g}_m \in L_2(\Omega)$.

Umgekehrt liegt jedes Funktional der obigen Gestalt im starken Dual von $K^m(\Omega)$.

Beweis:

Wir definieren $C_p := C(\Omega)$ für $0 \leq p \leq m-1$ und

$H_{p_J} := L_2(\Omega_{J^c})$ für $1 \leq |J| \leq n-1$, $0 \leq p_J \leq m_J-1$.

Dann betrachten wir den Produktraum

$$Z := \prod_{0 \leq p \leq m-1} C_p \times \prod_{\substack{0 \leq p_J \leq m_J-1 \\ 1 \leq |J| \leq n-1}} H_{p_J} \times L_2(\Omega)$$

versehen mit der kanonischen Produktnorm. Die Abbildung

$$\Phi : K^m(\Omega) \longrightarrow Z$$

mit

$$\Phi(F) := \Big(\prod_{0 \leq p \leq m-1} F^{(p)} \times \prod_{\substack{0 \leq p_J \leq m_J-1 \\ 1 \leq |J| \leq n-1}} f_{p_J} \times f_m \Big)$$

ist eine isometrische Einbettung von $K^m(\Omega)$ in Z. Wir betrachten nun ein Funktional $L \in K^m(\Omega)'$. Dann ist offensichtlich $L \circ \Phi^{-1}$ ein stetiges lineares Funktional auf

$$\Phi\,(K^m(\Omega)) \subset Z\ ,$$

welches nach dem Satz von Hahn-Banach eine stetige Fortsetzung auf ganz Z besitzzt. Nun haben aber die Funktionale auf Z bekanntlich die oben behauptete Darstellung, und man sieht daraus natürlich sofort die Behauptung. Die Umkehrung ist trivial.

q.e.d.

4. Spline-Interpolation auf $K^m(\Omega)$.

Wir betrachten nun den Produkthilbertraum

$$W_o := \prod_{\substack{0 \le p_J \le m_J - 1 \\ 1 \le |J| \le n-1}} H_{p_J} \times L_2(\Omega)$$

und die Abbildung $\varphi = \Phi \circ proj_{W_o}$. Man sieht nun sehr leicht, daß φ eine lineare und steige Abbildung ist, wenn W_o die kanonische Produkthilbertraumstruktur trägt. Weiterhin folgt sofort aus der Taylorformel, daß φ surjektiv ist mit

$$\text{Kern}(\varphi) = P^{m-1}(\Omega) \;.$$

Sei nun

$$N' := \prod_{1 \le j \le n} m_j$$

die Dimension von Kern(φ). Gibt man sich nun $N \geq N'$ stetige auf $K^m(\Omega)$ linear unabhängige Funktionale vor so, daß die ersten N' zusätzlich noch linear unabhängig auf Kern(φ) sind, so ist die Abbildung

$$F := \bigtimes_{1 \le i \le N} L_i$$

eine lineare steige Abbildung von $K^m(\Omega)$ auf $\mathbb{R}^N$ mit abgeschlossenem Bild. Somit gilt das folgende

<u>Theorem 5:</u>

Das System

$$(K^m(\Omega), W_o, R^N, \varphi, F)$$

bildet ein Sard System, das die "completness - condition" erfüllt.

Beweis:

Aufgrund der vorhergehenden Betrachtungen reicht es, die "completness condition" zu zeigen. Dazu betrachten wir die Abbildung

$$F_0 := \bigtimes_{1\leq i\leq N'} L_i$$

$$F_0 : K^m(\Omega) \longrightarrow \mathbb{R}^{N'}$$

und definieren die Sesquilinearform

$$(F,G)_H := (F_0(F),F_0(G))_{\mathbb{R}^{N'}} + (\gamma F,\gamma G)_{W_0}$$

Wegen der linearen Unabhängigkeit der L_i folgt sofort die positive Definitheit. Aufgrund der Vollständigkeit der Komponentenräume und der Surjektivität von F_0 und γ folgt natürlich sofort die Vollständigkeit des Raumes

$$(K^m(\Omega), \|\cdot\|_H) .$$

Wegen der Stetigkeit der Abbildung F_0 in der Banach-Norm gibt es eine Konstante $K>0$ mit:

$$\|F\|_H \leq K \|F\|_B \quad \text{für alle } F\in K^m(\Omega).$$

Also ist

$$id: (K^m(\Omega), \|\cdot\|_B) \longrightarrow (K^m(\Omega), \|\cdot\|_H)$$

stetig. Aufgrund des Homomorphiesatzes von Banach ist id^{-1} stetig. Also gilt:

$$\|F\|_B \leq K(\|\gamma(F)\|_{W_0} + \|F_0(F)\|_{\mathbb{R}^{N'}}) \quad \text{mit } K>0.$$

Daraus folgt aber nun sofort die Behauptung.

q.e.d.

Aufgrund der in 2. entwickelten Theorie gilt nun:

Korollar 1:

Zu jedem $F \in K^m(\Omega)$ existiert genau eine Splineapproximation $S_F := SP(F)$ mit:

$$L_i(F) = L_i(S_F) \quad 1 \leq i \leq N.$$

Korollar 2:

Für jedes $F_o \in K^m(\Omega)$ ist die Splineapproximierende

$$S_{F_o} := SP(F_o)$$

das eindeutig bestimmte Element unter allen $F \in K^m(\Omega)$ mit $L_i(F) = L_i(F_o)$ $1 \leq i \leq N$, welches das Funktional

$$F \longmapsto \|\varphi(F)\|_{W_o}$$

minimiert.

Sei nun $l \in K^m(\Omega)'$ ein stetiges lineares Funktional und

$$\mathrm{Appr}(l) = \left\{ a \in K^m(\Omega)' : a(p) = l(p) \text{ für alle } p \in P^{m-1}(\Omega) \right\}$$

die Menge der zulässigen Funktionale. Dann gilt für den nach dem 2. Kapitel existierenden verallgemeinerten Peano-Kern K_{l-a} das

Korollar 3:

Sei $l \in K^m(\Omega)'$ gegeben. Dann gilt:

(i) $l(F) = a(F) + (\varphi(F), K_{l-a})_{W_o}$ für alle $a \in \mathrm{Appr}(l)$.

(ii) $a_o := l \circ SP$ liegt in Appr(l).

(iii) a_o ist optimal im Sinne von Sard.

5. Zur Konstruktion der Spline-Approximation

Zunächst betrachten wir das Skalarprodukt

$$(F^1,F^2)_T := \sum_{p=0}^{m-1} c_p(F^1)c_p(F^2) + \sum_{\substack{p_J=0 \\ 1\le|J|\le n-1}}^{m_J-1} (f^1_{p_J},f^2_{p_J})_{H_{p_J}} + (f^1_m,f^2_m)_{L_2(\ )}$$

auf $K^m(\Omega)$. Man erkennt sofort, daß $K^m(\Omega)$ versehen mit diesem Skalarprodukt ein Hilbertraum, der homöomorph zum Produktraum $\mathbb{R}^{N'}$ x W_o ist. Weiterhin gilt jedoch:

Theorem 6:

Der Hilbertraum $(K^m(\Omega),(.,.)_T)$ besitzt den reproduzierenden Kern

$$K(X,S) = \sum_{p=0}^{m-1} e_p(X)\ e_p(S) + \sum_{\substack{p_J=0 \\ 1\le|J|\le n-1}}^{m_J-1} (g_{p_J}(X,.),g_{p_J}(S,.))_{H_{p_J}} + (g_m(X,.),g_m(S,.))_{L_2(\Omega)}$$

Beweis:

Offensichtlich gehört $K(.,S)$ zu $K^m(\Omega)$ für jeden Punkt $S\in\Omega$. Die reproduzierende Eigenschaft von K ist eine direkte Konsequenz aus den Relationen

$$D^p K(Z,S) = e_p(S) \qquad 0\leq p\leq m-1$$

$$D^{m_{J^c},p_J} K(x_{J^c},z_J;S) = g_{p_J}(X,s_{J^c}) \quad 1\leq |J| \leq n-1; 0\leq p_J\leq m_J-$$

$$D^m K(X,S) = g_m(S)$$

für jeden Punkt $S\in\Omega$.

q.e.d.

Für den Hauptsatz diese Paragraphen benötigen wir noch das folgende Lemma, dessen Beweis sich analog zu dem von Kösters-Schloßer [7] vorgeschlagenen durchführen läßt.

Lemma 3:

Sei $L\in K^m(\Omega)'$. Dann gilt für ein beliebiges $F\in K^m(\Omega)$ und $S\in\Omega$ die Gleichung:

$$\langle(\varphi F,\varphi K(.,S))_T, L_S\rangle = (\varphi F, \varphi \langle K(.,S), L_S\rangle)_{W_0} .$$

Bemerkung:

Mit Hilfe von üblichen Abschätzungsschlüssen läßt sich nun recht leicht zeigen, daß die auftretenden Skalarprodukte $(.,.)_T$ und $(.,.)_H$ zu äquivalenten Normierungen führen (Scheffolf-Schloßer [14]). Da der Spline-Projektor jedoch bezüglich $(.,.)_H$ definiert ist, benötigen wir zu dessen expliziter Darstellung den reproduzierenden Kern des Raumes $(K^m(\Omega),(.,.)_H)$ (vgl. Kösters-Schloßer [7]).

Sei (φ_i) $1 \leq i \leq N'$ die Dualbasis zu den Funktionalen L_i $1 \leq i \leq N'$, dann ist die Abbildung

$$P(F) = \sum_{i=1}^{N'} \langle F, L_i \rangle \varphi_i$$

ein orthogonaler Projektor auf $(K^m(\Omega), (.,.)_T)$ mit

$$\mathrm{Im}(P) = \mathrm{Kern}(\varphi).$$

Bezeichnen wir den zugehörigen supplementären Projektor mit Q so folgt mit den bekannten funktionalanalytischen Standardschlüssen (vgl. Mansfield [8] und Kösters-Schloßer [7]) das

Theorem 7:

Ist K der reproduzierende Kern von $(K^m(\Omega), (.,.)_T)$, dann ist

$$K_1(X,S) = Q_X \circ Q_S\, K(X,S) + \sum_{i=1}^{N'} \varphi_i(X) \varphi_i(S)$$

der reproduzierende Kern von $(K^m(\Omega), (.,.)_H)$.

Berücksichtigt man nun die üblichen Schlüsse der Theorie der reproduzierenden Kerne (Schempp-Tippenhauer [16]), so erhalten wir den Spline-Projektor explizit durch

$$SP(F)(X) = \sum_{i=1}^{N} a_i \langle K_1(X,S), L_{i,S} \rangle$$

mit

$$\sum_{i=1}^{N'} \langle\langle K_1(X,S), L_{i,S} \rangle, L_{j,X} \rangle = \langle F, L_j \rangle \quad 1 \leq j \leq N \quad .$$

Da die Matrix des obigen linearen Gleichungssystems positiv definit und symmetrisch ist, läßt sie sich mit üblichen Methoden der numerischen Mathematik lösen.

Literaturverzeichnis

[1] Birkhoff, G.: Piecewise bicubic interpolation and approximation in polygons. In: Approximation with special emphasis on spline functions. (I.J. Schoenberg ed.) pp. 185-221. New York-London: Academic Press 1969.

[2] De Boor, C.R., Lynch, R.E.: On splines and their minimum properties. J. Math. Mech. 15, 953-969(1966).

[3] Delvos, F.J.: Über die Konstruktion von Spline Systemen. Dissertation. Ruhr-Universität Bochum 1972.

[4] Delvos, F.J., Schempp, W.: Sard's method and the theory of spline systems. Erscheint in J. Approximation Theory.

[5] Delvos, F.J., Schempp, W.: On spline systems. Monatsh. Math. 34, 399-409(1970).

[6] Delvos, F.J., Schloßer, K.-H.: Das Tensorproduktschema von Spline Systemen. In: Spline-Funktionen. Vorträge und Aufsätze. (K. Böhmer, G. Meinardus, W. Schempp eds.) pp. 59-73. Mannheim-Wien-Zürich: Bibliographisches Institut 1974.

[7] Kösters, H.W., Schloßer, K.-H.: On spaces related to bivariate Spline Interpolation. (Erscheint demnächst).

[8] Mansfield, L.: On the variational characterization and convergence of bivariate splines. Numer. Math. 20, 99-114(1972).

[9] Nielson, G.M.: Bivariate spline functions and the approximation of linear functionals. Numer. Math. 21, 138-160(1973).

[10] Sard, A.: Optimal approximation. J. Functional Anal. 1. 224-244 (1967) and 2, 368-369 (1968).

[11] Sard, A.: Approximation based on nonscalar observations. J. Approximation Theory 8, 315-334 (1973).

[12] Sard, A.: Instances of generalized Splines. In: Spline-Funktionen. Vorträge und Aufsätze. (K. Böhmer, G. Meinardus, W. Schempp eds.) pp. 215-241. Mannheim-Wien-Zürich: Bibliographisches Institut 1974.

[13] Sard, A.: Linear approximation. Providence, Rhode Island, American Mathematical Society, 1963.

[14] Scheffold, E., Schloßer, K.-H.: Spline-Funktionen mehrerer Veränderlicher. (Erscheint demnächst).

[15] Schempp, W.: Zur Theorie der Spline Systeme. In: Spline-Funktionen. Vorträge und Aufsätze. (K. Böhmer, G. Meinardus, W. Schempp eds.) pp. 275-289. Mannheim-Wien-Zürich: Bibliographisches Institut 1974.

[16] Schempp, W., Tippenhauer, U.: Reprokerne zu Spline-Grundräumen. Math. Z. 136, 357-369(1974).

[17] Schloßer, K.-H.: Mehrdimensionale Spline-Interpolation mittels Spline-Systemen. Zeitr. Angew. Math. Mech. 55, T260-T262 (1975).

[18] Schloßer, K.-H.: Zur mehrdimensionalen Spline-Interpolation. Dissertation. 79pp. Bochum 1974.

[19] Schoenberg, I.J.: On best approximation of linear operators. Nederl. Akad. Wetensch. Indag. Math. 26, 155-163(1964).

[20] Tippenhauer, U.: Reproduzierende Kerne in Spline-Grundräumen. Dissertation 101pp. Bochum 1973.

Dr. Karl-Heinz Schloßer
Ruhr-Universität Bochum

D-4630 Bochum
Universitätsstraße 150 NA
Rechenzentrum
Bundesrepublik Deutschland

TOWARD A CONSTRUCTIVE THEORY OF GENERALIZED SPLINE FUNCTIONS

Larry L. Schumaker

This paper is divided into two parts. In the first part we define a rather general class of piecewise functions, and discuss some of its basic algebraic and structural properties. In the second part, we specialize to a (still quite general) class of Tchebycheffian splines and discuss their zero properties and approximation powers.

§1. Introduction.

Considerable effort has been expended over the past 10 to 15 years in building up an extensive heirarchy of generalized spline functions (e.g., the L-splines, g-splines, Lg-splines, pLg-splines, Λ-splines, M-splines, singular-splines, perfect-splines, etc., etc.). Most of these generalized splines arise as solutions of certain smoothest interpolation problems, and hence are the result of what might be called the variational approach to splines. This approach is quite abstract, relatively complete, and very general.

On the other hand, in the past few years, certain classes of polynomial spline functions have begun to play a very important role in applications, and in particular, in the design of algorithms for the numerical solution of a wide variety of problems in Applied Mathematics. In spite of the beauty of the variational approach and all of the attendent optimal properties, it is probably fair to say that it is really some of the more mundane algebraic, structural, and analytic properties of these splines which accounts for their extensive applications. For example, we may mention that 1) there exists a very convenient basis (B-splines) for the space of polynomial splines which makes them easy to store, evaluate, and manipulate on a computer; 2) polynomial splines have many structural and analytic properties such as sign change properties, zero properties, and determinental properties similar to polynomials; 3) polynomial spline functions possess excellent approximation powers. It is no surprise, then, that polynomial splines are indeed an ideal tool for most applications.

Still, there are some applications where other classes of piecewise functions (generalized splines) may be more suitable than polynomial splines. We mention just two examples: 1) in certain data fitting problems there may be a strong motivation to use spaces of piecewise trigonometric, exponential, logarithmic, or rational functions; 2) in some methods for the numerical solution of singular boundary-value problems, it is necessary to use spaces of nonpolynomial splines.

The study of the basic algebraic, structural, and analytic properties of spaces of piecewise functions might be called the <u>constructive approach</u> to splines. The constructive theory of polynomial splines is now reasonably complete (although interesting and important properties still continue to be discovered). While there are quite a number of papers dealing with constructive properties of various spaces of nonpolynomial splines, it seems to me that (compared with the variational theory) there has been disproportionately little effort devoted to the development of a more complete constructive theory of generalized spline functions.

The purpose of this paper is to take a step towards such a general constructive theory. In particular, in the first part of the paper (sections 2-5) we define a rather general class of piecewise (real-valued) functions (in essentially one dimension), and proceed to discuss some of its basic algebraic and structural properties. The remainder of the paper (sections 6- 10) is devoted to some classes of generalized splines for which we can handle more difficult questions on local support bases, zero properties, and approximation powers. Clearly, there remain many unsolved problems, and it is to be hoped that this paper will provide a stimulus for further work on the constructive approach.

§2. A class of splines.

As hinted at above, in this paper we regard the defining feature of splines to be the fact that they are piecewise functions with the pieces somehow tied together. This notion allows us to define a rather general class of splines. We shall consider only real-valued functions, although it is possible to define complex-valued splines. We need some notation. Let Ω be a well ordered set, and suppose that $\Delta = \{x_1 < x_2 < \dots < x_k\}$ is a set of distinct elements in Ω. The set Δ partitions Ω into k+1 "intervals" $I_0 = \{x < x_1\}$, $I_i = \{x_i \le x < x_{i+1}\}$, i = 1,2, ...,k-1 and $I_k = \{x_k \le x\}$. Suppose that S_i are linear spaces of real-valued functions defined on I_i with bases $u_{i1},..,u_{in_i}$, $i = 0,1,...,k$. For each $0 \le i < j \le k$, let

$$\Gamma_{ij} = \{(\gamma^-_{ijv}, \gamma^+_{ijv})\}_{v=1}^{r_{ij}},$$

where γ^-_{ijv} and γ^+_{ijv} are linear functionals defined on S_i and on S_j, respectively. We write Γ for the set $\{\Gamma_{ij}\}_{0\le i<j\le k}$. We now define

$$(2.1)\quad \mathcal{S}(S_0,..,S_k;\Gamma;\Delta) = \{s: s_i = s|_{I_i} \in S_i,\ i = 0,..,k,$$
$$\gamma^-_{ijv}s_i = \gamma^+_{ijv}s_j,\ v = 1,2,..,r_{ij},\ 0\le i<j\le k\}.$$

This is a class of real-valued functions defined on Ω such that in each subinterval I_i, the function s belongs to S_i, and moreover, the pieces in the ith and j^{th} intervals are tied together by the requirement that certain linear functionals on each of these pieces have a common value. Because of this characteristic piecewise structure, we call $\mathcal{S}$ a <u>space</u> <u>of</u> <u>spline</u> <u>functions</u>.

We shall see a number of interesting examples of $\mathcal{S}$ in the following section. In any case, clearly the class $\mathcal{S}$ is sufficiently general to include nearly all of the classes of functions which are presently known as splines.

We begin now the task of determining the algebraic properties of $\mathcal{S}$. Our first result shows that it is a linear space, and under appropriate assumptions on the spaces S_i and the sets of linear functionals Γ_{ij} (which assure their consistency), we identify its dimension.

THEOREM 2.1. Suppose that for each $j = 1,2,\ldots,k$, the set of linear functionals tying the piece s_j to all previous pieces is a linearly independent set over S_j. Specifically, we assume that the matrix A_j^+ defined by

$$A_j^+ = \begin{bmatrix} A_{0j}^+ \\ A_{1j}^+ \\ \cdot \\ A_{j-1,j}^+ \end{bmatrix}, \quad A_{ij}^+ = \begin{bmatrix} \gamma_{ij1}^+ u_{j1} & \cdots & \gamma_{ij1}^+ u_{jn_j} \\ \gamma_{ij2}^+ u_{j1} & \cdots & \gamma_{ij2}^+ u_{jn_j} \\ & & \\ \gamma_{ijr_{ij}}^+ u_{j1} & \cdots & \gamma_{ijr_{ij}}^+ u_{jn_j} \end{bmatrix}$$

is of full rank $r_j = r_{0j} + \cdots + r_{j-1,j}$ for $j = 0,1,\ldots,k$. (We recall that $u_{j1},\ldots,u_{jn_j}$ denotes a basis for S_j. This condition can be fulfilled, of course, only when $r_j \le n_j$, $j = 1,2,\ldots,k$). Then, $\mathcal{S}$ is a linear space, and

$$\text{(2.2)} \qquad \dim \mathcal{S} = n_0 + \sum_{j=1}^{k} (n_j - r_j).$$

Proof: It is clear that $\mathcal{S}$ is a linear space, and that each $s \in \mathcal{S}$ can be written in the form

$$s(x) = \left\{ s_j(x) = \sum_{p=1}^{n_j} c_{jp} u_{jp}(x) \text{ for } x \in I_j,\ j = 0,\ldots,k \right. .$$

The conditions tying the pieces together can be written as a linear system of equations on the coefficients of s. Indeed, if we write $c_j = (c_{j1},\ldots,c_{jn_j})^T$ for the coefficient vector of the jth piece, and define $r_{ij} \times n_j$ matrices A_{ij}^- exactly as for the A_{ij}^+, but using the linear functionals γ_{ijv}^-, the system of equations is

$$\begin{bmatrix} A_{01}^- & -A_{01}^+ & 0 & \cdot & & 0 \\ A_{02}^- & 0 & A_{02}^+ & \cdot & & 0 \\ \cdot & & & & & \\ A_{0k}^- & & & & & -A_{0k}^+ \\ 0 & A_{12}^- & -A_{12}^+ & \cdot & & \cdot \\ \cdot & & & & & \\ 0 & 0 & \cdot & \cdot & A_{k-1,k}^- & -A_{k-1,k}^+ \end{bmatrix} \begin{bmatrix} c_0 \\ c_1 \\ \\ \\ \\ \\ c_k \end{bmatrix} = 0.$$

The matrix of this system has $\sum_0^k n_j$ columns and $\sum_1^k r_j$ rows. As it is easily seen that the hypotheses guarantee it is of full rank (equal to the number of rows), it follows that the dimensionality of its null space is $\sum_0^k n_j - \sum_1^k r_j$, which, is of course, also the dimensionality of $\mathcal{S}$. ■

We turn now to the question of constructing a basis for $\mathcal{S}$. Our method of approach will be to mimic the construction of the one-sided basis for the polynomial splines. In particular, we shall first construct n_0 splines with support on all of Ω. Then associated with each knot x_i, we shall construct $m_i = n_i - r_i$ splines which vanish identically for $x < x_i$, $i = 1,2,\ldots,k$. We shall define each basis spline in terms of its coefficients in each subinterval. In particular, for $1 \le j \le m_i$ and $0 \le i \le k$, (where we set $m_0 = n_0$ for convenience), let

$$B_{ij}(x) = \left\{ \sum_{q=1}^{n_p} \alpha_{ijpq} u_{pq}(x) \text{ for } x \in I_p,\ p = 0,\ldots,k. \right.$$

To get the first m_0 basis splines, we simply extend each of the functions $u_{01},\ldots,u_{0n_0}$ to be a spline. That this is possible is the content of the following lemma.

<u>LEMMA 2.2.</u> <u>There</u> <u>exist</u> <u>splines</u> $B_{01},\ldots,B_{0m_0}$ <u>so</u> <u>that</u>

$$B_{0j}|_{I_0} = u_{0j} \quad , \; j = 1,2,\ldots,m_0 = n_0 \; .$$

<u>Proof</u>: Fix $1 \le j \le m_0$. We define B_{0j} to be u_{0j} on I_0. Now to extend it to I_1, we need to choose coefficients $\alpha_{0j11},\ldots,\alpha_{0j1n_1}$ so that the pieces of B_{0j} in I_0 and I_1 are properly tied together. This requires that the α's satisfy the system of equations

$$A_{01}^{+}\begin{bmatrix}\alpha_{0j11}\\ \cdot\\ \alpha_{0j1n_1}\end{bmatrix} = \begin{bmatrix}\bar{\gamma}_{011}u_{0j}\\ \cdot\\ \bar{\gamma}_{01r_{01}}u_{0j}\end{bmatrix} .$$

Since this is a system of $r_1 \le n_1$ equations, it can always be solved. Clearly, this process can now be continued to extend B_{0j} into the intervals $I_2,\ldots,I_k$. ■

Now, associated with each knot x_i, we want to construct m_i splines (which, hopefully, will turn out to be linearly independent). We have

<u>LEMMA 2.3</u>. <u>There</u> <u>exist</u> <u>splines</u> $B_{i1},\ldots,B_{im_i}$ <u>which</u> <u>vanish</u> <u>identically</u> <u>to</u> <u>the</u> <u>left</u> <u>of</u> x_i <u>and</u> <u>are</u> <u>linearly</u> <u>independent</u> <u>on</u> I_i , $i = 1,2,\ldots,k$.

<u>Proof</u>: Fix $1 \le i \le k$ and $1 \le j \le m_i$. Define $B_{ij}(x) = 0$ for $x < x_i$. To define B_{ij} in the interval I_i, we need to describe its coefficients there. Let A_i^{+} be the matrix defined in Theorem 2.1 describing the ties to previous intervals. Let $\alpha = (\alpha_{iji1},\ldots,\alpha_{ijin_i})^T$ be a solution of the system of equations $A_i^{+}\,\alpha = \delta(n_i,r_i+j)$, where

$\delta(n_i, r_i+j)$ is an n_i vector with all 0 components except for the r_i+j^{th}, which is to be a 1. Then, defining B_{ij} in the interval I_i with the coefficients $\alpha_{iji1}, \ldots, \alpha_{ijin_i}$, it is clear that B_{ij} satisfies the correct ties with all previous intervals. Finally, it is also evident that just as in the proof of Lemma 2.2, B_{ij} can be further extended to all of Ω to be an element of $\mathfrak{S}$.

It remains to check that the $B_{i1}, \ldots, B_{im_i}$ are linearly independent on I_i. Suppose that $\beta_1 B_{i1} + \ldots + \beta_{m_i} B_{im_i} \equiv 0$ on I_i. We may regard the vector α to be a solution of some augmented (nonsingular) system $\tilde{A}\alpha = \delta(n_i, r_i+j)$, obtained by adding m_i rows to A_i^+. Each of these additional rows of $\tilde{A}$ defines a linear functional on S_i; viz., if $(\tilde{A}_{r_i+\nu,1}, \ldots, \tilde{A}_{r_i+\nu,n_i})$ is the νth additional row, then

$$\lambda_\nu u = \lambda_\nu\Big(\sum_{p=1}^{n_i} c_p u_{ip}\Big) = \sum_{p=1}^{n_i} c_p \tilde{A}_{r_i+\nu,p} \quad .$$

Moreover, by the construction of the B_{ij}, it is clear the linear functionals $\lambda_1, \ldots, \lambda_{m_i}$ are dual to the $B_{i1}, \ldots, B_{i,m_i}$; that is,

$$\lambda_p B_{ij} = \delta_{pj} \quad , \quad j,p = 1,2,\ldots,m_i \, ,$$

(where here δ is the Kronecker delta). Now, for each $p = 1,2,\ldots,m_i$ we have

$$\beta_p = \lambda_p \Big(\sum_{j=1}^{m_i} \beta_j B_{ij}\Big) = 0,$$

and the required linear independence is established. ■

It is clear that there is some arbitrariness in the construction of the basis elements in Lemmas 2.2 and 2.3. However, it is also clear that whatever choices are made, the <u>span</u> of the resulting functions on I_i is unaffected.

THEOREM 2.4. The set of splines $\{B_{ij}\}_{j=1,i=0}^{m_i \ k}$ defined in Lemmas 2.2 and 2.3 form a basis for $\mathcal{S}$.

Proof: By construction, each of these functions belongs to $\mathcal{S}$, which we know by Theorem 2.1 is of dimension $m_0+\ldots+m_k$. It remains to check that they are linearly independent. Suppose that

$$\sum_{i=0}^{k} \sum_{j=1}^{m_i} \beta_{ij} B_{ij} \equiv 0 \quad \text{on } \Omega .$$

Restricting attention to I_0, we see that $\beta_{01}B_{01} + \ldots + \beta_{0m_0} \cdot B_{0m_0} \equiv 0$ there. But these are linearly independent on I_0, and we conclude the corresponding coefficients must be 0. Now we consider the interval I_1. We now have that $\beta_{11}B_{11} + \ldots + \beta_{1m_1} \cdot B_{1m_1} \equiv 0$ on I_1. By the linear independence established in Lemma 2.3, we conclude these coefficients are also zero. This process can be continued, moving one interval to the right at a time, and the theorem is proved. ■

The construction of a one-sided basis described here is, of course, considerably simpler in the case where $S_i = S$, $i = 0,1,\ldots,k$, since in this case, there is no problem of extending the functions to be in $\mathcal{S}$.

§3. A periodic class of splines.

In some cases it is desirable to work with classes of functions which are periodic. To define a general class of periodic splines, suppose now that Ω is an ordered set with a first element a and a last element b. Then if we associate a with b, we may think of Ω as being circular. Suppose now that Ω is partitioned into k subintervals by a set $\Delta = \{ x_1 < x_2 \ldots < x_k \}$. (The last of the "intervals" in this case is $I_k = \{x_k \le x < x_1\}$. With a set $\mathring{\Gamma} = \{\Gamma_{ij}\}_{1\le i<j\le k}$ of pairs of linear functionals and with linear spaces $S_1,\ldots,S_k$, we define the space of periodic splines as

$$(3.1)\quad \mathring{\mathcal{S}}(S_1,\ldots,S_k;\mathring{\Gamma};\Delta) = \{s : s_i = s|_{I_i} \in S_i,\ i = 1,\ldots,k,$$
$$\gamma^-_{ijv}s_i = \gamma^+_{ijv}s_j,\ v = 1,\ldots,r_{ij},\ 1\le i<j\le k\}.$$

Arguing as in the proof of Theorem 2.1, it is not hard to show that $\mathring{\mathcal{S}}$ is a linear space of dimension $m_1 + \ldots + m_k$. The construction of a one-sided basis can also be carried out exactly as in the non-periodic case.

§4. Examples.

The most common choice for Ω would be an interval $[a,b]$. If we take all of the spaces S_i to be the same, then with appropriate choices of the linear functionals, we obtain all of the usual splines in the literature. As these spaces are well-known, there is no need to go into any further detail on them.

In most of the spaces of splines in the literature, the tie conditions take the form of continuity of certain derivatives or combinations of derivatives at the knots $x_1,\ldots,x_k$. Our first example in this section is a linear space of piecewise polynomials (used already by actuaries in the late 1800's) where the pieces tie together at several points and over several intervals.

EXAMPLE 4.1. Let $\Omega = [a,b]$, and suppose $S_i = \mathcal{P}_4$, the space of cubic polynomials, $i = 0,1,\ldots,k$. For each $i = 0,1,\ldots,k-1$, let $\Gamma_{i,i+1} = \{(e_i,e_i),(e_{i+1},e_{i+1}),(e_{i+2},e_{i+2})\}$ and $\Gamma = \{\Gamma_{i,i+1}\}_{0\le i\le k-1}$, where $e_i f = f(x_i)$.

<u>Analysis</u>: In this case, the space $\mathcal{S}$ defined in (2.1) consists of piecewise cubic polynomials with considerable ties between the various pieces. In particular, s_i and s_{i+1} agree at the points x_i, x_{i+1}, x_{i+2} ; s_i and s_{i+2} agree at the points x_{i+1}, x_{i+2} ; and s_i and s_{i+3} agree at the point x_{i+2}. By Theorem 2.1 the space is of dimension $k + 4$. As a basis for $\mathcal{S}$, we may take $B_{01}, B_{02}, B_{03}, B_{04}$ to be the powers $1, x, x^2, x^3$, along with the one-sided splines

$$B_{i1}(x) = \begin{cases} 0 & , \quad a \le x < x_i, \\ (x-x_{i-1})(x-x_i)(x-x_{i+1}) & , \quad x_i \le x \le x_k. \end{cases}$$

for $i = 1,2,\ldots,k$. ■

It might not be unreasonable in certain data fitting problems, for example, to use a space of splines with a different structure in each subinterval. To illustrate the results of section 2 for such a class we consider the following example.

<u>EXAMPLE 4.2</u>. Let $\Omega = [-1,2]$ and $\Delta = \{0,1\}$. Let $S_0 = \mathcal{P}_2$, $S_1 = \text{span}\ \{e^x\}$, and $S_2 = \text{span}\ \{\cos(x), \sin(x)\}$. Suppose that Γ consists of $\Gamma_{01} = \{(e_0,e_0)\}$ and $\Gamma_{12} = \{(e_1,e_1)\}$.

<u>Analysis</u>: This space consists of functions which are linear polynomials in the first interval, exponential in the second, and trigonometric in the third. The ties require continuity at the points $x_1 = 0$ and $x_2 = 1$. It is easy to compute the dimension from Theorem 2.1 as 3. As a basis we may take the following 3 functions:

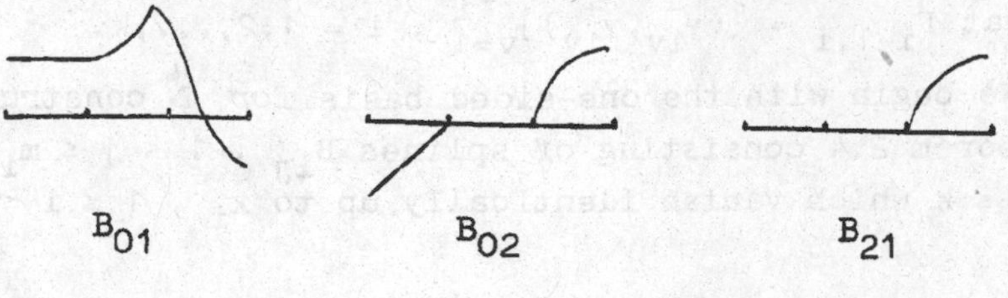

whose analytic formulae can be easily worked out. In particular, we may take B_{01} to consist of the three pieces $1, e^x$, $e\cos(x)/\cos(1)$; B_{02} to consist of $x, 0$, $\sin(x)/\sin(1) - \cos(x)/\cos(1)$; and B_{21} to consist of $0, 0$, $\sin(x)/\sin(1) - \cos(x)/\cos(1)$. ■

If Ω is taken as some discrete set , we obtain generalizations of the discrete splines (see § 11 for references). If Ω is taken to be a curve in the plane, (and if we work with complex valued functions on this curve), we obtain complex and analytic splines. We do not bother to give explicit examples.

§ 5. Local support bases.

For actual numerical applications, it is known that one-sided bases for piecewise functions are usually not satisfactory. It is much more desirable to have a basis of functions each with a relatively small local support. In this section we give some general results which are useful for constructing a local support basis by taking linear combinations of the one-sided bases constructed previously.

While it is possible to give results in the general case, the notation is considerably simplified if we restrict our attention to the case where all of the spaces S_i are the same and where $\Gamma = \{\Gamma_{0,1}, \Gamma_{1,2}, .., \Gamma_{k-1,k}\}$, (so that only adjoining pieces must satisfy some tying conditions). In the interest of saving subscripts, we assume henceforth that $S_i = S$ is spanned by $u_1, .., u_m$, and that $\Gamma_{i-1,i} = \{(\gamma_{iv}^-, \gamma_{iv}^+)\}_{v=1}^{r_i}$, $i = 1,2,...,k$.

We begin with the one-sided basis for $\mathcal{S}$ constructed in Theorem 2.4 consisting of splines B_{ij} , $1 \le j \le m_i$, $0 \le i \le k$ which vanish identically up to x_i , $1 \le i \le k$.

It is convenient to renumber these splines with a single subscript. We number them lexicographically as

(5.1) $\rho_1,\ldots,\rho_{m+K} = B_{01},\ldots,B_{0m},B_{11},\ldots,B_{1m_1},\ldots,B_{k1},\ldots,B_{km_k}$,

where $K = m_1 + \ldots + m_k$. If we introduce the notation

(5.2) $y_{m+1} \le y_{m+2} \le \ldots \le y_{m+K} = x_1,\ldots,x_1,\ldots,x_k,\ldots,x_k$,

where each x_i is repeated exactly m_i times, then it follows that each of the splines $\rho_i(x)$ vanishes identically for $x < y_i$, $i = m+1,\ldots,m+K$.

Since $S_i = \text{span}\,\{u_1,\ldots,u_m\}$ for all i, we know that for $i = 1,2,\ldots,m$,

(5.3) $$\rho_i(x) = \sum_{j=1}^{m} c_{ij}u_j(x)$$

while for $i = m+1,\ldots,m+K$

(5.4) $$\rho_i(x) = \begin{cases} 0 & , x < y_i \\ \sum_{j=1}^{m} c_{ij}u_j(x) & , x \ge y_i \ . \end{cases}$$

It follows that the basis $\{\rho_i\}_1^{m+K}$ for $\mathcal{S}$ is completely determined by the points $y_{m+1},\ldots,y_{m+K}$ and the matrix $C = (c_{ji})_{i=1,j=1}^{m+K\ \ m}$. It will be convenient to use the notation $C\langle i_1,\ldots,i_r\rangle$ for the submatrix of C obtained by taking only the columns $i_1,\ldots,i_r$.

The following lemma is a rather general result on when local support splines can be constructed as linear combinations of one-sided splines as in (5.3)-(5.4).

LEMMA 5.1. *Suppose* $1 \le i_1 < i_2 < \ldots < i_r \le m+K$, *and suppose that* $\delta = (\delta_1,\ldots,\delta_r)^T$ *is a solution of*

$$(5.5) \qquad C\langle i_1,\ldots,i_r\rangle\delta = 0.$$

Then

$$(5.6) \qquad B(x) = \sum_{j=1}^{r} \delta_j \, \rho_{i_j}(x)$$

is a spline which vanishes for $x < y_{i_1}$ (*if* $i_1 > m$) *and also for* $x > y_{i_r}$.

Proof: Since each of the ρ's in the sum (5.6) vanish for $x < y_{i_1}$ when $i_1 > m$, it follows that B has the same property. Now for $x > y_{i_r}$ we know that for all $j = 1,\ldots,r$,

$$\rho_{i_j}(x) = (u_1(x),\ldots,u_m(x))\, C\langle i_j\rangle ,$$

and hence,

$$B(x) = (u_1(x),\ldots,u_m(x))\, C\langle i_1,\ldots,i_r\rangle\, \delta = 0. \ \blacksquare$$

In applying Lemma 5.1 to the construction of local support splines, we may choose any value of r for which a corresponding nontrivial δ satisfying (5.5) can be found. In some cases, local support bases can be constructed from just two one-sided splines. In most cases, however, it is necessary to take r somewhat larger, and usually at least m + 1.

Lemma 5.1 may be used to construct a number of local support splines in $\mathcal{S}$. An important associated question is when are the resulting splines linearly independent? The following lemma gives a simple algebraic condition which assures linear independence.

<u>LEMMA 5.2. Suppose</u> $\{\beta_v = (\beta_{v1},\ldots,\beta_{v,m+K})\}_{v=1}^{q}$ <u>is a set of</u> q <u>linearly independent vectors in</u> R^{m+K}, <u>and that</u>

$$(5.7)\qquad B_v(x) = \sum_{j=1}^{m+K} \beta_{vj}\,\rho_j\ ,\qquad v = 1,2,\ldots,q.$$

<u>Then</u> $B_1,\ldots,B_q$ <u>are linearly independent splines in</u> $\mathcal{S}$.

<u>Proof</u>: Suppose $d_1B_1 + \ldots + d_qB_q \equiv 0$ on Ω. Then,

$$\sum_{v=1}^{q} d_v \sum_{j=1}^{m+K} \beta_{vj}\,\rho_j = \sum_{j=1}^{m+K} \rho_j \sum_{v=1}^{q} d_v\,\beta_{vj} = 0\ .$$

By the linear independence of the ρ's, it follows that $d_1\beta_1 + \ldots + d_q\beta_q = 0$. Now, by the linear independence of the β's, we conclude that $d_1 = \ldots = d_q = 0$, and the desired linear independence is established. ■

Lemmas 5.1 and 5.2 show how a local support basis for $\mathcal{S}$ could be constructed. We must find m+K linearly independent vectors in R^{m+K} which at the same time are such that the corresponding splines have small supports. How successful an attempt to choose such vectors will be depends heavily on the properties of the matrix C. As Example 5.4 below shows, in some cases there will be no local support basis at all. In other cases it is possible to construct local support bases with each element of minimal support. The following result gives one set of sufficient conditions on C to guarantee the existence of a local support basis with support intervals not too long.

It will be convenient to introduce some further notation. We need to know more precisely how the x's and the y's introduced in (5.2) are related to each other. For i = 1,2,...,k , let $\epsilon_i + 1$ be the index of the first y which is equal to x_i. Explicitly, we set $\epsilon_0 = 0$ and $\epsilon_i = \epsilon_{i-1} + m_{i-1}$, i = 1,2,...,k+1. With this notation, we note that the last y which is equal to x_i must have the index ε_{i+1}.

<u>THEOREM 5.3. Suppose that the matrix</u> C <u>describing the one-sided basis in</u> (5.3)-(5.4) <u>has the property that</u>

(5.8) $C\langle \varepsilon_{i+1}+1,\ldots,\varepsilon_{i+m+1}\rangle$ <u>is of full rank</u> m

<u>for</u> i = 0,1,...,k-m. <u>Then there exists a basis</u> $\{B_i\}_1^{m+K}$ <u>for</u> $\mathcal{S}$ <u>with the properties that</u>

(5.9) $B_{\varepsilon_i+1},\ldots,B_{\varepsilon_i+m_i}$ <u>have support on</u> $\{x_i \le x \le x_{i+m}\}$

<u>for</u> i = 0,1,...,k-m, <u>while</u>, <u>for</u> i = k-m+1,....,k

(5.10) $B_{\varepsilon_i+1},\ldots,B_{\varepsilon_i+m_i}$ <u>have support on</u> $\{x_i \le x \le b\}$.

<u>Proof</u>: We construct vectors $\beta_1,\ldots,\beta_{m+K}$ so that the B's defined by (5.7) have the desired properties. First, for i = 0,1,....,k-m and j = 1,2,....,m_i, choose β_{ε_i+j} to be an m+K vector with the ε_i+j component equal to 1; the $\varepsilon_{i+1}+1,\ldots,\varepsilon_{i+m+1}$ components equal to δ, where δ is any solution of

$$C\langle \varepsilon_{i+1}+1,\ldots.,\varepsilon_{i+m+1}\rangle\ \delta = -\ C\langle \varepsilon_i+j\rangle\ ;$$

and the remaining components equal to 0. By Lemma 5.1 the corresponding splines have the stated support properties. Now, for i = k-m+1,...,k and j = 1,2,...,m_i, let β_{ε_i+j} be an m+K vector with its ε_i+j th component equal to 1 and the other components equal to 0 ; i.e.,

$$B_{\varepsilon_i+j} = \rho_{\varepsilon_i+j}\ ,\quad j = 1,..,m_i \text{ and } i = k-m+1,\ldots,k.$$

These splines also clearly have the stated support properties. By the construction, it is clear that the vectors $\beta_1,\ldots,\beta_{m+K}$ are linearly independent , and by Lemma 5.2, we know that the B's form a basis for $\mathcal{S}$. ∎

We should emphasize once again that in many cases there exist local support bases with considerably smaller supports. We close this section with an example where no local support basis exists.

EXAMPLE 5.4. Let $m = 2$, $S = \text{span}\,\{1,x\}$, and $\Gamma_{i-1,i} = \{(e_i',e_i')\}$ for $i = 1,2,\ldots,k$, where $e_i'f = f'(x_i)$, and where $a < x_1 < \ldots < x_k < b$ with $\Omega = [a,b]$.

Analysis: The space $\mathcal{S}$ of splines in this case consists of piecewise linear functions with equal slopes in each subinterval (the linear pieces are not required to match continuously at the knots). Its dimension is k+2. The one-sided basis constructed in section 2 in this case is given by $1,x,\ (x-x_1)_+^0,\ldots,(x-x_k)_+^0$. Lemma 5.1 can be used to construct a number of local support splines; e.g.,

$$B_i(x) = (x-x_i)_+^0 - (x-x_{i+1})_+^0 \quad , \; i = 1,\ldots,k-1.$$

Still, there can be no local support basis for $\mathcal{S}$, since clearly $\mathcal{S}$ contains the function x, and every local support spline must have a zero slope everywhere. The matrix C in this case is given by

$$C = \begin{bmatrix} 1 & 0 & 1 & \cdots & 1 \\ 0 & 1 & 0 & \cdots & 0 \end{bmatrix} .$$

The condition (5.8) is clearly not satisfied for this matrix.

§6. Tchebycheffian splines.

In the remainder of the paper we intend to concentrate on some of the finer structural properties of splines, such as zeros, sign changes, determinants, etc. It is clear that we cannot expect to obtain such results for the general class of splines $\mathcal{S}$ discussed in §2 without some kind of further assumptions on the nature of the spaces S_i and of the linear functionals describing the ties between the pieces. In this section, we introduce a class of generalized splines for which many of these finer properties can be established, and which, at the same time, is still sufficiently general to include the bulk of the splines in the literature. We also take the opportunity to illustrate some of the results of the previous sections for this class of splines.

Let $\Omega = [a,b]$, and suppose Δ is a partition defined by points $a = x_0 < x_1 < \ldots < x_k < x_{k+1} = b$. Suppose that u_1 is a positive continuous function on $[a,b]$, and that

$$u_2(x) = u_1(x) \int_a^x d\sigma_2(s_2)$$

(6.1) $$\ldots$$

$$u_m(x) = u_1(x) \int_a^x \ldots \int_a^{s_{m-1}} d\sigma_m(s_m) \ldots d\sigma_2(s_2),$$

for some absolutely continuous monotone increasing functions $\sigma_2, \ldots, \sigma_m$ on $[a,b]$. The set of functions $u_1, \ldots, u_m$ form a Complete Tchebycheff (CT-) system. (This also holds if the σ's are right continuous, see Remarks 10,11). Our main interest in using CT- systems is the fact that any linear combination of $u_1, \ldots, u_m$ can have at most m-1 zeros, counting multiplicities appropriately - see [43]).

We are going to consider a class of splines where $S_i = S = U$, $U = \text{span}\,\{u_1,\ldots,u_m\}$. To define the linear functionals describing the ties between the pieces, we need to introduce certain linear operators associated with the CT- system U. For any $a \le x < b$ and any $\varphi \in U$, we define

$$(6.2) \qquad D_0\varphi(x) = \varphi(x) \,/\, u_1(x) \,,$$

and

$$(6.3) \qquad D_j^+\varphi(x) = \lim_{\delta \downarrow 0} \frac{\varphi(x+\delta) - \varphi(x)}{\sigma_{j+1}(x+\delta) - \sigma_{j+1}(x)} \,,$$

for $j = 1,2,\ldots,m-1$. For $a < x \le b$, we define operators $D_1^-,\ldots,D_{m-1}^-$ similarly, except that we take $-\delta$ in the above. It will be convenient to use the notation

$$(6.4) \qquad \begin{array}{l} L_j^+ = D_j^+ \cdots D_1^+ D_0 \\ L_j^- = D_j^- \cdots D_1^- D_0 \end{array} \,, \quad j = 0,1,\ldots,m-1.$$

Let $M = (m_1,\ldots,m_k)$ be a vector of integers with $1 \le m_i \le m$, $i = 1,2,\ldots,k$.

We define the space of <u>Tchebycheffian</u> <u>spline</u> <u>functions</u> <u>with</u> <u>knots</u> $x_1,\ldots,x_k$ <u>of</u> <u>multiplicities</u> $m_1,\ldots,m_k$ to be

$$(6.5) \qquad \mathcal{S}(U;M;\Delta) = \{s : s_i = s|_{I_i} \in U \,,\ i = 0,1,\ldots,k, \text{ and } L_j^- s_{i-1}(x_i) = L_j^+ s_i(x_i),\ j = 0,1,\ldots,m-1-m_i \text{ and } i = 1,2,\ldots,k\}.$$

The space of Tchebycheffian splines was studied in considerable detail in [43]. Still, in preparation for the development of new results in the following sections, and to help illustrate the general results of the previous sections, we will recall some of the algebraic facts about them. Theorem 2.1 implies that $\mathcal{S}(U;M;\Delta)$ is a linear space of dimension $m + K$, where $K = m_1 + \ldots + m_k$. The one-sided basis constructed in §2 can be given explicitly for Tchebycheffian splines. In order to do so, we need to introduce some further notation. For each $j = 1,\ldots,m-1$ we define the jth reduced system corresponding to U by

$$(6.6)\qquad \begin{aligned} v_{j,1}(x) &= 1 \\ v_{j,2}(x) &= \int_a^x d\sigma_{j+2}(s_{j+2}) \\ &\cdots \\ v_{j,m-j}(x) &= \int_a^x \cdots \int_a^{s_{m-1}} d\sigma_m(s_m) \ldots d\sigma_{j+2}(s_{j+2}) . \end{aligned}$$

We write $U^j = \{v_{j,1},\ldots,v_{j,m-j}\}$. Each of these sets is also a CT- system, of course. Moreover, it is clear that

$$(6.7)\qquad L_j^+ u_i = \begin{cases} v_{j,i-j} , & j = 1,\ldots,i-1 \\ 0 , & j = i,\ldots,m-1, \end{cases}$$

for $i = 1,2,\ldots,m$. We also have need for certain adjunct functions defined by

$$(6.8)\qquad \begin{aligned} v_{j,1}^*(y) &= 1 \\ v_{j,2}^*(y) &= \int_a^y d\sigma_2(t_2) \\ &\cdots \\ v_{j,m-j}^*(y) &= \int_a^y \cdots \int_a^{t_3} d\sigma_2(t_2) \ldots d\sigma_{m-j}(t_{m-j}). \end{aligned}$$

Now let

$$(6.9)\quad g_j(x,y) = \begin{cases} \sum_{i=1}^{j} u_i(x)v^*_{m-j,j-i+1}(y)(-1)^{j-1}, & y \le x, \\ 0 & , x < y, \end{cases}$$

for $j = 1,2,\ldots,m$.

It is shown in [43] that a basis for $\mathcal{S}(U;M;\Delta)$ is given by

$$(6.10)\qquad \{B_{ij}(x) = g_{m-j+1}(x,x_i)\}_{j=1,i=0}^{m_i\ \ k}, \quad (m_0=m).$$

In fact, all of these one-sided splines come from the basic Green's function g_m by application of appropriate "differentiation" operators. In particular, if we define

$$E_j^-\psi = \lim_{\delta\uparrow 0} \frac{\psi(x) - \psi(x-\delta)}{\sigma_{m-j+1}(x) - \sigma_{m-j+1}(x-\delta)}, \quad j = 1,..,m-1$$

and $R_j^- = E_j^- \ldots E_1^-$, then

$$(6.11)\qquad B_{ij}(x) = (-1)^{j-1}R_{j-1}^- g_m(x,x_i), \quad j = 1,\ldots,m_i,$$

where here the operator R_j^- is applied to the second variable in g_m. It can be shown (cf. [43]) that this Green's function has strong total positivity properties.

The results of section 5 can now be applied to construct a local support basis for $\mathcal{S}$, and indeed, one with splines of very small support. To describe the basis, we need one further piece of notation. Given points $a \le t_1 < t_2 < \ldots < t_{m+1} \le b$ and functions $\varphi_1,\ldots,\varphi_{m+1}$ on $[a,b]$, we define

$$(6.12)\qquad D\binom{\varphi_1,\ldots,\varphi_{m+1}}{t_1,\ldots,t_{m+1}} = \det(\varphi_j(t_i))_1^{m+1}.$$

This definition can be extended to allow coincident t's if we agree on a convention for how to treat successive rows when repeated t's appear (cf. e.g. [16,43]). For functions $\varphi \in V = \text{span}\ \{v^*_{0,1},\dots,v^*_{0,m}\}$, we define (6.12) for all $a \le t_1 \le t_2 \le \dots \le t_{m+1} \le b$ by using the convention that when t's are repeated, then successive rows should be replaced by the "derivatives" $E^-_1,\dots,E^-_{m-1}$ of the previous rows, (where at the point a we use the corresponding right derivatives E^+_i. Thus, for example, if $t_1 = t_2 = t_3$, then we replace the second row by $(R_1\varphi_1(t_1),\dots, R_1\varphi_{m+1}(t_1))$ and the third row by $(R_2\varphi_1(t_1),\dots,R_2\varphi_{m+1}(t_1))$. The following theorem is proved in [43].

THEOREM 6.1. *Let $y_{m+1} \le y_{m+2} \le \dots \le y_{m+K}$ be an enumeration of the sequence $x_1,\dots,x_1,\dots,x_k,\dots,x_k$, where each x_i is repeated exactly m_i times*, $i = 1,2,\dots, k$. *In addition, let* $c = y_1 = \dots = y_m$ *and* $d = y_{m+K+1} = \dots = y_{2m+K}$. *Then the functions*

$$(6.13)\qquad B_i(x) = D\begin{pmatrix} v^*_{0,1},\dots,v^*_{0,m}, & g_m(x,\cdot) \\ y_i,\dots\dots & , y_{i+m} \end{pmatrix},\ i = 1,\dots,m+K$$

form a basis for $\mathcal{S}$. *Moreover*,

$$(6.14)\qquad B_i(x) > 0 \text{ on } (y_i,y_{i+m}),$$

and

$$(6.15)\qquad B_i(x) = 0 \text{ for } a \le x < y_i \text{ and } y_{i+m} < x \le b,$$

$i = 1,2,\dots,m+K.$

The expression (6.13) is, of course, a kind of generalized divided difference, and the definition here is a direct analog of the usual definition of B-splines in the polynomial spline case. In fact, the reader will be quick to note that the entire development of Tchebycheffian splines parallels that for polynomial splines, which are, of course, a special case. The generalized B-splines in (6.13) also have many other properties similar to those for the classical B-splines. In particular, the basis in Theorem 6.1 has strong total positivity properties which we shall discuss in some detail in the following section.

We close this section with some observations about the derivatives of Tchebycheffian splines. We define $L_j\, s$ for a spline $s \in \mathcal{S}(U;M;\Delta)$ by the rule

$$(6.16) \qquad L_j s(x) \;=\; \begin{cases} L_j^+ s(x) & \text{for } a \le x < b\,, \\ L_j^- s(b) & \text{for } x = b\,. \end{cases}$$

It is clear that $L_j s$ will also be a piecewise function with pieces in the span of the set $U^j = \{v_{j,1},\ldots,v_{j,m-j}\}$, cf. (6.7). In fact, if we examine (6.5) more closely, we see that

$$(6.17) \qquad L_j s \in \mathcal{S}(U^j;M^j;\Delta)\,,$$

where $M^j = (m_1^j,\ldots,m_k^j)$ with $m_i^j = \min(m_i, m-j)$, $i = 1,2,\ldots,k$. We may note that with $j = m-1$, this is just the class of piecewise constants with (possible) jumps at the knots.

The assumption that the σ's in the canonical representation of the u's in (6.1) are absolutely continuous implies that each of the u's as well each of the $L_j u$ (i.e. each of the reduced system functions) is continuous, and if u_1 is absolutely continuous, even absolutely continuous. (See also Remark 10).

§7. Bounds on zeros and applications.

In this section we give a bound on the number of zeros nontrivial Tchebycheffian splines can have, and quote a number of important corollaries. The feasibility of doing this stems from the fact that nonzero elements in the span of the CT system U cannot have more than m-1 zeros, counting multiplicities appropriately. In particular, it is shown in [43] that if $u \in U = \text{span}\{u_i\}$, $u \neq 0$, then $Z(u) \leq m-1$, where Z counts all zeros of u in [a,b] according to the following rule: we say u has a zero of multiplicity z at a point $t \in [a,b]$, provided that

$$(7.1) \quad u(t) = L_1 u(t) = \ldots = L_{z-1} s(t) = 0 \neq L_z s(t) \ ,$$

where $L_1, \ldots, L_{m-1}$ are the operators defined in (6.16); i.e. to be equal to the L^+ values everywhere except at the point b where L^-'s are used.

We now want to state the analogous result for Tchebycheffian splines. First, however, we must agree on how to count the zeros of such a spline. Suppose that $s \in \mathcal{S}(U;M;\Delta)$ and $s \neq 0$. Since between any two knots, s is an element of U, by the above mentioned result, it must either vanish identically throughout this interval, or it can be zero at only a finite number (at most m-1) of isolated points in the interval. If t is isolated, $t \notin \Delta$, we say that s has a zero of multiplicity $1 \leq z \leq m-1$ provided the corresponding element $s_i \in U$ has a zero of multiplicity z at t (counted as in (7.1) above).

When s vanishes identically on an interval between two knots, we count the entire interval as a zero of order either m or m+1 according to the following rules:

(7.2) Suppose $s(t) = 0$ on $[a,x_j)$, but $s(t) \neq 0$ for all $x_j < t < x_j+\epsilon$, for some $\epsilon > 0$. Then $[a,x_j]$ is counted as an interval zero of multiplicity $z = m$. A similar count is used if s vanishes identically on an interval ending at b,

or if s vanishes on an interval interior to (a,b),

(7.3) Suppose $s(t) = 0$ on $[x_i,x_j)$, but s is not zero on $(x_j,x_j+\epsilon)$ or $(x_i-\epsilon,x_i)$ for some $\epsilon > 0$. Then we say $[x_i,x_j)$ is an interval zero of multiplicity

$$z = \begin{cases} m+1, & \text{if } m \text{ is even and } s(x_i-\epsilon/2)s(x_j+\epsilon/2) < 0, \\ m+1, & \text{if } m \text{ is odd and } s(x_i-\epsilon/2)s(x_j+\epsilon/2) > 0, \\ m, & \text{otherwise.} \end{cases}$$

It remains to consider the case where s is zero at a knot, but not in an interval containing the knot, or where s jumps through zero at a knot. If $t \in \Delta$ and s does not vanish identically on any interval containing t, then we define the multiplicity of t as follows:

(7.4) Suppose $t = x_i$, and that s_{i-1} and s_i are the elements of U to the left and right of x_i. Let $\alpha = \max(\ell,r)$, where ℓ and r are the number of times s_{i-1} and s_i vanish at x_i, respectively. Then we say s has a zero at t of multiplicity

$$z = \begin{cases} \alpha+1, & \text{if } \alpha \text{ is even and s changes sign at t} \\ \alpha+1, & \text{if } \alpha \text{ is odd and s doesn't change sign at t} \\ \alpha, & \text{otherwise}. \end{cases}$$

This rule counts a jump through 0 at a knot as a zero of multiplicity 1. It can be shown that the multiplicities produced by these rules coincide with the limits of the number of zeros of appropriate sequences of splines converging to s, with simple zeros; i.e., the definitions are the natural ones. Moreover, it is easily checked that s has a sign change at a zero of odd multiplicity, and no sign change at a zero of even multiplicity.

The key to obtaining a bound on the number of zeros of a Tchebycheffian spline function is the following analog of Rolle's theorem : If $L_j s$ has no jumps at knots, then

$$(7.5) \qquad Z(L_j s) \le Z(L_{j+1} s) + 1 \quad , \quad j = 0,1,\dots,m-2,$$

(cf. [43]). Using (7.5), it was shown in [43] that the following result holds.

THEOREM 7.1. *For every* $s \in \mathcal{S}(U;M;\Delta)$ *with* $s \neq 0$,

$$(7.6) \qquad Z(s) \le m+K-1 \ ,$$

where Z *counts the number of zeros of* s *in* [a,b], *with multiplicities, as in* (7.1) - (7.4).

Theorem 7.1 has a number of important corollaries relating to the sign properties of Tchebycheffian splines, and to determinants formed from the B-splines. For example, one can show (cf. [43]) that for $1 \le i \le m+K$ and $0 \le j \le m-1$, the B-spline B_i defined in (6.13) satisfies

$$(7.7) \qquad Z_{(y_i, y_{i+m})}(L_j B_i) \le j \ .$$

To give a more important application, define

$$(7.8) \qquad M\begin{pmatrix} B_1,\dots,B_{m+K} \\ t_1,\dots,t_{m+K} \end{pmatrix} = (B_j(t_i))_1^{m+K}$$

for $a \leq t_1 < t_2 < \ldots < t_{m+K} \leq b$. We also define this matrix in the case of equalities among the t's by agreeing to replace successive rows by the "derivatives" $L_1, \ldots$ of the previous ones. We shall use the notation D for the determinant of this matrix. Then using Theorem 7.1, it can be shown [43] that the following result holds.

THEOREM 7.2. Let $m > 1$. Then for any $1 \leq \nu_1 < \ldots < \nu_p \leq m+K$,

$$(7.9) \qquad D\begin{pmatrix} B_{\nu_1}, \ldots, B_{\nu_p} \\ t_1, \ldots, t_p \end{pmatrix} \geq 0$$

for all $a \leq t_1 \leq \ldots \leq t_p \leq b$, (with at most m of the t's equal to any given value), and moreover, strict inequality holds if and only if

$$(7.10) \qquad t_i \in \text{int} (\text{supp} (B_{\nu_i})) = (y_{\nu_i}, y_{\nu_i+m})$$

for $i = 1,2,\ldots,p$.

Theorem 7.2 asserts that the basis $\{B_i\}_1^{m+K}$ of B-splines for the space of Tchebycheffian splines is in fact a Weak Tchebycheff (WT-) system. This fact in turn allows the development of a more or less classical theory of best approximation with respect to the class $\mathcal{S}$. We do not bother to recapitulate the details of such a theory, as they are well known (cf. [37,38,39,40]). Theorem 7.2 actually asserts considerably more; namely, that the Gram matrix M in (7.8) is totally positive. This puts the extensive theory of total positivity (see [16]) at our disposal.

To give one example of the kind of result which follows from total positivity, we prove a relation between the number of sign changes of a Tchebycheffian spline, and the number of sign changes of its coefficient vector (relative to the B-spline basis.

Let $S^-(v)$ be the number of sign changes in the vector $v = (v_1,\ldots,v_p)$, where 0 components are discarded. For any function f defined on $[a,b]$, we define

$$(7.11)\quad S^-(f) = \sup\ \{S^-(v) : v = (f(t_1),\ldots,f(t_p))\ \},$$

where the supremum is taken over all p and all $a \le t_1 < \ldots < t_p \le b$.

THEOREM 7.3. Let $s = \sum_1^{m+K} c_i B_i$, where $B_1,\ldots,B_{m+K}$ are the B-splines in (6.13). Then

$$(7.12)\quad S^-(s) \le S^-(c) ,$$

where $c = (c_1,\ldots,c_{m+K})$.

Proof: By convolving each of the B-splines B_i with the standard kernel (cf. [20 , p. 15]), we obtain a Descartes system $B_{1,\sigma},\ldots,B_{m+K,\sigma}$. Then, by Theorem 4.4 of [20], it follows that

$$Z(s_\sigma) \le S^-(c) \ , \quad \text{for} \quad s_\sigma = \sum_1^{m+K} c_i B_{i,\sigma} \ ,$$

where here Z counts simple zeros. Now, since $s_\sigma \to s$ pointwise as $\sigma \to 0$, (7.12) follows. ∎

§8. A Budan-Fourier Theorem.

The main result of section 7 was based on Rolle's theorem for splines, as stated in (7.5). It may be repeatedly applied as long as the successive derivative splines $L_j s$ do not have jumps at the knots. In particular, suppose we define the space of Tchebycheffian splines with simple knots to be

(8.1) $\mathcal{S}(U;\Delta) = \mathcal{S}(U;M_s;\Delta)$, $M_s = (1,\ldots,1)$.

If $s \in \mathcal{S}(U;\Delta)$, then $L_j s \in \mathcal{S}(U^j;\Delta)$ for $j = 0,1,\ldots,m-1$, where L_j are the operators defined in (6.16). We write $Z(L_j s)$ for the number of zeros of $L_j s$, counting multiplicities as explained in §7 (e.g., $L_j s$ has a double zero at t if $L_j s(t) = L_{j+1} s(t) = 0$, etc. We note that the spline $L_{m-1}s$ is a piecewise constant, and $Z(L_{m-1}s)$ simply counts the number of jumps of $L_{m-1}s$ to or through 0). Now, repeated application of (7.5) yields

(8.2) $$Z(L_j s) \le Z(L_{m-1}s) + m-j-1 ,$$

$j = 0,1,\ldots,m-1$.

The inequality (8.2) involves zeros on the entire interval [a,b]. If we restrict our attention to (a,b), then (8.2) can be considerably sharpened if we take into account the exact behavior of s and its derivatives at a and b. In this section we prove an extension of the classical Budan-Fourier Theorem for polynomials which gives a result of this type. We first need some additional notation.

If $v = (v_1,\ldots,v_p)$ is a vector of real numbers, we denote the number of sign changes in the sequence $v_1,\ldots, v_p$, where 0's are counted as either + or −, by the symbol $S^+(v)$. To save subscripts, we write

(8.3) $$A_i = L_i^+ s(a) , \quad B_i = L_i^- s(b) , \quad i = 0,1,\ldots,m-1,$$

and set

(8.4) $$\alpha_j = S^+(A_j,-A_{j+1},\ldots,(-1)^{m-j-1}A_{m-1})$$
$$\beta_j = S^+(B_j,B_{j+1},\ldots,B_{m-1}) , \quad j = 0,1,\ldots,m-1.$$

<u>THEOREM 8.1. For</u> $j = 0,1,\ldots,m-1$, <u>let</u> $Z_j = \tilde{Z}(L_j s)$, <u>where</u> $\tilde{Z}$ <u>counts the number of zeros of</u> $L_j s$ <u>inside the interval</u> (a,b), <u>with multiplicities, exactly as in section</u> 7 <u>except that if</u> $L_j s$ <u>vanishes on an interval ending at</u> a <u>or</u> b, <u>we count such an interval as</u> 0 <u>rather than as</u> m <u>as in</u> (7.2). <u>For</u> $j = 0,1,\ldots,m-1$, <u>let</u> $A_j, B_j, \alpha_j, \beta_j$ <u>be the quantities defined in</u> (8.3)-(8.4), <u>and define</u>

$$(8.5)\qquad \begin{aligned} \ell &= \max_{1\le i\le m} \{i : A_{i-1} \neq 0\} \\ r &= \max_{1\le i\le m} \{i : B_{i-1} \neq 0\}. \end{aligned}$$

<u>Then</u>, <u>if</u> $L_{m-1}s \neq 0$,

$$(8.6)\qquad Z_j \le Z_{m-1} + m-j-3 - \alpha_j - \beta_j$$

<u>for</u> $j = \max(\ell,r),\ldots,m-2$;

$$(8.7)\qquad Z_j \le Z_{m-1} + m-j-2 - \alpha_j - \beta_j$$

<u>for</u> $j = \min(\ell,r),\ldots, \min(m-2, \max(\ell,r)-1)$;

$$(8.8)\qquad Z_j \le Z_{m-1} + m-j-1 - \alpha_j - \beta_j$$

<u>for</u> $j = 0,1,\ldots, \min(\ell,r) - 1$.

<u>Proof</u>: We begin with some preliminary remarks and some additional notation. We shall use the abbreviation $s_j = L_j s$. We say that the point a is a Rolle's point for s_j provided that either $s_j(a) = 0$, or for all $\epsilon > 0$ sufficiently small, there exists a point $a < t < a+\epsilon$ with $s_j(t)s_{j+1}(t) > 0$. Similarly, we say that b is a Rolle's point for s_j provided that either $s_j(b) = 0$, or for all $\epsilon > 0$ sufficiently small, $s_j(t)s_{j+1}(t) < 0$ for some $b-\epsilon < t < b$. If we refer to points $a < t < b$ where

$s_j(t) = 0$ also as Rolle's points, then it is clear that between any two Rolles points of s_j, the spline s_{j+1} must have at least one zero . (In the case $j = m-2$ where $s_{j+1} = s_{m-1}$ is a piecewise constant, this may be a jump type zero).

Next, we need some observations about the vectors $A_0,\ldots,A_{m-1}$ and $B_0,\ldots,B_{m-1}$. Suppose that

$$(8.9)\quad A_{j+1} = \ldots = A_{m-1} = 0 \quad \text{for some } 0 \le j \le m-2.$$

Then, since $s_{m-1} \ne 0$, there exists a knot x_L with s_{m-1} identically zero on $[a,x_L)$, but $s_{m-1}(x_L) \ne 0$. It follows then that $s_{j+1},\ldots,s_{m-2}$ are also identically zero on this interval. Similarly, if

$$(8.10)\quad B_{j+1} = \ldots = B_{m-1} = 0 \quad \text{for some } 0 \le j \le m-2,$$

then $s_{j+1},\ldots,s_{m-1}$ vanish identically on some interval of the form $[x_R,b]$, while $s_{m-1}(\tilde{x}_R) \ne 0$, $\tilde{x}_R = x_R-\epsilon$.

We define I_j and J_j so that $\alpha_j = \alpha_{j+1} + I_j$ and $\beta_j = \beta_{j+1} + J_j$, respectively. The I_j and J_j take on either the value 0 or 1, depending on whether the augmented A or B vector has one more sign change or not. When $I_j = 1$, we can say considerable about the vector A. In particular, $I_j = 1$ if and only if either (8.9) holds or

$$(8.11)\quad A_{j+1} = \ldots = A_{j+p-1} = 0 \quad \text{for some } 1 \le p \le m-j-1,$$
$$A_{j+p} \ne 0 \text{ , and } A_j \cdot A_{j+p} \ge 0 \text{ .}$$

In case (8.11) holds, by Lemma 8.2 below, it follows that a is a Rolle's point for s_j. Similarly, $J_j = 1$ if and only if either (8.10) persists, or

$$(8.12)\quad B_{j+1} = \dots = B_{j+q-1} = 0 \text{ for some } 1 \le q \le m-j-1,$$
$$B_{j+q} \neq 0, \text{ and } B_j \cdot B_{j+q} (-1)^q \ge 0.$$

Again, Lemma 8.2 implies that if (8.12) holds, then b is a Rolle's point for s_j.

Finally, we observe that if s_j has no zeros, then $Z_j \le Z_{j+1}$, trivially. If s_j does have some zeros in (a,b), say at points $z_1,\dots,z_w$, then Rolle's theorem asserts that

$$(8.13)\quad Z_j \le Z_{j+1} + 1 ,$$

where the zeros of s_{j+1} counted in (8.13) lie in (z_1,z_w).

We are now ready to proceed with the proof of the theorem. The analysis will be divided into several cases depending on the values of ℓ and r. Each of these will in turn be settled by a series of simple (if somewhat tiresome) steps. We may suppose $1 \le \ell \le r \le m$.

<u>Case 1</u>: Suppose $\ell = r = m$. Then $A_{m-1} \neq 0$, $B_{m-1} \neq 0$, and by the above remarks, this means that if $I_j = 1$, then a is a Rolle's point for s_j, and similarly, if $J_j = 1$, then b is a Rolle's point for s_j. We now prove (8.8) by induction. For j = m-1 there is nothing to prove. Suppose now that the result is established for $0 < j+1 \le m-1$; i.e.,

$$(8.14)\quad Z_{j+1} \le Z_{m-1} + m-j-2 - \alpha_{j+1} - \beta_{j+1} .$$

Then, (8.8) follows for j if we can show that

$$(8.15)\quad Z_j \le Z_{j+1} - I_j - J_j + 1,$$

since then

$$Z_j \le Z_{m-1} + m-j-1 - \alpha_{j+1} - \beta_{j+1} - I_j - J_j$$
$$= Z_{m-1} + m-j-1 - \alpha_j - \beta_j.$$

We have three subcases:

(1.A) If $I_j = J_j = 0$, then (8.15) is trivial by (8.13).

(1.B) If $I_j = 1$ and $J_j = 0$, then (8.15) reads $Z_j \le Z_{j+1}$. This is trivially true if $Z_j = 0$. If $Z_j > 0$, then s_{j+1} satisfies (8.13), and has an additional zero not counted in (8.13), namely, between the Rolle's points a and z_1. If $I_j = 0$ and $J_j = 1$, the analysis is identical.

(1.C) Suppose $I_j = J_j = 1$. Now we need to prove $Z_j \le Z_{j+1} -1$. This is true if $Z_j = 0$ since s_{j+1} must have a zero somewhere between the two Rolle's points a and b. If $Z_j > 0$, then in addition to the zeros guaranteed in (8.13), we also have zeros of s_{j+1} in (a,z_1) and (z_w,b). The proof of (8.8) is complete in this case.

<u>Case 2</u>. Suppose $\ell = r = m-1$. Now we have $A_{m-2} \ne 0$ and $B_{m-2} \ne 0$, but s_{m-1} vanishes identically on intervals $[a,x_L)$ and $[x_R,b]$. First we prove (8.8) for $j = m-2$.

(2.A) We must prove $Z_{m-2} \le Z_{m-1} - 1$. But this follows from (8.13) and the fact that s_{m-1} has the two jumps.

Now (8.8) can be established for $j = m-3,..,0$ by induction just as in case 1.

<u>Case 3</u>. Suppose $\ell < m-1, r = m$. We have $s_\ell,\dots,s_{m-1}$ vanishing identically on $[a,x_L)$ while $B_{m-1} \ne 0$. (Thus, b is a Rolle's point whenever $J_j = 1$, $j = 0,1,\dots,m-1$. We begin now by proving (8.7) for $j = m-2$. There are two subcases.

(3.A) If $J_{m-2} = 0$, we must show $Z_{m-2} \le Z_{m-1} -1$. When $Z_{m-2} = 0$, this holds since s_{m-1} has a jump zero at x_L. If $Z_{m-2} > 0$, then in addition to the zeros in (8.13), s_{m-1}

has the jump zero at x_L and another zero in (x_L,z_1).

(3.B) If $J_{m-2} = 1$, we need to show $Z_{m-2} \le Z_{m-1} - 2$. When $Z_{m-2} = 0$, this holds since s_{m-1} has the jump at x_L and another zero in (x_L,b). When $Z_{m-2} > 0$, then s_{m-1} has this jump zero, plus zeros in (x_L,z_1) and (z_w,b) in addition to those in (8.13).

Our next task is to prove (8.7) for $\ell \le j < m-2$ by induction. The result for j will follow from that for $j+1 : Z_{j+1} \le Z_{m-1} + m-j-3 - \alpha_{j+1} - \beta_{j+1}$, provided that we can show $Z_j \le Z_{j+1} - J_j$. Again, there are two cases.

(3.C) If $J_j = 0$, we need $Z_j \le Z_{j+1}$. If $Z_j = 0$, this is trivial. Otherwise, s_{j+1} has an extra zero in (x_L,z_1).

(3.D) If $J_j = 1$, we need $Z_j \le Z_{j+1} - 1$. If $Z_j = 0$, this is true since s_{j+1} has a zero in (x_L,b). If $Z_j > 0$, then s_{j+1} has extra zeros in (x_L,z_1) and (z_w,b) not counted in (8.13).

We now claim that (8.8) holds for $j = \ell-1$. Indeed, we already have $Z_\ell \le Z_{m-1} + m-\ell-2 - \alpha_\ell - \beta_\ell$. Again, there are two cases.

(3.E) If $J_j = 0$, it suffices to show $Z_{\ell-1} \le Z_\ell + 1$, which is true by (8.13).

(3.F) If $J_j = 1$, we need $Z_{\ell-1} \le Z_\ell$. When $Z_{\ell-1} = 0$, this is trivial. When $Z_{\ell-1} > 0$, then s has an extra zero in (z_w,b) not counted in (8.13).

If $\ell = 1$, the theorem is proved in this case. If not, we can now prove (8.8) for $j = \ell - 2,\ldots,0$ by induction exactly as in Case 1, since both a and b are Rolle's points in view of the fact that $A_{\ell-1} \ne 0$, $B_{\ell-1} \ne 0$.

<u>Case 4</u>. Suppose $\ell < r = m-1$. We now have s_{m-1} vanishing on $[a,x_L)$ and $[x_R,b]$ and $B_{m-2} \ne 0$. Now we must start with (8.7) for $j = m-2$.

(4.A) We need to show $Z_{m-2} \le Z_{m-1} - 2$. If $Z_{m-2} = 0$ this follows since s_{m-1} has the two jump zeros. If $Z_{m-2} > 0$, then s_{m-1} has the two jumps plus a zero in (x_L, z_1) on top of (8.13).

Now that we have got (8.7) started, the remainder of the proof is identical with case 3 since $B_{m-2} \ne 0$.

Case 5. Suppose $1 \le \ell \le r \le m-2$. We now have $s_\ell, \ldots, s_{m-1}$ and $s_r, \ldots, s_{m-1}$ vanishing identically on $[a, x_L)$ and $[x_R, b]$, respectively. This case can be started with (8.6) for $j = m-2$.

(5.A) We must show $Z_{m-2} \le Z_{m-1} - 3$. If $Z_{m-2} = 0$, this follows since s_{m-1} has the two jump zeros, plus a zero in (x_L, x_R). If $Z_{m-2} > 0$, then we have the two jumps, plus zeros in (x_L, z_1) and (z_w, x_R).

We next prove (8.6) by induction for $r \le j \le m-2$. We suppose (8.6) holds for $j+1$; i.e., $Z_{j+1} \le Z_{m-1} + m-j-4 - \alpha_{j+1} - \beta_{j+1}$, and prove it for j.

(5.B) Since $I_j = J_j = 1$, (8.6) will follow for j if $Z_j \le Z_{j+1} - 1$. This holds for $Z_j = 0$ since s_{j+1} has a zero in (x_L, x_R). If $Z_j > 0$, then s_{j+1} has zeros in (x_L, z_1) and (z_w, x_R) not counted in (8.13).

We now come to a transition. We already have $Z_r \le Z_{m-1} + m-r-3 - \alpha_r - \beta_r$. There are two cases.

(5.C) Suppose $\ell < r$. Now we can prove (8.7) for $j = r-1$. This follows if $Z_{r-1} \le Z_r$. For $Z_{r-1} = 0$ this is trivial. When $Z_r > 0$, it follows since s_r has an extra zero in (x_L, z_1).

(5.D) Suppose $\ell = r$. Now (8.8) follows for $j = r-1$ from our result for r and (8.13).

If (5.D) was the case, we now have $A_{\ell-1} \ne 0$, $B_{\ell-1} \ne 0$, and can prove (8.8) holds for $j = \ell-1, \ldots, 0$ by induction exactly as in Case 1. If (3.C) was the case, we can prove (8.7) holds for $j = \ell,\ldots,r-1$ by induction, exactly as in Case 2. We now come to another transition step in the latter situation. We have $Z_\ell \leq Z_{m-1} + m-\ell-2 - \alpha_\ell - \beta_\ell$ and want to establish (8.8) for $\ell-1$. We have the final two cases.

(5.E) If $I_r = 0$, (8.8) follows for $j = \ell-1$ from the ℓ result and (8.13).

(5.F) If $I_r = 1$, (8.8) follows if $Z_{\ell-1} \leq Z_\ell$. When $Z_\ell = 0$, this is trivial, while if $Z_{\ell-1} > 0$, then s_ℓ has an extra zero in (z_w,b).

The theorem will be completely proved if we now establish (8.8) for $j = \ell-1,\ldots,0$ by induction just as in Case 1. ■

The following lemma is obvious for polynomials. Although the proof for CT-systems is practically identical, we give it anyway for completeness.

LEMMA 8.2. *Suppose that* $u_1,\ldots,u_m$ *is a CT-system in the canonical form* (6.1) *with absolutely continuous* σ's. *Suppose also that* u *is a linear combination of the* $u_1,\ldots,u_m$ *with*

$$L_{j+1}u(a) = \ldots = L_{j+p-1}u(a) = 0 \quad , \quad 0 \leq j < j+p \leq m-1.$$

Then, $L_{j+p}u(a) > 0$ *implies that for some* $\epsilon > 0$,

$$(8.16) \quad L_iu(t) > 0 \quad , \quad a < t < a+\epsilon \ , \ i = j,\ldots,j+p.$$

If $L_{j+p}u(b) < 0$, *then the* $L_iu(t) < 0$ *in* $(a,a+\epsilon)$. *Similarly, if*

$$L_{j+1}u(b) = \ldots = L_{j+q-1}u(b) = 0 \ , \ 0 \leq j < j+q \leq m-1,$$

then $L_{j+q}u(b) > 0$ *implies that for some* $\varepsilon > 0$,

$$(8.17) \qquad (-1)^{j+q-i} L_i u(t) > 0, \; b-\varepsilon < t < b, \quad i = j,\ldots,j+q.$$

If $L_{j+q}u(b) < 0$, *then* (8.17) *holds with the opposite sign.*

Proof: We consider only the first case at a with $L_{j+p}u(a) > 0$. The other cases are similar. By the continuity of $L_{j+p}u$, we know that it is positive throughout some interval of the form $(a,a+\varepsilon)$. Now, since for $i = j,\ldots,j+p-1$,

$$L_i u(t) = \int_a^t L_{i+1}u(x)dx ,$$

(8.16) follows. ■

It is easy to show with an example that Theorem 8.1 cannot hold if we try to count end intervals as in (7.2). We conclude this section with an example to show that Rolle's theorem and the refined Budan-Fourier theorem do not hold if we admit splines with multiple knots.

EXAMPLE 8.3. Let $[a,b] = [0,4]$, $\Delta = \{2\}$, and $M = (2)$. Let $u_1 = 1$ and $u_2 = x$. Thus, we are considering linear splines with a double knot at 2. Evidently $s(x) = x - 1 - 2(x-2)_+^0$ is such a spline, and $s'(x) \equiv 1$. Thus, $Z_1 = 0$ but $Z_0 = 3$, (as s has zeros at 1,2, and 3). We see that neither Rolle's theorem (8.13) nor the Budan-Fourier result (8.8) holds in this case.

§9. Tchebycheffian splines with HB ties.

In the previous three sections we have discussed properties of Tchebycheffian splines where the pieces are tied together by the requirement that a sequence of derivatives $L_0,\ldots,L_{m-1-m_i}$ be continuous at each knot x_i, $i = 1,2,\ldots,k$. In this section we shall examine to what extent similar results can be established for a somewhat larger class of Tchebycheffian splines in which the ties are defined by only a selection of these derivatives.

We begin with a precise definition of the class of interest in this section. Let

$$(9.1) \qquad E = (E_{ij})_{i=1,j=0}^{k \quad m-1}$$

be a prescribed matrix of 0's or 1's with exactly K entries having the value 1. We define γ_{ij}^{+} by

$$(9.2) \qquad \gamma_{ij}^{+}\varphi = L_{m-j-1}^{+}\varphi(x_i) \quad ,$$

and γ_{ij}^{-} similarly using L_{m-j-1}^{-}.

Let U be a CT-system as in (6.1). We call

$$(9.3) \qquad \mathcal{S}(U;E;\Delta) = \{s : s_i = s|_{I_i} \in U \ , \ i = 0,1,..,k \text{ and } \gamma_{ij}^{-}s_{i-1} = \gamma_{ij}^{+}s_i \ , \text{ all } i,j \text{ with } E_{ij} = 0 \ \}$$

the <u>space of Tchebycheffian splines with</u> HB- <u>ties at the knots</u> $x_1,\dots,x_k$. The space $\mathcal{S}(U;M;\Delta)$ defined in (6.5) is obtained as a special case if we choose E with the property that

$$E_{ij} = \begin{cases} 1 \ , & j = 0,1,..,m_i-1 \\ 0 \ , & j = m_i,\dots,m-1 \ , \end{cases}$$

for $i = 1,2,\dots,k$.

The general results of sections 2 - 5 can now be applied to derive some of the basic algebraic properties of the space $\mathcal{S}(U;E;\Delta)$.

THEOREM 9.1. The space $\mathcal{S}(U;E;\Delta)$ is of dimension $m + K$, where K is the number of 1's in the incidence matrix E. Moreover, a one-sided basis for $\mathcal{S}$ is given by

$$(9.4) \qquad \{u_1,\ldots,u_m\} \cup \{g_{m-j}(x,x_i) : E_{ij} = 1\},$$

where g_{m-j} is the Green's function defined in (6.9).

Proof: The dimensionality statement follows immediately from Theorem 2.1 as the required rank conditions are easily checked. We recall from section 6 that

$$g_{m-j}(x,x_i) = \begin{cases} 0 & , a \le x < x_i , \\ u_{m-j} + \ldots & , x_i \le x \le b , \end{cases}$$

and that

$$L_i^+ g_{m-j}(x,x_i)\big|_{x=x_i} = \delta_{i,m-j-1} , \; i = 0,1,\ldots,m-1.$$

It follows that whenever $E_{ij} = 1$, g_{m-j} has the correct piecewise structure to be a spline, and also the correct ties at the knot x_i, (all derivatives at x_i are zero except for the $m-j-1^{th}$ which has a jump there). It is clear that the g's associated with a particular knot are linearly independent since each starts with a different u with coefficient 1. The linear independence of the entire set (9.4) then follows just as in Theorem 2.4. For later use, we note

$$g'_{m-j}(x,x_i) = (-1)^j R_j^- g_m(x,x_i) \quad . \blacksquare$$

We can also give conditions under which a local support basis exists.

THEOREM 9.2. Suppose that the incidence matrix E is such that each row of E contains a 0 in the last column. (This assumption assures that every spline s is at least contin-

<u>uous. Then there exists a basis</u> $\{B_i\}_1^{m+K}$ <u>for</u> $\mathcal{S}$ <u>consisting of splines with support over no more than</u> m <u>of the subintervals defined by</u> Δ; <u>i.e., with support as described in</u> (5.9) <u>and</u> (5.10).

<u>Proof:</u> We may apply Theorem 5.3. The hypothesis (5.8) is verified since we observe that the matrix

$$C\langle \epsilon_{i+1}+1,\ldots,\epsilon_{i+m+1}\rangle$$

contains the matrix

$$C\langle \epsilon_{i+1}+1,\epsilon_{i+2}+1,\ldots,\epsilon_{i+m}+1\rangle \;=\; (v^*_{0,\nu}(x_{i+\mu}))_{\nu,\mu=1}^{m} \;,$$

where the $v^*_{0,\nu}$ are the adjunct functions defined in (6.8). Since they form a CT-system, the matrix is nonsingular. ∎

Example 5.4 serves to illustrate the fact that local support bases do not always exist for Tchebycheffian splines with HB-ties. We shall come back to the question of constructing splines with smaller support than those in Theorem 9.2 later in this section, and in particular, we will look at analogs of the B-splines constructed in Theorem 6.1. Before doing so, however, we want to give some results on zeros of splines in $\mathcal{S}(U;E;\Delta)$.

We recall that if $u \in U$ and if L_j is the operator in (6.16), then we say that $L_j u$ has a zero of multiplicity $1 \le z \le m-j-1$ at the point t provided that

$$(9.5) \qquad L_j u(t) = L_{j+1}u(t) = \ldots = L_{j+z-1}u(t) = 0 \;,$$
$$L_{j+z}u(t) \neq 0.$$

Now if s is a spline in $\mathcal{S}(U;E;\Delta)$, then we count a zero of $L_j s$ at a point $t \notin \Delta$ which is isolated (i.e. $L_j s$ does not vanish in any interval containing t) as in (9.5).

For isolated zeros at knots we use

(9.6) Let $t = x_i$, and suppose that s_{i-1} and s_i are the elements of U representing s to the left and right of x_i. Let $0 \le \ell \le m-2-j$ and $0 \le r \le m-2-j$ be the number of times that $L_j s_{i-1}$ and $L_j s_i$ vanish at x_i, respectively, and let $\alpha = \min(\ell, r)$. Then we say that $L_j s$ has a zero at t of multiplicity

$$z = \begin{cases} \alpha+1, & \text{if } \alpha \text{ is even and s changes sign at t,} \\ \alpha+1, & \text{if } \alpha \text{ is odd and s doesn't change sign at t,} \\ \alpha \quad, & \text{otherwise.} \end{cases}$$

This rule counts a jump through 0 at a knot as a zero of multiplicity 1. Moreover, $L_j s$ changes sign at odd order zeros, and does not change sign at even order ones. We count the zeros of $L_{m-1}s$ by simply counting the jumps. Now we have to explain how to count zeros when $L_j s$ vanishes identically on intervals, $j < m-1$. First, if $L_j s$ vanishes on an interval ending at a, we use

(9.7) Suppose that $L_j s(t) = 0$ for $a \le t < x_i$ and that $L_j s(x_i) = \dots = L_{j+z-1}s(x_i) = 0$, $L_{z+j}s(x_i) \ne 0$. Then we count the interval $[a, x_i)$ as a zero of $L_j s$ of multiplicity z. (If $L_j s(x_i) \ne 0$, we count the jump as $z = 1$).

An analogous rule can be used on an interval of the form $[x_i, b]$ where $L_j s$ vanishes identically. Now we suppose that $L_j s$ vanishes identically on an interior interval.

(9.8) Suppose that $L_j s$ is identically zero on $[x_p, x_q)$ in (a,b). Suppose the adjoining pieces s_{p-1} and s_q are such that $L_j s_{p-1}$ has a zero of multiplicity ℓ at x_p and $L_j s_q$ has a zero of multiplicity r at x_q. Let $\alpha = \min(\ell, r)$. Then, we count the interval $[x_p, x_q)$ as a zero of multiplicity α or $\alpha+1$ just as in (9.6).

Thus, if $L_j s$ vanishes on an interval in (a,b) and has a jump at either end of it, then the interval is counted as a zero of $L_j s$ of order 1.

We can now state our zero result for Tchebycheffian splines with HB-ties. The result involves the structure of the incidence matrix E. If $1 \le i \le k$, and $E_{i,\nu},\ldots,E_{i,\mu}$ are consecutive elements in the ith row of E with

$$(9.9) \qquad E_{i,\nu} = \cdots = E_{i,\mu} = 1 \quad \text{and} \quad \begin{cases} \nu = 1 \text{ or } E_{i,\nu-1} = 0, \\ \mu = m-1 \text{ or } E_{i,\mu+1} = 0, \end{cases}$$

then we call $E_{i,\nu},\ldots,E_{i,\mu}$ a sequence of 1's in E. We call such a sequence odd (even) if it has an odd (even) number of elements.

<u>THEOREM 9.3.</u> <u>Let</u> $s \in \mathcal{S}(U;E;\Delta)$ <u>be such that</u> $s \not\equiv 0$. <u>Let</u> $Z^*(s) = Z^*(L_0 s)$ <u>denote the number of zeros of</u> s, <u>counting multiplicities as in</u> (9.5)-(9.8). <u>Let</u> K <u>be the number of</u> 1's <u>in the incidence matrix</u> E, <u>and let</u> p <u>be the number of odd sequences in</u> E <u>which do not start in the first column. Then</u>

$$(9.10) \qquad Z^*(s) \le m + K + p - 1.$$

<u>Proof:</u> The idea of the proof is to relate the number of zeros of $L_0 s$ to the number of zeros of $L_1 s$, then to $L_2 s$, etc. until $L_{m-1}s$. First, we need to introduce a classification scheme for zeros. Consider $L_j s$, and suppose that τ denotes either an isolated zero at a point c = d, or a zero on an interval [c,d) in (a,b). Let s_L and s_R denote the elements of U which represent s to the left and right of c and d, respectively. We say τ is a <u>continuous zero</u> of $L_j s$ provided

$$(9.11) \qquad \begin{cases} L_{j+1}s_L(c)L_{j+1}s_R(d) > 0 \quad \text{and} \\ L_j s_R(d)\, L_{j+1}s_R(d) \ge 0 \ , \ L_j s_L(c) L_{j+1} s_L(c) \le 0 \end{cases}$$

or if

(9.12) $L_j s_L(c) = L_j s_R(d) = 0.$

If τ is not a continuous zero, we say it is a <u>discontinuous</u> one, and further type it according to the following classification scheme:

(9.13) <u>type +</u>, if $L_{j+1}s_L(c)L_{j+1}s_R(d) < 0$

(9.14) <u>type -</u>, if $L_{j+1}s_L(c)L_{j+1}s_R(d) > 0$ and

$$L_j s_L(c)L_{j+1}s_L(c) > 0 \ , \ L_j s_R(d)L_{j+1}s_R(d) < 0.$$

(9.15) <u>type 0</u>, otherwise.

We also need to classify the zero intervals τ of the form $[a,d)$ or $[c,b]$. Consider the first case. We say such a τ is a continuous zero of $L_j s$ provided that

(9.16) $L_j s_R(d) = 0$.

If $L_j s_R(d) \neq 0$, we call $[a,d)$ a discontinuous zero, and further type it as follows:

(9.17) <u>type +</u> , if $L_{j+1}s_R(d) \neq 0$

(9.18) <u>type 0</u> , otherwise.

A similar definition applies if $\tau = [c,b]$.

Let θ_j^+, θ_j^-, and θ_j^o denote the number of discontinuous zeros of $L_j s$ of types +,-, and 0. We now compare the number of zeros of $L_j s$ with the number of zeros of $L_{j+1}s$. For convenience, set $Z_j^* = Z^*(L_j s)$. If $Z_j^* = 0$, then it is trivially true that

(9.19) $Z_j^* \leq Z_{j+1}^* + 1 + \theta_j^o + 2\theta_j^-$.

We claim that this inequality also holds if $Z_j^* > 0$. Indeed, suppose that $L_j s$ has zeros at the points or intervals $\tau_1, .., \tau_n$ with multiplicities $\omega_1, ..., \omega_n$, so that $Z_j^* = \omega_1 + ... + \omega_n$. Then, from the definition of multiple zeros, it follows (cf. [43] for a similar consideration) that $L_{j+1}s$ has zeros at the same places of multiplicities $\omega_1 - 1, ..., \omega_n - 1$.

In addition to these zeros of $L_{j+1}s$, we can find additional zeros. There are at least $n - 1 - \theta_j^- - \theta_j^o - \theta_j^+$ intervals for which both endpoints are continuous type zeros of $L_j s$. If we throw out all such intervals which contain a type - discontinuous zero, we are left with at least $n - 1 - 2\theta_j^- - \theta_j^o - \theta_j^+$ intervals on which it is easily checked that $L_{j+1}s$ must have an additional sign change. In addition, by the definition of type + zeros, it follows that $L_{j+1}s$ must have a jump zero at each such zero of $L_j s$. Hence, we conclude that

$$Z_{j+1}^* \geq \omega_1 - 1 + ... + \omega_n - 1 + n-1 - 2\theta_j^- - \theta_j^o ,$$

which is the statement (9.19).

It is clear that discontinuous zeros of $L_j s$ are always associated with knots, and can occur for a zero involving the knot x_i if and only if $E_{i,m-j-1} = 0$. It is convenient to introduce a companion matrix to E which describes the zero structure of the discontinuous zeros of s and its derivatives. We define the matrix

$$E^* = (E_{i,j}^*)_{i=1,j=0}^{k \quad m-1}$$

by

(9.20) $$E_{i,j}^* = \begin{cases} 2 \text{ if } L_{m-1-j}s \text{ has a type - zero involving } x_i, \\ 1 \text{ if the zero of } L_{m-1-j} \text{ is type } 0, \\ 0, \text{ otherwise.} \end{cases}$$

We can now restate (9.19) as

$$Z_j^* \le Z_{j+1}^* + \sum_{i=1}^{k} E_{i,m-j-1}^* + 1, \; j = 0,1,\ldots,m-2.$$

Since $L_{m-1}s$ is a piecewise constant function whose only zeros are at those knots x_i where $E_{i0} = 1$, we conclude that

$$Z_{m-1}^* \le \sum_{i=1}^{k} E_{i0}^* .$$

Combining these inequalities, we finally have

$$(9.21) \quad Z_0^* \le m - 1 + \sum_{i=1}^{k} \sum_{j=0}^{m-1} E_{ij}^* .$$

It remains to relate the double sum in (9.21) to the original incidence matrix. First, we observe that $E_{i,0}^*$ are always either 0 or 1. Moreover, if $E_{i,j}^* = 2$ for some $j > 1$, then $E_{i,j-1}^*$ must be 0, and if $j < m-1$, $E_{i,j+1}^*$ must also be zero. Indeed, $E_{i,j}^* = 2$ means that at a zero associated with x_i we have $L_{m-j-1}s$ has a type - discontinuous zero. This implies $L_{m-j}s$ has a jump at x_i, but not through 0, and $E_{i,j-1}^* = 0$. Similarly, if $L_{m-j-2}s$ has a discontinuous zero associated with x_i while its derivative $L_{m-j-1}s$ has a type - discontinuous zero there, then $L_{m-j-1}s$ has a strong sign change there, so L_{m-j-2} has a type + zero and $E_{i,j+1}^* = 0$.

Now we can estimate the double sum in (9.21). Suppose that the incidence matrix has a sequence of 1's in the ith row. If it has an even number in it, then the corresponding string in E* has a sum no larger than the sum of the 1's. (The only way it could be larger would be if some 2's appeared, but then they would have to have the form 0,2,0,2,.., etc. whose sum is no larger). If a sequence of 1's starts in the second column or later and has an odd number of 1's in it, then the corresponding sequence in E^* could have the form 2,0,2,...,0,2, which has a sum one greater than the number of 1's in the E sequence. (If it starts in the first column, then the first element in the E^* sequence is always a 0 or a 1, and the sum cannot increase). In summary, we see that the sum of the entries in E^* can be no larger than the number of 1's in E plus the number of consecutive sequences of 1-s in E which begin in the second column or later, and which have an odd number in them. The theorem is proved. ■

It is of interest to have a version of this theorem for splines which vanish identically outside of (x_1,x_k), and where only zeros inside of (x_1,x_k) are counted.

<u>THEOREM 9.4</u>. <u>Suppose that</u> $s \in \mathcal{S}(U;E;\Delta)$ <u>is such that</u> $s \not\equiv 0$ <u>but</u> s <u>vanishes identically outside of</u> (x_1,x_k). <u>Suppose that</u> s <u>is of exact order</u> m; <u>i.e.</u>, $L_{m-1}s(t) \neq 0$ <u>for some</u> t <u>in</u> (a,b). <u>Then</u>

$$(9.22) \qquad Z^*_{(x_1,x_k)} \leq K + p - m - 1.$$

<u>Proof:</u> Since the proof is nearly identical with that of Theorem 9.3, we do not need to go into great detail. The important difference is that now instead of (9.19), we can show that

$$Z^*_j \leq Z^*_{j+1} - 1 + \theta^o_j + 2\theta^-_j ,$$

if we count only zeros in (x_1,x_k). Indeed, the two end intervals $[a,x_1)$ and $[x_k,b]$ are not counted in Z^*_j, but Z^*_{j+1} has the same number of zeros as in Theorem 9.3. ■

The number p in Theorem 9.4 can be related to a concept of importance in the theory of poisedness for HB-interpolation problems. We say that a sequence of 1's in the incidence matrix E is an <u>odd supported sequence</u>

if its first element E_{ij} starts in the second column or later, and if there exist $i' < i < i''$ and $0 \le j', j'' < j$ with $E_{i'j'} = E_{i''j''} = 1$. Now, we observe that if $L_{m-j-1}s$ has a type - discontinuous zero, then since it vanishes identically on $[a,x_1)$ and on $[x_k,b]$, $L_{m-j}s$ must have a jump somewhere to the left and somewhere to the right of this zero of L_{m-j-1}. This means that the corresponding sequence must be an odd supported sequence when the count in E* is 1 larger than the number of 1's in the E sequence.

Theorems 9.3 and 9.4 are not correct if we try to use the stronger zero count of section 7, as the following example shows.

EXAMPLE 9.5. Let $U = \{1,x,\ldots,x^8\}$, $[a,b] = [-1,2]$, $\Delta = \{0,1\}$, and

$$E = \begin{bmatrix} 1&0&0&0&0&0&0&0&1 \\ 1&0&0&0&0&0&0&0&1 \end{bmatrix} .$$

Consider the spline

$$s(x) = \begin{cases} x^8 , & -1 \le x < 0 \\ 1 , & 0 \le x < 1 \\ -(x-1)^8 , & 1 \le x \le 2 . \end{cases}$$

Analysis: Here $m = 9$, $K = 4$, $p = 2$. Hence Theorem 9.3 would require that $Z(s) \le 14$. On the other hand, if we use the count of section 7, we see that s has 8 zeros at 0 and 8 zeros at 1. ■

We return now to the question of constructing local support splines in $\$(U;E;\Delta)$. First, we define an extended incidence matrix $\tilde{E}$ associated with E. Let

$$(9.23) \qquad \tilde{E}_{ij} = \begin{cases} E_{ij} , & j = 0,1,..,m-1 , \quad 1 \le i \le k \\ 1 , & j = 0,1,..,m-1 , \quad i = 0,k+1. \end{cases}$$

If $F = (F_{ij})_{i=\ell, j=0}^{r \quad m-1}$ is a matrix of 0's and 1's, we define

$$(9.24) \qquad F \subset \tilde{E} \quad \text{iff} \quad F_{ij} = 1 \text{ implies } \tilde{E}_{ij} = 1 .$$

We are going to show how to construct splines with local support corresponding to matrices $F \subset \tilde{E}$ with m+1 entries equal to 1. Suppose F is such a matrix, and consider the set

$$(9.25) \qquad \{(-1)^j e_{x_i} R_j\}_{F_{ij}=1} ,$$

where e_t denotes the point evaluator (defined by $e_t f = f(t)$) and R_j is the operator defined by

$$R_j f(x) = \begin{cases} R_j^- f(x) & , \quad a < x \le b \\ R_j^+ f(x) & , \quad x = a , \end{cases}$$

(cf. § 6).

The set (9.25) contains m+1 linear functionals. Let $\eta_1, \ldots, \eta_m$ be the reverse lexicographical ordering of this set; i.e.,

$$e_{x_i} R_j < e_{x_{i'}} R_{j'} \quad \text{provided} \quad \begin{cases} i < i' & \text{or} \\ i = i' \text{ and } j' < j . \end{cases}$$

Given any functions $\psi_1, \ldots, \psi_{m+1}$ for which the η's can be applied, we define

$$D\binom{\eta_1, \ldots\ldots, \eta_{m+1}}{\psi_1, \ldots\ldots, \psi_{m+1}} = \det (\eta_i \psi_j)_{i,j=1}^{m+1} .$$

We define $u_i^* = v_{0,1}^*$, $i = 1,2,\ldots\ldots,m$, where the v^*'s are defined in (6.8). Let g_m be the Green's function defined in (6.9). We are now ready to define local support splines in $\mathfrak{S}(U;E\Delta)$ with quite small support.

LEMMA 9.6. Let $F \subset \tilde{E}$ be a matrix with m+1 entries equal to 1, and let $\eta_1, \ldots, \eta_{m+1}$ be the associated linear functionals defined above. Define

$$(9.26) \qquad B_F(x) = D\begin{pmatrix} \eta_1, \ldots\ldots, & \eta_{m+1} \\ u_1^*, \ldots, u_m^*, & g_m(x,\cdot) \end{pmatrix}.$$

Then

$$(9.27) \qquad B_F(x) = \sum_{i=1}^{m+1} (-1)^i D_i \, \eta_i g_m(x, x_i),$$

where

$$(9.28) \qquad D_{ij} = D\begin{pmatrix} \eta_1^i, \ldots, \eta_m^i \\ u_1^*, \ldots, u_m^* \end{pmatrix}$$

η_i applies to the second variable of g_m, and $\eta_1^i, \ldots, \eta_m^i$ is the set $\{\eta_1, \ldots, \eta_{m+1}\} \setminus \{\eta_i\}$. Thus, $B_F \not\equiv 0$ whenever some $D_i \neq 0$. A necessary condition for this to happen is that

$$(9.29) \qquad M_v(F) = \sum_{i=\ell}^{r} \sum_{j=0}^{v} F_{ij} \geq v+1, \quad v = 0, 1, \ldots, m-1.$$

(ie., F satisfies the Pólya condition). The function B_F is a spline in $\mathcal{S}(U;E;\Delta)$, and

$$(9.30) \qquad B_F(x) = 0 \quad \text{for} \quad x < x_\ell \quad \text{and} \quad x_r < x.$$

Finally,

$$(9.31) \qquad Z^*_{(x_\ell, x_r)}(B_F) \leq p + m - m',$$

where m' is the exact order of B_F; i.e. $m' = \max\{m-j : D_{ij} \neq 0, \ \ell \leq i \leq r, \ 0 \leq j \leq m-1\}$.

Proof: If we apply Laplace's expansion to (9.26), we obtain (9.27)-(9.28). If the Pólya condition is not satisfied, suppose $M_v \leq v$. Then in the first v+1 rows of the determinant defining B_F, we have only v nonzero entries. But then the Laplace expansion gives 0. In view of (9.27), it is clear that B_F is a spline. The assertion (9.30) for $x < x_\ell$ is trivial since all of the g_j's are identically 0 there. For $x_r < x$, we recall from section 6 that g_m is a linear combination of the u^*'s, and hence the determinant is again 0. Finally, (9.31) is just an immediate application of Theorem 9.4. ∎

Whether Lemma 9.6 can be used to construct a basis for $\mathcal{S}(U;E;\Delta)$ of splines of the form (9.26) depends on whether it is possible to construct m+K matrices $F_1,\ldots,F_{m+K}$ involving as few x's as possible, but such that the resulting B-splines are linearly independent. It is possible to give conditions under which such a basis is constructable, but it requires going into the theory of Hermite-Birkhoff interpolation with respect to a CT-system, and this paper is long enough already.

The function B_F in (9.26) has been called the Birkhoff kernel. It is, of course, the kernel in the representation of a certain linear functional, and as such plays exactly the same role as the usual B-splines. (See also Remark 15).

§10. Approximation powers of Tchebycheffian splines.

In this section we want to briefly describe how well certain classes of smooth functions can be approximated by some linear spaces of generalized splines. We begin by introducing some additional notation. Assuming $\mathcal{S}$ is a linear space of splines, and f is an appropriately smooth function, we define

$$(10.1)\qquad d_{j,q}(f,\mathcal{S}) = \inf_{s\in\mathcal{S}} \|D^j(f-s)\|_q ,$$

where D^j is the usual derivative operator and $\|.\|_q$ is the usual q-norm defined on the interval [a,b]. We are interested in relating the behavior of $d_{j,q}(f,\mathcal{S})$ to the smoothness of f. This means that we want upper bounds, lower bounds, and inverse results -- in short, a constructive theory of spline approximation.

While it would be possible to develop a constructive theory for Tchebycheffian splines as defined in section 6 (but using the L_j derivatives) there is probably more interest in having results involving the usual derivatives. Clearly, to prove such results we have to further restrict the class of Tchebycheffian splines to be considered.

Let $a_j \in C^j[a,b]$, $j = 0,1,\ldots,m-1$, and consider the linear differential operator

$$(10.2)\qquad L = D^m + \sum_{0}^{m-1} a_j D^j \quad .$$

Then, there exist functions $u_i \in C^m[a,b]$, $i = 1,2,\ldots,m$ which span the null space N_L of L. We do not assume that the u's form a CT-system.

Let $M = (m_1,\ldots,m_k)$ be a multiplicity vector ($1\le m_i \le m$) and let Δ be a partition of [a,b]. Then we define

$$(10.3)\qquad \mathcal{P}(L;\Delta) = \{s : s_i = s|_{I_i} \in N_L ,\ i = 0,1,..,k\},$$

and

$$(10.4)\qquad \mathcal{S}(L;M;\Delta) = \{s \in \mathcal{P}(L;\Delta) : D^j s_{i-1}(x_i) = D^j s_i(x_i),\ 0\le j\le m-m_i-1 ,\ 1\le i\le k \}.$$

The latter class is similar to the class of Tchebycheffian

splines discussed in §6, and when N_L is spanned by an Extended Complete Tchebycheff (ECT-) system, then they are in fact the same.

It is clear that for any multiplicity vector M,

$$(10.5) \qquad \mathcal{S}(L;M_s;\Delta) \subseteq \mathcal{S}(L;M;\Delta) \subseteq \mathcal{P}(L;\Delta)$$

where $M_s = (1,1,..,1)$. Hence, in proving direct theorems (upper bounds on $d_{j,q}(f,\mathcal{S})$), it suffices to work with the space $\mathcal{S}(L;M_s;\Delta)$ of splines with simple knots. On the other hand, if we want lower bounds or inverse theorems, we should work with the space of piecewise N_L functions $\mathcal{P}(L;\Delta)$. It will often turn out that the upper and lower bounds are of the same order, so that all of the spline spaces have the same approximation power, independent of M.

Before stating our first direct theorem, we need a little more notation. Let

$$(10.6) \qquad L_q^m[a,b] = \{f : f^{(m-1)} \text{is absolutely continuous}, \; f^{(m)} \in L_q[a,b] \}$$

be the usual Sobolev space with the norm

$$(10.7) \qquad \|f\|_{L_q^m[a,b]} = \sum_0^{m-1} \|D^j f\|_{L_q[a,b]} .$$

Associated with L_q, L_q^m, and the differential operator L, we define the Peetre K-functional

$$(10.8) \qquad K(t,f,q,L) = \inf \{\|f-g\|_q + t^m \|Lg\|_q : g \in L_q^m \}$$

for $0 < t < \infty$. K gives a measure of the smoothness of f, and as we shall see below, is closely related to various moduli of smoothness.

Given a partition Δ of $[a,b]$, we define

$$(10.9)\quad \bar{\Delta} = \max_{0\le i\le k} (x_{i+1}-x_i) \quad\text{and}\quad \underline{\Delta} = \min_{0\le i\le k} (x_{i+1}-x_i).$$

<u>THEOREM 10.1.</u> <u>Let</u> $1 \le q < \infty$. <u>Then, there exists a constant</u> C_q (<u>which may also depend on</u> a,b,<u>and</u> L), <u>such that</u>

$$(10.10)\quad d_{0,q}(f, \mathcal{S}(L;M_s;\Delta)) \le C_q \cdot K(\bar{\Delta},f,q,L)$$

<u>for all</u> $f \in L_q[a,b]$ <u>and all</u> Δ <u>with</u> $\bar{\Delta}$ <u>sufficiently small</u>. <u>Moreover</u>, <u>if</u> $q = \infty$, <u>then</u> (10.10) <u>also holds for all</u> $f \in C[a,b]$ <u>and</u> $\bar{\Delta}$ <u>sufficiently small</u>.

<u>Proof:</u> It is clear that $d_{0,q}(f, \mathcal{S}) \le d_{0,q}(f, \mathcal{S}^*)$ for any $\mathcal{S}^* \subset \mathcal{S}$. Now, given the partition Δ, we define

$$\Delta^* = \{a = y_0 < y_1 < \dots < y_{k^*+1} = b\},$$

where $y_1,\dots,y_{k^*}$ are defined recursively by

$$y_j = \min\{x_i : y_{j-1} + \bar{\Delta} \le x_i < y_{j-1} + 2\bar{\Delta}\},$$

and where k^* is the maximum integer for which this set is nonempty. It is easily checked that $\underline{\Delta} \le \bar{\Delta} \le \underline{\Delta}^* \le \bar{\Delta}^* \le 3\bar{\Delta}$, so that $\bar{\Delta}^*/\underline{\Delta}^* \le 3$. We define $\mathcal{S}^* = \mathcal{S}(L;M_s;\Delta^*)$. Since the theorem is proved in [15] for this class of splines with bounded mesh ratio, the general result stated here also follows. ■

Although Theorem 10.1 is stated in terms of the K functional, it is relatively easy to translate it into a statement involving classical moduli of smoothness. Let $\Delta_u^r f(x)$ denote the usual r^{th} forward difference of the function f, and define

$$(10.11)\quad \omega_r(f,t)_\infty = \sup_{\substack{|u|\le t\\ x,x+ru\in[a,b]}} |\Delta_u^r f(x)|$$

For $1\le q<\infty$, we define

$$(10.12)\quad \omega_r(f,t)_q = \sup_{|u|\le t} \left(\int_a^b |\Delta_u^r f(x)|^q dx\right)^{1/q} ,$$

(where either only those u such that s+ru stay in [a,b] are considered, or where f is extended to [2a-b,2b-a], see [14]). We write $\mathcal{S}$ for $\mathcal{S}(L;M_s;\Delta)$ and ω for ω_0.

<u>THEOREM 10.2.</u> <u>For</u> <u>all</u> $f\in L_q[a,b]$, $1\le q<\infty$,

$$(10.13)\quad d_{0,q}(f,\mathcal{S}) \le C_q[\bar{\Delta}^m\|f\|_q + \omega_m(f,\bar{\Delta})_q] \quad .$$

<u>The</u> <u>same</u> <u>result</u> <u>holds</u> <u>for</u> $q=\infty$ <u>for</u> <u>all</u> $f\in C[a,b]$. <u>Moreover</u>, <u>if</u> $1\le r\le m$, <u>then</u> <u>for</u> <u>all</u> $1\le q\le\infty$,

$$(10.14)\quad d_{j,q}(f,\mathcal{S}) \le C_{j,q}(\bar{\Delta})^{r-j-1}[\omega(D^{r-1}f,\bar{\Delta}) + (\bar{\Delta})^{m-r}\|f\|_{L_q^{r-1}[a,b]}]$$

<u>for</u> <u>all</u> $f\in C^{r-1}[a,b]$, <u>and</u> <u>all</u> $j=0,1,\ldots,r-1$. <u>Also</u>,

$$(10.15)\quad d_{j,q}(f,\mathcal{S}) \le C_{j,q}(\bar{\Delta})^{r-j}[\omega(D^r f,\bar{\Delta})_q + (\bar{\Delta})^{m-r}\|f\|_{L_q^r[a,b]}]$$

<u>for</u> <u>all</u> $f\in L_q^r[a,b]$, <u>and</u> $j=0,1,\ldots,r$. <u>Finally</u>, <u>if</u> $1\le p\le q\le\infty$, <u>then</u>

$$(10.16)\quad d_{j,q}(f,\mathcal{S}) \le C_{j,q,p}(\bar{\Delta})^{r-j+1/q-1/p}[\|D^r f\|_{L_p[a,b]} + (\bar{\Delta})^{m-r}\|f\|_{L_p^{r-1}[a,b]}]$$

for all $f \in L_p^r[a,b]$ and $j = 0,1,\dots,\min(r,m-1)$.

Proof: These results can be obtained as a corollary of Theorem 10.1 by applying estimates of the K functional in terms of moduli of smoothness (cf. [14]), or can be proved directly using a certain spline approximation scheme (cf. [12]). ■

It is straightforward to apply the theory of interpolation spaces to the results of Theorem 10.2 to obtain direct theorems for functions in intermediate spaces; e.g. in the Besov spaces (cf. [9]). We also note that it is possible to give upper bounds in terms of a modulus of continuity defined with generalized divided differences with respect to $u_1,\dots,u_m$, provided that they form a T-system (see [15,45]).

In the following theorem we quote lower bounds on approximation with $\mathcal{S}$ which follow from certain results on n-widths (see [12]). These results are the exact companions of those in Theorem 10.2, and serve to show that the upper bounds there were optimal order, at least if the partitions are such that $\bar{\Delta} / \underline{\Delta}$ is bounded.

THEOREM 10.3. Fix $1 \le r \le m$ and $1 \le q \le \infty$. Then, there exists a function $F \in C^{r-1}[a,b]$ with $\|F\|_{L_q^{r-1}} \le 1$ so that

$$(10.17)\qquad c_{j,q}(\underline{\Delta})^{r-j-1}\,\omega(D^{r-1}F,\underline{\Delta}) \;\le\; d_{j,q}(F,\mathcal{S}(L;\Delta)),$$

where $c_{j,q}$ is a constant depending on F, but independent of Δ, $j = 0,1,\dots,r-1$. Furthermore, there exists a function $F \in SL_q^r[a,b] = \{f \in L_q^r[a,b] : \|f\|_{L_q^r} \le 1\}$ with

$$(10.18)\qquad c_{j,q}(\underline{\Delta})^{r-j} \;\le\; d_{j,q}(F,\mathcal{S}(L;\Delta)),$$

where $c_{j,q}$ is a constant independent of Δ.

It is also possible to establish inverse results which show that if a particular f can be approximated well with Tchebycheffian splines, then it must be appropriately smooth. We quote just one example of this type of result. Suppose Δ_ν is a sequence of partitions of [a,b]. We say this sequence is <u>mixed</u> provided that for some k_0 and $d > 0$

$$(10.19) \qquad \sup_{\nu \geq k} \text{dist}\,(t,\Delta_\nu) \geq d\,\underline{\Delta}_k\,,\ \text{all } k \geq k_0,\ t \in \Delta_k.$$

The sequence is called <u>strongly mixed</u> if there exist some $\nu_1,\ldots,\nu_r \geq k$ so that

$$(10.20) \qquad \max_{1 \leq i \leq r} \text{dist}\,(t,\Delta_{\nu_i}) \geq d\,\underline{\Delta}_k\,,\ \text{all } k \geq k_0,\ t \in \Delta_k.$$

<u>THEOREM 10.4.</u> <u>Suppose</u> Δ_ν <u>is a sequence of mixed partitions of</u> [a,b]. <u>Then there exists a constant</u> c <u>and a</u> k_0 <u>so that for all</u> $k \geq k_0$ <u>and</u> $f \in C[a,b]$,

$$(10.21) \qquad K(\overline{\Delta}_k,f,\infty,L) \leq c \sup_{\nu \geq k} d_{0,\infty}(f,\ \mathcal{S}(L;\Delta_\nu))\ .$$

<u>If</u> Δ_ν <u>is strongly mixed, then a similar result holds for</u> $1 \leq q < \infty$ <u>and all</u> $f \in L_q[a,b]$.

<u>Proof:</u> See [15]. Since K is closely related to classical moduli of smoothness, (10.21) gives classical smoothness results when f can be well-approximated from $\mathcal{S}(L;\Delta)$. ∎

The inverse theorem 10.4 can be coupled with the direct theorem 10.1 to give complete characterizations of various classes of smooth functions (generalized Lipschitz classes) in terms of their approximability with Tchebycheffian splines (see [34,35]). Theorem 10.4 also leads immediately to the following saturation result which gives a limit on how well one can approximate with such splines.

THEOREM 10.5. <u>Suppose that Δ_ν is a sequence of partitions of $[a,b]$ which is strongly mixed with $\bar{\Delta}_\nu \to 0$ as $\nu \to \infty$. Suppose that ψ is a positive function on $(0,\epsilon)$ with $\psi(t) \to 0$ as $t \to 0$. Then, if $f \in L_q[a,b]$ is such that</u>

$$(10.22) \quad d_{0,q}(f, \mathcal{S}(L;\Delta)) \leq c\,(\bar{\Delta}_\nu)^m\,\psi(\bar{\Delta}_\nu) ,$$

<u>f must belong to N_L. A similar result holds for $q = \infty$ and $f \in C[a,b]$ under the milder assumption that Δ_ν is mixed.</u>

This result asserts that we cannot get higher than m^{th} order convergence with m^{th} order splines, no matter how smooth f may be (unless it is already in N_L).

§11. <u>Notes and remarks</u>.

1. The first paper to deal with a class of nonpolynomial splines seems to be that of Schoenberg[36], where certain trigonometric splines were introduced. At almost the same time, Greville[8] took the first step towards a constructive theory of generalized splines by studying a linear space of functions which belonged piecewise to a given m dimensional linear space U_m, and which had global smoothness $C^{m-2}[a,b]$. In particular, he showed that the spline space was finite dimensional, and with the help of a certain Green's function, constructed a one-sided basis for it. Unfortunately, the paper was never published, and it is a pleasure to acknowledge its considerable influence on the developments in the first 5 sections of this paper.

2. It is also possible to define a more general class of splines than that considered in (2.1). In particular, we may take the S_i to simply be spaces of functions depending on n_i parameters (possibly nonlinearly) and we may relax the assumption that the functionals describing the ties

be linear. One must, however, expect that considerably less can be established without linearity.

3. As an example of the nonlinear splines mentioned in the previous remark, we may mention the series of papers on rational splines by Werner, Schaback, Braess, Arndt, and others at Münster (see [33,46] for further references).

4. When Ω in (2.1) is a discrete set, it has become customary to call the resulting splines discrete splines. They were introduced in Mangasarian and Schumaker [25,26]. Some constructive properties were given in [41], and further developed in Lyche[23].

5. When Ω is a curve in the plane and the functions are taken to be complex-valued, one obtains the complex and analytic splines introduced and studied in the papers of Ahlberg, Nilson, and Walsh (e.g., see [1,3] and references therein).

6. Example 4.1 essentially deals with the classical 4-point central interpolation scheme. Since our definition of splines (2.1) permits ties between non-adjacent intervals, other osculatory interpolation methods used by actuaries in the late 1800's and early 1900's may now be regarded as splines (see Greville[7] for references to some of this literature).

7. The development in §5 concerning construction of local support bases is a slight extension of the ideas in Jerome and Schumaker[13]. Example 5.4 is taken directly from that paper.

8. Spaces of splines consisting of pieces in the null-space of L^*L, where L is a differential operator as in (10.2) were studied by Ahlberg, Nilson, and Walsh[2] and by Schultz and Varga[44], but primarily as solutions of an appropriate best interpolation problem. Under the

assumption that L satisfies property W of Pólya (so that its null space is spanned by an ECT-system), Karlin and Ziegler also studied such spaces, which they called Tchebycheffian splines. Since they also studied a certain Green's function, a one-sided basis, and certain determinants, we may regard it as the second paper dealing with constructive properties of generalized splines. A local support basis (B-splines) and its total positivity properties were later elaborated in Karlin's book on Total Positivity [16],(see also Karlin and Karon [18], and Karlin and Schumaker[19]).

9. Section 6 is a survey of results from my recent paper on generalized Tchebycheffian splines. The main difference as compared with the Tchebycheffian splines of Karlin and Studden is the fact that the splines are based on a broad class of CT-systems rather than on ECT systems. This difference is of more than theoretical interest, however, as there are applications where the extended notion must be used (cf. Reddien and Schumaker[31] where an applica- involving singular differential equations is given).

10. The system of functions in (6.1) remain a CT-system even if we assume only that the σ's are right continuous and monotone increasing (rather than absolutely continuous), cf. Mühlbach[30]. However, in this case the u's themselves may not even be continuous. For example, the functions 1 and $x + (x-1)_+^0$ form a CT-system on [0,2], but the second function has a jump at 1. We should also warn the reader than with the weaker assumptions on σ's, integration is not the inverse operation of D^+ differentiation. For example, $D^+(x + (x-1)_+^0) \equiv 1$, whose integral is x not $x + (x-1)_+^0$. It is for this reason that we have decided to restrict our attention to CT-systems with absolutely continuous σ's, although most of the results can be established in a wider setting.

11. At this writing, it is not completely clear to what extent a general CT-system can be put in the canonical form (6.1). In this connection, we mention Rutman[32] and recent work of Zielke[47,48].

12. Surprisingly, strong zero properties for spline functions have only recently been developed, although such results for monosplines have long been known. In [42] I proved Theorem 7.1 for polynomial splines, and showed how it could be used to obtain a wide variety of results on total positivity properties of certain Green's functions and of the B-spline basis. The result for Tchebycheffian splines was established in [43].

13. Budan-Fourier theorems for splines have also only very recently been considered (although, again, they had been studied earlier for monosplines). Indeed, Professor deBoor first acquainted me with such results at Karlsruhe by showing me a copy of the manuscript for his contribution to this volume, and a manuscript of Melkman[28]. Theorem 8.1 is somewhat stronger than the versions I saw at that time in that a stronger zero count is used, the two finer cases (8.6) and (8.7) are delineated, and, of course, it is stated for Tchebycheffian splines rather than polynomial splines. I must, however, apologize for the intricacy of the proof of Theorem 8.1, and I expect that a more elegant proof (and probably a more elegant theorem) must exist. The present proof represents the wages paid for the sin of proving the result at the last minute before the deadline for this paper.

14. The results of section 9 are an extension and improvement of results of Birkhoff[4], Ferguson[6], and Lorentz[22], (and also Jetter, according to a manuscript he gave me in Karlsruhe). The extension is, of course, to Tchebycheffian splines, and the improvement is to a

stronger zero count. Example 9.5 is taken from [42].

15. For other results on the Birkhoff kernel in (9.26), (in the polynomial spline case) see the references in Remark 14. The polynomial spline results can all be readily carried over to the Tchebycheffian spline framework. For a survey of results on Hermite-Birkhoff interpolation with ECT-systems and further references, see Karlin and Karon[18]. These results can also be carried over to the CT-system setting.

16. In §10 we have surveyed only a part of the available results on approximation powers of generalized spline functions. Although error bounds were available for interpolation with L-splines [44], (piecewise in the null space of L*L) the first step towards getting estimates of the approximation power of more general classes of splines was taken by Jerome[10], where a basic approximation scheme was introduced, and a bound for C[a,b] functions was given. Direct theorems in q-norms were later obtained by Scherer[34] and Jerome and Schumaker[12]. The direct result in Theorem 10.1 with the K functional is due to Johnen and Scherer[15], although the elimination of the mesh ratio dependence follows Jerome and Schumaker[12], (who in turn acted on a suggestion of Marsden to follow Meir and Sharma[27]). Connections between the K functional and classical moduli of smoothness are thoroughly investigated in Johnen[14]. The lower bounds in Theorem 10.3 were proved on the basis of n-width results by Jerome and Schumaker[12]. It should be noted that the lower bound corresponding to (10.16) is missing--a consequence of the lack of an appropriate n-width result. For more on inverse theorems, saturation, and characterization of classes of generalized Lipschitz functions, see Scherer [34], Johnen and Scherer[15], and deVore and Richards[45].

References.

1. Ahlberg, J.H., Splines in the complex plane, in Approximations with Special Emphasis on Spline Functions, I.J. Schoenberg, ed., Academic Press, N.Y., 1969, 1-27.

2. Ahlberg, J.H., E.N. Nilson, and J.L. Walsh, Fundamental properties of generalized splines, Proc. Nat. Acad. Science USA 52(1964), 1412-1419.

3. Ahlberg, E.H., E.N. Nilson, and J.L. Walsh, Complex polynomial splines on the unit circle, J. Math. Anal. Appl. 33(1971), 234-257.

4. Birkhoff, G.D., General mean value and remainder theorems, Trans. Amer. Math. Soc.7(1906), 107-136.

5. deBoor, C. and I.J. Schoenberg, Cardinal interpolation and spline functions VIII. The Budan-Fourier Theorem for splines and applications, MRC Rpt. 1546, Univ. of Wis., 1975 (see also these proceedings).

6. Ferguson, D.R., Sign changes and minimal support properties of Hermite-Birkhoff splines with compact support, SIAM J. Numer. Anal. 11(1974), 769-779.

7. Greville, T.N.E., The general theory of osculatory interpolation, Trans. Actuar. Soc. Amer. 45(1944), 202-265.

8. Greville, T.N.E., Interpolation by generalized spline functions, MRC Rpt. 476, Univ. Wis., 1964.

9. Hedstrom, G.W. and R.S. Varga, Application of Besov spaces to spline approximation, J. Approx. Th. 4 (1971), 295-327.

10. Jerome, J.W., On uniform approximation by certain generalized spline functions, J. Approx. Th. 7 (1973), 143-154.

11. Jerome, J.W. and L.L. Schumaker, Local bases and computation of g-splines, Meth. und Verfahren der Math. Physik 5(1971), 171-199.

12. Jerome, J.W. and L.L. Schumaker, On the distance to a class of generalized spline functions, in Linear Operators and Approximation Theory II., P.L. Butzer and B. Sz.-Nagy, Eds., Birkhäuser, ISNM 25, Basel, 503-517.

13. Jerome, J.W. and L.L. Schumaker, Local support bases for a class of spline functions, J. Approx. Th., to appear.

14. Johnen, H., Inequalities connected with the moduli of smoothness, Mat. Vesnik 9(1972), 289-303.

15. Johnen, H. and K. Scherer, Direct and inverse theorems for best approximation by Λ-splines, this proceedings

16. Karlin, S., Total Positivity, Stanford Univ. Press, 1968.

17. Karlin, S. and J. Karon, A remark on B-splines, J. Approx. Th. 3(1970), 455.

18. Karlin, S. and J. Karon, On Hermite-Birkhoff interpolation, J. Approx. Th. 6(1972), 90-114.

19. Karlin, S. and L.L. Schumaker, The fundamental theorem of Algebra for Tchebycheffian monosplines, J. Anal. Math. 20(1967), 233-270.

20. Karlin, S. and W.J. Studden, Tchebycheff Systems: With Applications in Analysis and Statistics, Interscience, New York, 1966.

21. Karlin, S. and Z. Ziegler, Chebychevian spline functions, SIAM J. Numer. Anal. 3(1966), 514-543.

22. Lorentz, G.G., Zeros of splines and Birkhoff's kernel, to appear.

23. Lyche, T., Discrete Polynomial Spline Approximation Methods, dissertation, Univ. of Texas, 1975 (also Rpt. ISBN 82-553-0202-6, No.2, Univ. Oslo.

24. Lyche, T. and L.L. Schumaker, Local spline approximation methods, J. Approx. Th., to appear.

25. Mangasarian, O.L. and L.L. Schumaker, Discrete splines via mathematical programming, SIAM J. Control 9(1971), 174-183.

26. Mangasarian, O.L. and L.L. Schumaker, Best summation formulae and discrete splines, SIAM J. Numer. Anal. 10(1973), 448-459.

27. Meir, A. and A Sharma, On uniform approximation by cubic splines, J. Approx. Th. 2(1969), 270-274.

28. Melkman, A.A., The Budan-Fourier theorem for splines, Israel J. Math., to appear.

29. Mühlbach, G., A recurrence formula for generalized divided differences and some applications, J. Approx. Th. 9(1973), 165-172.

30. Mühlbach, G., Chebysev-Systeme und Lipschitzklassen, J. Approx. Th. 9(1973), 192-203.

31. Reddien, G.W. and L.L. Schumaker, On a collocation method for singular two point boundary value problems, submitted.

32. Rutman, M.A., Integral representation of functions forming a Markov series, Doklady 164(1965), 1340-1343.

33. Schaback, R., Interpolationen mit nichtlinearen Klassen von Spline-Funktionen, J. Approx. Th. 8 (1973), 173-188.

34. Scherer, K., Best approximation by Chebychevian splines and generalized Lipschitz spaces, Proc. Conf. on Constructive Theory of Functions, Cluj, 1973.

35. Scherer, K., Characterization of generalized Lipschitz classes by best approximation with splines, SIAM J. Numer. Anal. 11(1974), 283-304.

36. Schoenberg, I.J., On trigonometric spline interpolation, J. Math. Mech 13(1964), 795-825.

37. Schumaker, L.L., Uniform approximation by Tchebycheffian spline functions, I., J Math. Mech. 18(1968), 369-378.

38. Schumaker, L.L., Uniform approximation by Tchebycheffian spline functions, II., SIAM J. Numer. Anal. 5(1968), 647-656.

39. Schumaker, L.L., On the smoothness of best spline approximation, J. Approx. Th. 2(1969), 410-418.

40. Schumaker, L.L., Approximation by splines, in Theory and Applications of Spline Functions, T.N.E. Greville, ed., Academic Press, N.Y., 1969, 65-85.

41. Schumaker, L.L., Constructive aspects of discrete polynomial spline functions, in Approximation Theory, G.G. Lorentz, ed., Academic Press, N.Y., 1973, 469-476.

42. Schumaker, L.L., Zeros of spline functions and applications, J. Approx. Th., to appear.

43. Schumaker, L.L., On Tchebycheffian spline functions, J. Approx. Th., to appear.

44. Schultz, M.H. and R. S. Varga, L-splines, Numer. Math. 10(1967), 345-369.

45. deVore, R. and F. Richards, The degree of approximation by Chebyshevian splines, to appear.

46. Werner, H., Tshebyscheff-Approximation mit nichtlinearen Spline-Funktionen, in Spline-Funktionen, K. Böhmer, G. Meinardus, W. Schempp, eds., B.I., Mannheim, 1974, 303-314.

47. Zielke, R., On transforming a Tchebyshev-System into a Markov-System, J. Approx. Th. 9(1973), 357-366.

48. Zielke, R., Alternation properties of Tchebyshev-systems and the existence of adjoined functions, J. Approx. Th. 10(1974), 172-184.

Larry L. Schumaker*
Department of Mathematics
The University of Texas
Austin, Texas 78712

* Supported in part by the Deutsche Forschungsgemeinschaft at the University of Munich, and by the United States Air Force under Grant AFOSR-74-2895A.

Discussion of Prof. Meinardus' paper

Larry L. Schumaker

The existence of interpolating periodic splines in the case where the number of knots is odd follows immediately from certain strong zero properties for splines which I recently obtained [1]. As the result holds also for Hermite interpolation, perhaps it is best to begin with notation. Let $[a,b]$ be an interval, and identify b with a (so we are thinking of it as a circle). Let $\Delta = \{x_1<x_2<..<x_k\}$ be a partition of $[a,b]$ into subintervals $I_i = [x_i,x_{i+1})$, $i = 1,2,..,k-1$ and $I_k = [x_k,x_1)$. Given a vector $M = (m_1,...,m_k)$ of integers with $1 \le m_i \le m$, we define the class of <u>periodic polynomial splines with knots at</u> $x_1,...,x_k$ <u>of multiplicities</u> $m_1,...,m_k$ <u>by</u>

$$\mathring{\mathcal{S}}(\mathcal{P}_m;M;\Delta) = \{s : s_i = s|_{I_i} \in \mathcal{P}_m,\ i = 1,2,...,k \text{ and}$$
$$s^{(j)}_{i-1}(x_i) = s^{(j)}_i(x_i),\ j = 0,1,...,m-1-m_i,$$
$$i = 1,2,...,k\}.$$

It is shown in [1] that $\mathring{\mathcal{S}}$ is a linear space of dimension $K = \sum_i^k m_i$. Moreover, Theorem 3.1 of [1] asserts that for all nontrivial s ,

$$Z(s) \le K-1 \qquad \text{if } K \text{ is odd},$$

where Z counts the number of zeros of s with multiplicities (for the exact definition, see [1]). As the dimension of $\mathring{\mathcal{S}}$ is K, we would expect to be able to interpolate K pieces of data. We have

<u>THEOREM</u>. <u>Suppose</u> $\{y_i^j\}_{j=0}^{m_i-1}{}_{,i=1}^{k}$ <u>are given real numbers</u>. <u>Then if</u> K <u>is odd</u>, <u>there exists a unique</u> $s \in \mathring{\mathcal{S}}$ <u>such that</u>

$$s^{(j)}(x_i) = y_i^j \quad , \quad j = 0,1,\ldots,m_i-1 \ , \ i = 1,\ldots,k.$$

<u>Proof:</u> If $\{B_i\}_1^K$ is some basis for $\mathring{\mathcal{S}}$, then writing $s = \sum_i^K c_i B_i$, the interpolation conditions lead immediately to K equations for the determination of the coefficients. This system of equations will have a unique solution if and only if the corresponding homogeneous system has a unique solution; i.e. if the corresponding matrix is nonsingular. Now, if the matrix were singular, then there would exist a nontrivial $s_0 \in \mathring{\mathcal{S}}$ with

$$s_0^{(j)}(x_i) = 0 \quad , \quad j = 0,1,\ldots,m_i-1 \ , \ i = 1,2,\ldots,k.$$

But then s_0 would have K zeros, counting multiplicities, a contradiction of the theorem mentioned above.

This result also holds for a certain class of generalized Tchebycheffian splines (see [2]).

1. Schumaker, L.L., Zeros of spline functions and applications, J. Approx.Th., to appear.

2. Schumaker, L.L., On Tchebycheffian spline functions, J. Approx. Th., to appear.

ON AN INEQUALITY OF LORENTZ IN THE THEORY OF BERNSTEIN POLYNOMIALS

by

F. Schurer and F.W. Steutel

§ 1. *Introduction and summary*

For a function f defined on [0,1], the Bernstein polynomial of order n is defined by

$$B_n(f;x) = \sum_{k=0}^{n} f(\tfrac{k}{n})p_{nk}(x),$$

where

$$p_{nk}(x) = \binom{n}{k}x^k(1-x)^{n-k} \quad , \qquad (n = 1,2,\ldots;\ k = 0,1,\ldots,n).$$

In [2] and [3] Sikkema studies the degree of approximation of functions in C[0,1] by Bernstein polynomials. This note deals with similar problems for functions in $C^1[0,1]$. The following result in this respect is due to Lorentz and can be found in his well-known book on Bernstein polynomials.

THEOREM I (Lorentz [1], p. 21). *If* $f \in C^1[0,1]$ *and if* $\omega_1(.)$ *is the modulus of continuity of* f', *then for* n = 1,2,... *one has*

$$\|B_n(f) - f\| \le C \frac{1}{\sqrt{n}}\, \omega_1(\frac{1}{\sqrt{n}}) \tag{1}$$

with $C = \frac{3}{4}$. *Here* $\|\cdot\|$ *denotes the Chebyshev norm.*

The object of this communication can be stated as follows: what is the best constant in the right-hand side of (1)? To be more precise, for each fixed n let c_n be the infimum of the set of all numbers C_n for which the inequality

$$\|B_n(f) - f\| \le C_n \frac{1}{\sqrt{n}}\, \omega_1\ (\frac{1}{\sqrt{n}})$$

holds for all*) $f \in C^1[0,1]$. Obviously we have for n = 1,2,...

$$c_n = \sup_{f\in C^1[0,1]} \frac{\sqrt{n}\,\|B_n(f) - f\|}{\omega_1(\frac{1}{\sqrt{n}})}\ . \tag{2}$$

---

*) Clearly, linear functions are of no interest in this respect, as for these both sides of (1) vanish. Therefore we shall disregard them in the sequel of this communication.

It is our aim to determine

(3) $$c = \sup_n c_n .$$

Section 2 of this note contains some preliminary results that will be needed in the sequel. In section 3 we prove that $c < \frac{1}{2}$, and in the last section we give a simple proof of our main result $c = \frac{1}{4}$.

A detailed exposition of the stated results will be published elsewhere. Moreover, we shall deal there with $\sup_{n\geq 2} c_n$ and with the asymptotic behaviour of c_n as $n \to \infty$.

§ 2. *Preliminary results*

In the course of section 3 we shall need the following two lemmas.

LEMMA 1 (Lorentz [1], p. 14). *Defining*

$$T_{ns}(x) = \sum_{k=0}^{n} (k - nx)^s p_{nk}(x) , \qquad (n = 1,2,\ldots;\ s = 0,1,\ldots)$$

one has

(4) $$T_{n0}(x) = 1 \quad , \quad T_{n1}(x) = 0 ,$$

(5) $$T_{n2}(x) = nx(1-x), \quad T_{n4}(x) = 3n^2x^2(1-x)^2 + nx(1-x) - 6nx^2(1-x)^2 .$$

LEMMA 2. *Let*

$$S_n(x) = \sqrt{n} \sum_{k=0}^{n} \left|\frac{k}{n} - x\right| p_{nk}(x) , \qquad (n = 1,2,\ldots) .$$

The S_n *have the following properties:*

(i) $$S_n(x) = S_n(1-x) , \qquad (x \in [0,1])$$

(ii) $$S_n(x) = \frac{2}{\sqrt{n}} (n-\ell)\binom{n}{\ell} x^{\ell+1}(1-x)^{n-\ell} , \qquad \left(0 \leq \frac{\ell}{n} \leq x < \frac{\ell+1}{n} \leq \frac{1}{2}\right)$$

(iii) $$\frac{1}{2} = \|S_1\| > \|S_3\| > \|S_5\| > \ldots ,$$

$$\frac{8}{27}\sqrt{2} = \|S_2\| > \|S_4\| > \|S_6\| > \ldots .$$

As the proof of lemma 2 is tedious, though elementary, it is omitted here.

LEMMA 3. *If* c_n *is defined as in* (2), *then*

$$c_1 = \frac{1}{4} .$$

Proof. Using the mean value theorem we have

$$|B_1(f;x) - f(x)| = |x(1-x)f'(\xi_0) - x(1-x)f'(\xi_1)| = x(1-x)|f'(\xi_0) - f'(\xi_1)| \le \frac{1}{4}\,\omega_1(1).$$

Taking $f(x) = \frac{1}{2}|x - \frac{1}{2}|$, $0 \le x \le 1$, it follows that $c_1 = \frac{1}{4}$. The fact that f is not differentiable at $x = \frac{1}{2}$ does not, of course, affect the argument. In fact, for our purposes, it is allowed to replace the space $C^1[0,1]$ by the set of continuous functions that have a bounded derivative that is continuous in [0,1] with the possible exception of finitely many points. This set of functions will be denoted by K.

DEFINITION.

$$K_\delta = \{f \mid f \in K,\ \omega_1(\delta) \le 1\}, \ \delta > 0 .$$

The next lemma is fundamental in determining $c = \sup_n c_n$.

LEMMA 4. *Let* n *be a positive integer and let* $x_0 \in [0,1]$ *be arbitrary and fixed. Then*

$$\sup_{f \in K_\delta} |B_n(f;x_0) - f(x_0)| = B_n(\tilde{f};x_0) - \tilde{f}(x_0) ,$$

where $\tilde{f}$ *is determined by*

$$\tilde{f}'(x) = j + \frac{1}{2} , \qquad (j\delta < x - x_0 \le (j+1)\delta;\ j = 0, \pm 1, \pm 2, \ldots), \tag{6}$$

up to an additive constant that does not affect the value of $B_n(\tilde{f};x_0) - \tilde{f}(x_0)$.

Proof. In what follows we only give a rough sketch of the proof. It can be shown that it is sufficient to consider only those functions in K_δ that are convex. Furthermore it can be assumed without loss of generality that $f(x_0) = 0$ and $f'(x_0 + 0) = \frac{1}{2}$. Given a convex function $f \in K_\delta$ we associate with it a function f^* that is completely determined by the following conditions:

$$\begin{cases} f^*(x_0 + j\delta) = f(x_0 + j\delta) , & (j = 0, \pm 1, \pm 2, \ldots;\ 0 \le x_0 + j\delta \le 1) \\ f^*(0) = f(0) ,\ f^*(1) = f(1) , & \\ f^* \text{ is linear in between} . & \end{cases}$$

As a consequence of this construction f^* is a convex function in K_δ with a derivative that is a step function with steps ≤ 1.
We now introduce a class of piece-wise linear functions K_δ^* defined as follows:

$$(7) \qquad K_\delta^* = \{f \mid f \in K_\delta,\ f \text{ convex},\ f \equiv f^*,\ f(x_0) = 0,\ f'(x) = \tfrac{1}{2} \text{ if } x_0 < x < x_0 + \delta\} .$$

Using the fact that B_n is a positive linear operator it can be shown that

$$\sup_{f \in K_\delta} |B_n(f;x_0) - f(x_0)| = \max_{f \in K_\delta^*} (B_n(f;x_0) - f(x_0)) .$$

It now follows from the representation

$$B_n(f;x_0) - f(x_0) = \sum_{k=0}^{n} p_{nk}(x_0) \int_{x_0}^{\frac{k}{n}} f'(t)dt$$

that the maximum is attained by taking $f = \tilde{f}$, i.e. by taking the steps of f' as large as possible.

Remark. The function $\tilde{f}$ will be called extremal; it contains δ and x_0 as parameters. As for the extremal functions $\omega_1(\delta)$ is equal to 1, we shall from now on restrict ourselves to functions with $\omega_1(\delta) = 1$.

§ 3. *An upperbound for c*

The object of this section is to prove the following theorem.

THEOREM 1. *If* c *is defined as in* (3), *then*

$$c < \tfrac{1}{2} .$$

Proof. Let $n \geq 2$, let $x \in [0,1]$ be arbitrary but fixed, and let δ be an arbitrary positive number. In view of (4) and using a well-known property of the modulus of continuity it is easily verified that one has

$$|B_n(f;x) - f(x)| = |\sum_{k=0}^{n} \{f(\tfrac{k}{n}) - f(x)\} p_{nk}(x)| = |\sum_{k=0}^{n} p_{nk}(x) \int_x^{\frac{k}{n}} \{f'(t) - f'(x)\} dt| \le$$

$$\le \sum_{|\frac{k}{n}-x|\le\delta} |\int_x^{\frac{k}{n}} (f'(t)-f'(x))dt| p_{nk}(x) + \sum_{|\frac{k}{n}-x|>\delta} |\int_x^{\frac{k}{n}} (f'(t)-f'(x))dt| p_{nk}(x) \le$$

$$\le \omega_1(\delta) \sum_{|\frac{k}{n}-x|\le\delta} |\tfrac{k}{n}-x| p_{nk}(x) + \omega_1(\delta) \sum_{|\frac{k}{n}-x|>\delta} |\int_x^{\frac{k}{n}} (\frac{|t-x|^3}{\delta^3}+1)dt| p_{nk}(x) \le$$

$$\le \omega_1(\delta) \sum_{k=0}^{n} |\tfrac{k}{n}-x| p_{nk}(x) + \frac{\omega_1(\delta)}{\delta^3} \sum_{|\frac{k}{n}-x|>\delta} |\int_x^{\frac{k}{n}} |t-x|^3 dt| p_{nk}(x) =$$

$$= \omega_1(\delta) \left\{ \sum_{k=0}^{n} |\tfrac{k}{n}-x| p_{nk}(x) + \frac{1}{4\delta^3} \sum_{|\frac{k}{n}-x|>\delta} (\tfrac{k}{n}-x)^4 p_{nk}(x) \right\} \le$$

$$\le \omega_1(\delta) \left\{ \sum_{k=0}^{n} |\tfrac{k}{n}-x| p_{nk}(x) + \frac{1}{4\delta^3} \sum_{k=0}^{n} (\tfrac{k}{n}-x)^4 p_{nk}(x) \right\} .$$

Taking $\delta = \frac{1}{\sqrt{n}}$ and taking into account lemmas 1 and 2 we obtain

$$(8) \qquad |B_n(f;x) - f(x)| \le \frac{1}{\sqrt{n}} \omega_1(\tfrac{1}{\sqrt{n}}) \{S_n(x) + \frac{1}{4n^2} T_{n4}(x)\} .$$

The expression between brackets can be evaluated by means of the second part of (5) and lemma 2, (ii), (iii). Using these results, by straightforward calculation one has

$$\|S_n\| \le \|S_3\| = \tfrac{1}{4}\sqrt{3}, \qquad (n = 2,3,\ldots)$$

$$\frac{1}{4n^2}\|T_{n4}\| \le \frac{3}{64}, \qquad (n = 1,2,\ldots) .$$

In view of (8) and lemma 3 it follows that $c < \frac{1}{2}$. This proves theorem 1.

§ 4. *A simple proof of* $c = \frac{1}{4}$

In lemma 4 of section 2 we obtained the extremal function $\tilde{f}$, δ being an arbitrary positive number. As we wish to sharpen Lorentz' theorem I, we shall from now on assume that $\delta = \frac{1}{\sqrt{n}}$ and we write $\tilde{f}_n$ instead of $\tilde{f}$. We recall (cf. lemma 4) that, assuming n and x_0 being fixed, the extremal function $\tilde{f}_n$ is uniquely determined up to an additive constant. In view of this we may take $\tilde{f}_n \in K^*_\delta$ (cf. (7)), and one has

$$\tilde{f}_n(x) = \frac{1}{2}|x-x_0| + \sum_{j=1}^{\infty} \left(|x-x_0| - \frac{j}{\sqrt{n}}\right)_+ , \tag{9}$$

where, as usual, $x_+ = \max(0,x)$.

Using the functions $\tilde{f}_n$ we shall prove in an elementary way that $c_n \le \frac{1}{4}$ for all $n = 1,2,\ldots$. To this end we introduce, n and $x_0 \in [0,1]$ being fixed, a quadratic function q_n defined by

$$q_n(x) = \frac{1}{8\sqrt{n}} + \frac{1}{2}\sqrt{n}\,(x-x_0)^2 . \tag{10}$$

The properties of the function q_n are given in the following lemma.

LEMMA 5. *Let* q_n *be defined by* (10) *and let* $\tilde{f}_n$ *be the extremal function defined by* (9), *then we have*

(i) $$q_n\left(x_0 + \frac{2k+1}{2\sqrt{n}}\right) = \tilde{f}_n\left(x_0 + \frac{2k+1}{2\sqrt{n}}\right) = \frac{2k^2+2k+1}{4\sqrt{n}}, \qquad (k = 0, \pm 1, \pm 2, \ldots)$$

(ii) $$q'_n\left(x_0 + \frac{2k+1}{2\sqrt{n}}\right) = \tilde{f}'_n\left(x_0 + \frac{2k+1}{2\sqrt{n}}\right) = k + \frac{1}{2}, \qquad (k = 0, \pm 1, \pm 2, \ldots)$$

(iii) $$q_n(x) \ge \tilde{f}_n(x), \qquad (x \in [0,1])$$

(iv) $$\sqrt{n}\, B_n(q_n;x_0) = \frac{1}{8} + \frac{x_0(1-x_0)}{2} .$$

Proof. In view of (6) it follows by integration from x_0 to $x_0 + \frac{2k+1}{2\sqrt{n}}$ that for $k \ge 0$ we have

$$\tilde{f}_n\left(x_0 + \frac{2k+1}{2\sqrt{n}}\right) = \frac{1}{\sqrt{n}}\left(\frac{1}{2} + \frac{3}{2} + \ldots + \frac{2k-1}{2} + \frac{2k+1}{4}\right) = \frac{2k^2+2k+1}{4\sqrt{n}} = q_n\left(x_0 + \frac{2k+1}{2\sqrt{n}}\right).$$

By symmetry we obtain (i) for $k < 0$. From (9) and (10) we immediately have (ii). Taking into account that $q_n(x_0) > \tilde{f}_n(x_0)$ and the fact that q_n is convex, property (iii) now follows from (i) and (ii). Finally, (iv) is an easy consequence of the first part of (5). This completely proves the lemma.

We are now in a position to prove the main result of this communication.

THEOREM 2. *For* n = 1,2,... *we have*

$$c = \sup_{n} \sup_{f \in C^1[0,1]} \frac{\sqrt{n}\,\| B_n(f) - f \|}{\omega_1(\frac{1}{\sqrt{n}})} = \frac{1}{4}.$$

Proof. Noting that B_n is a positive linear operator, it follows from properties (iii) and (iv) of lemma 5 that for all $x_0 \in [0,1]$ one has

$$\sqrt{n}(B_n(\tilde{f}_n;x_0) - \tilde{f}_n(x_0)) = \sqrt{n}\, B_n(\tilde{f}_n;x_0) \le \sqrt{n}\, B_n(q_n;x_0) = \frac{1}{8} + \frac{1}{2} x_0(1-x_0) \le \frac{1}{4}.$$

Hence, by lemma 4, $c_n \le \frac{1}{4}$ for n = 1,2,... . Taking into account lemma 3 and observing definition (3) of c we obtain $c = \frac{1}{4}$.

Acknowledgement. The authors are greatly indebted to Prof. Dr. P.C. Sikkema, Technological University Delft, who suggested the problem, and who as early as 1960 conjectured that $c = \frac{1}{4}$ (unpublished note). In that year he communicated his conjecture at a meeting of the GAMM-Tagung in Hannover.

References .

[1] Lorentz, G.G., Bernstein polynomials. University of Toronto Press, Toronto, 1953.

[2] Sikkema, P.C., Über den Grad der Approximation mit Bernstein-Polynomen. Num. Math. 1 (1959), 221-239.

[3] Sikkema, P.C., Der Wert einiger Konstanten in der Theorie der Approximation mit Bernstein-Polynomen. Num. Math. 3 (1961), 107-116.

Department of Mathematics,
Technological University Eindhoven,
Eindhoven, The Netherlands.

ALTERNANTEN BEI GLEICHMÄSSIGER APPROXIMATION MIT ZWEIDIMENSIONALEN SPLINEFUNKTIONEN

MANFRED SOMMER

In this paper the problem of approximating on special subspaces of $\mathbb{R}^2$ a given continous real function f in the uniform norm by spline functions with fixed knots is considered. The spline functions are tensor products of B-splines. Using alternation lattices one gets sufficient conditions for the existence and uniqueness of a minimal solution. On halfdiscrete subspaces of $\mathbb{R}^2$ also necessary conditions are given.

0. Einleitung

Die gleichmäßige Approximation von stetigen Funktionen durch Splines mit festen Knoten auf endlichen reellen Intervallen wurde von RICE [12] und SCHUMAKER [13] behandelt. Sie beweisen, daß Minimallösungen durch Alternantenbedingungen charakterisiert werden können.

In dieser Arbeit approximieren wir stetige reelle Funktionen auf Rechtecken im $\mathbb{R}^2$ gleichmäßig durch zweidimensionale Splinefunktionen, die als Tensorprodukte von B-Splines gebildet werden. Wir erhalten in Kapitel 1 hinreichende Bedingungen für Existenz und Eindeutigkeit von Minimallösungen. Dabei verwenden wir die von EHLICH, HAUSSMANN [5] eingeführten Alternantengitter. Diese beiden Autoren haben die Approximation durch Mengen, deren Elemente sich als Tensorprodukte von Elementen aus vollständigen Tschebyscheff-Systemen bilden lassen, untersucht. Sie beweisen, daß das Vorliegen eines

Alternantengitters für eine Fehlerfunktion f - s eine hinreichende Bedingung dafür ist, daß s die einzige Minimallösung für f ist (Satz 1.1). Wir benutzen eine Methode, die von BARRAR, LOEB [1] bei der Approximation durch Splines auf endlichen reellen Intervallen verwendet wurde, und approximieren zunächst durch von einem Parameter t > 0 abhängige Funktionen, die sich als Tensorprodukte von Elementen aus vollständigen Tschebyscheff-Systemen darstellen lassen und die für $t \to 0$ gleichmäßig gegen Elemente aus der gegebenen Klasse der zweidimensionalen Splines konvergieren. Für $t \to 0$ beweisen wir aus dem obigen Resultat von EHLICH, HAUSSMANN, daß die Existenz eines Alternantengitters eine hinreichende Bedingung für eine Minimallösung darstellt (Satz 1.2). Unter welchen Bedingungen an das Alternantengitter die Eindeutigkeit folgt, die beim Übergang für $t \to 0$ verloren geht, wird in Satz 1.3 untersucht. Schließlich wird in Satz 1.4 eine untere Abschätzung des Fehlers angegeben.

In Kapitel 2 wählen wir als spezielle Teilmengen des R^2 halbdiskrete Mengen. Für unser Approximationsproblem erhalten wir auf solchen Mengen - wie die Ergebnisse von EHLICH, HAUSSMANN [5] erwarten lassen - auch notwendige Bedingungen. Wir beweisen (wieder für $t \to 0$), daß stets eine Minimallösung existiert, deren zugehörige Fehlerfunktion entweder ein Alternantengitter oder eine Alternante auf einer achsenparallelen Geraden besitzt (Satz 2.2). Außerdem zeigen wir, daß es zu jeder Minimallösung ein minimales Alternantengitter oder eine minimale Alternante gibt (Satz 2.3). Schließlich verwenden wir ein Ergebnis von STRAUSS [16] und beweisen, daß die Eindeutigkeit einer Minimallösung eine notwendige Bedingung dafür ist, daß in Teilmengen des gegebenen halbdiskreten Bereichs spezielle Alternantenbedingungen gelten (Satz 2.4).

Die Ergebnisse dieser Arbeit lassen sich auch auf höherdimensionale Approximation verallgemeinern.

1. Hinreichende Bedingungen bei der Approximation auf Rechtecken durch zweidimensionale Splinefunktionen

Es sei $R = [\alpha,\beta]\times[\gamma,\delta]$ ein Rechteck im $\mathbb{R}^2$. Außerdem seien 2 natürliche Zahlen $m \geq 2$ und $n \geq 2$ gegeben. R sei durch das Knotengitter

$$\alpha = x_0 \leq x_1 \leq \ldots \leq x_k = \beta$$

$$\gamma = y_0 \leq y_1 \leq \ldots \leq y_l = \delta,$$

wobei höchstens $m - 1$ der x_i und höchstens $n - 1$ der y_j zusammenfallen, unterteilt. Es sei $m_0 - 1$ die Anzahl der Knoten x_i $(i \neq 0)$, die mit x_0 übereinstimmen und $m_k - 1$ die Anzahl der Knoten x_i $(i \neq k)$, die mit x_k übereinstimmen. Analog seien n_0 und n_l festgelegt.
Wir führen die Hilfsknoten

$$x_{m_0-m} < \ldots < x_{-1} < x_0, \quad x_k < x_{k+1} < \ldots < x_{k-m_k+m}$$

$$y_{n_0-n} < \ldots < y_{-1} < y_0, \quad y_l < y_{l+1} < \ldots < y_{l-n_l+n}$$

ein. Nun sei ein System von Funktionen $\{h_\nu\}_{\nu=1,\ldots,m}$ auf $(-\infty,\infty)$ mit $h_\nu \in C^{m-\nu}(-\infty,\infty)$, $\nu = 1,\ldots,m$ und $h_\nu(x) > 0$ auf $(-\infty,\infty)$ gegeben. Dann bildet das System von Funktionen

$$g_1(x) = h_1(x)$$

$$g_2(x) = h_1(x) \int_a^x h_2(\xi_1)d\xi_1$$

.

.

.

$$g_m(x) = h_1(x) \int_a^x h_2(\xi_1) \int_a^{\xi_1} h_3(\xi_2) \ldots \int_a^{\xi_{m-2}} h_m(\xi_{m-1})d\xi_{m-1} \ldots d\xi_1$$

nach KARLIN [9], S. 276 ein erweitertes vollständiges Tschebyscheff-System auf $(-\infty,\infty)$ für jedes $a \in \mathbb{R}$.

Wir definieren die Funktion

$$\phi_m(x,t) := \begin{cases} h_1(x) \int_t^x h_2(\xi_1) \ldots \int_t^{\xi_{m-2}} h_m(\xi_{m-1})d\xi_{m-1} \ldots d\xi_1 & t \leq x \\ 0 & t > x \end{cases}$$

Dann ist $\phi_m(x,t) \in C^{m-2}(-\infty,\infty)$ in x und in t. Für $a = 0$, $h_\nu(x) \equiv \nu$, $\nu = 1,\ldots,m$ erhält man die polynomialen Splines.

Mit Hilfe der Funktionen ϕ_m führen wir zu der gegebenen Knotenverteilung die B-Splines ein. Dazu benötigen wir folgende Abkürzungen (sh. KARLIN [9], S. 14):

Gegeben sei eine Funktion $F(x,y)$, definiert in $[a,b]\times[c,d]$.

Dann sei

$$F\begin{pmatrix} \xi_1 & \ldots & \xi_n \\ \eta_1 & \ldots & \eta_n \end{pmatrix} := \begin{vmatrix} F(\xi_1,\eta_1) & \ldots & F(\xi_1,\eta_n) \\ \vdots & & \vdots \\ F(\xi_n,\eta_1) & \ldots & F(\xi_n,\eta_n) \end{vmatrix}$$

für $a \leq \xi_1 < \ldots < \xi_n \leq b$, $c \leq \eta_1 < \ldots < \eta_n \leq d$.

Dies benutzen wir im folgenden auch für den Fall

$F(i,x) = F_i(x)$, $i \in \mathbb{N}$.

Besitzt F $p - 1$ partielle Ableitungen nach x und $q - 1$

partielle Ableitungen nach y, so erweitert man den Definitionsbereich der eben definierten Determinanten dadurch, daß man unter den ξ_i bzw den η_j Gleichheit, jedoch jeweils bei höchstens p bzw q aufeinanderfolgenden, zuläßt. Für jede Menge gleicher ξ_i ersetzt man die entsprechenden Zeilen in der obigen Determinanten durch die entsprechenden Ableitungen von F nach x an der Stelle ξ_i und für jede Menge gleicher η_j ersetzt man die entsprechenden Spalten in der obigen Determinanten durch die entsprechenden Ableitungen von F nach y an der Stelle η_j. Für Funktionen F(i,x) lassen wir natürlich nur Gleichheit unter den Werten zu, die für x eingesetzt werden. Man bezeichnet diese so entstehende Determinante mit

$$F^{*}\begin{pmatrix}\xi_1 & \cdots & \xi_n\\ \eta_1 & \cdots & \eta_n\end{pmatrix}.$$

1.1 <u>DEFINITION</u>: Die Funktion

$$B_{m,\mu}(\xi) := \frac{\Phi_m^{*}\begin{pmatrix}x_\mu & \cdots & & x_{\mu+m}\\ a & \cdots & a & \xi\end{pmatrix}}{\Phi_{m+1}^{*}\begin{pmatrix}x_\mu & \cdots & x_{\mu+m}\\ a & \cdots & a\end{pmatrix}} \qquad (a < x_\mu \text{ beliebig})$$

heißt <u>B-Spline</u> der Ordnung m mit den Knoten $x_\mu,\ldots,x_{\mu+m}$.

Von den Eigenschaften der B-Splines, die KARLIN [9], S.521 ff. bewies, benötigen wir den folgenden Hilfssatz für Interpolationsaussagen.

1.1 <u>HILFSSATZ</u>: Für alle reellen Zahlen ξ_j mit $a<\xi_1<\ldots<\xi_p$ gilt:

$$\tilde{B}_m\begin{pmatrix}\mu & \cdots & \mu+p-1\\ \xi_1 & \cdots & \xi_p\end{pmatrix} =$$

$$\Phi_m^*\begin{pmatrix} x_\mu & \cdot & \cdot & \cdot & x_{\mu+p+m-1} \\ a & \ldots & a & \xi_1 & \ldots & \xi_p \end{pmatrix} \cdot \prod_{j=1}^{p-1} \Phi_m^*\begin{pmatrix} x_{\mu+j} & \ldots & x_{\mu+j+m-1} \\ a & \ldots & a \end{pmatrix},$$

wobei $\hat{B}_{m,\mu}(\xi) := \Phi_m^*\begin{pmatrix} x_\mu & \cdot & \cdot & \cdot & x_{\mu+m} \\ a & \ldots & a & \xi \end{pmatrix}$ ist.

Da wir für Eindeutigkeitsaussagen auch zusammenfallende Punkte ξ_j benötigen, bewiesen wir in [15] die Aussage dieses Satzes auch für alle reellen Zahlen $a<\xi_1 \leq \ldots \leq \xi_p$, wobei jedes ξ_j höchstens $(m - \max_{\nu=\mu,\ldots,\mu+p-1} m_\nu)$ - mal auftritt.

Dabei ist m_ν die größte Zahl gleicher x_ρ in $[x_\nu, x_{\nu+m}]$. Die Zahlen m_ν wurden so gewählt, daß alle Ableitungen von $B_{m,\nu}$, die auftreten können, stetig sind.

Dann folgt unter Verwendung eines Satzes von KARLIN [9], S.503:

1.1 <u>KOROLLAR</u>: Es gilt für $a < \xi_1 \leq \cdot \cdot \cdot \leq \xi_p$, $v(\xi_j) \leq m - \max_{\nu=\mu,\ldots,\mu+p-1} m_\nu$, wobei $v(\xi_j)$ die Vielfachheit von ξ_j ist:

$$B_m^*\begin{pmatrix} \mu & \ldots & \mu+p-1 \\ \xi_1 & \ldots & \xi_p \end{pmatrix} > 0 \iff x_{\mu+j-1} < \xi_j < x_{\mu+m+j-1}, \; j = 1,\ldots,p$$

Wir untersuchen im folgenden die Approximation einer stetigen reellen Funktion f auf R durch Elemente aus

$$V_{k,l}^{m,n} := \{s \mid s(x,y) = \sum_{\mu=m_o-m}^{k-m_k} \sum_{\lambda=n_o-n}^{l-n_l} \alpha_{\mu\lambda} B_{m,\mu}(x) B_{n,\lambda}(y), \; \alpha_{k-m_k, l-n_l} = 0\}$$

bzgl der Supremumsnorm.

Die Funktionen $\{B_{m,m_o-m}, \ldots, B_{m,k-m_k}\}$ bilden kein Tschebyscheff-System; aber jeder B-Spline $B_{m,\mu}$ läßt sich als Grenzwert einer Familie von einem Parameter $t > 0$ abhängiger Funktionen

$$u_{m,\mu}(x,t) := \frac{1}{\sqrt{2\pi}\,t} \int_{-\infty}^{\infty} e^{-\frac{(x-\xi)^2}{2t^2}} B_{m,\mu}(\xi)\, d\xi$$

darstellen. Es gilt nach KARLIN [9], S.528:

1.2 HILFSSATZ:

(i) $\| u_{m,\mu}(\ ,t) - B_{m,\mu} \|_{[-\infty,\infty]} = o(1)$ für $t \to 0$

(ii) $\mathfrak{U}_t = \{u_{m,m_o-m}(\ ,t), \ldots, u_{m,k-m_k}(\ ,t)\}$ ist ein vollständiges Tschebyscheff-System

Wir untersuchen nun zunächst die gleichmäßige Approximation von stetigen Funktionen auf R durch Elemente aus

$$V^{m,n}_{k,l;t} := \{s \mid s(x,y;t) = \sum_{\mu=m_o-m}^{k-m_k} \sum_{\lambda=n_o-n}^{l-n_l} \alpha_{\mu\lambda} u_{m,\mu}(x,t) u_{n,\lambda}(y,t),\ \alpha_{k-m_k,l-n_l} = 0\}$$

und prüfen, welche Aussagen beim Übergang zu $V^{m,n}_{k,l}$ für $t \to 0$ erhalten bleiben. Dabei ist offensichtlich, daß für jedes $s(\ ,\ ;t) \in V^{m,n}_{k,l;t}$ $\lim_{t\to 0} s(\ ,\ ;t) = s$ für $s \in V^{m,n}_{k,l}$ gilt. Der Übergang $t \to 0$ entspricht einer Idee von BARRAR, LOEB [1], die dies bei der Approximation von stetigen Funktionen durch Splines auf endlichen reellen Intervallen verwendet haben. EHLICH, HAUSSMANN [5] haben die Approximationsaufgabe bzgl solcher Mengen, deren Elemente als Tensorprodukte von Elementen aus vollständigen Tschebyscheff-Systemen gebildet werden, behandelt. Um hinreichende Bedingungen zu erhalten, führen sie Alternantengitter folgender Gestalt ein:

1.2 DEFINITION: Ein Punktegitter W mit (u+1)(v+1) Punkten in R mit der Gestalt

(1.1) $W = \bigcup_{\nu=0}^{v} \{\xi_{0\nu},\dots,\xi_{u\nu}\} \times \{\eta_\nu\}$

$(\eta_\mu < \eta_\nu$ für $\mu < \nu$, $\xi_{\rho\nu} < \xi_{\sigma\nu}$ für $\rho < \sigma)$

bzw

(1.2) $W = \bigcup_{\mu=0}^{u} \{\xi_\mu\} \times \{\eta_{\mu 0},\dots,\eta_{\mu v}\}$

$(\xi_\mu < \xi_\nu$ für $\mu < \nu$, $\eta_{\mu\rho} < \eta_{\mu\sigma}$ für $\rho < \sigma)$

heißt <u>Alternantengitter</u> für f bzgl R, wenn

$\varepsilon(-1)^{\mu+\nu} f(\xi_{\mu\nu},\eta_\nu) = ||f||_R$, $\mu = 0,\dots,u$, $\nu = 0,\dots,v$, $\varepsilon = \pm 1$

bzw

$\bar{\varepsilon}(-1)^{\mu+\nu} f(\xi_\mu,\eta_{\mu\nu}) = ||f||_R$, $\mu = 0,\dots,u$, $\nu = 0,\dots,v$, $\bar{\varepsilon} = \pm 1$

gilt.

Aus einem Satz von EHLICH, HAUSSMANN [5], der sich auf $V^{m,n}_{k,l;t}$ anwenden läßt, erhalten wir folgende hinreichende Bedingung für eine Minimallösung.

Wir setzen zuvor fest: $\tilde{m} = m - m_0 + k - m_k$

$\tilde{n} = n - n_o + l - n_l$

1.1 <u>SATZ</u>: Sei $s_0 \in V^{m,n}_{k,l;t}$. Besitzt $f - s_0$ ein Alternantengitter mit $(\tilde{m}+1)(\tilde{n}+1)$ Punkten in R, so ist s_0 eindeutige Minimallösung für f bzgl $V^{m,n}_{k,l;t}$ auf R.

Zum Beweis des Satzes benötigen EHLICH, HAUSSMANN folgenden Hilfssatz:

1.3 <u>HILFSSATZ</u>: Es sei W ein Punktegitter der Gestalt (1.1) bzw (1.2) mit $(\tilde{m}+1)(\tilde{n}+1)$ Punkten in R. Gilt dann für ein $s \in V^{m,n}_{k,l;t}$

$\varepsilon(-1)^{\mu+\nu} s(\xi_{\mu\nu},\eta_\nu) \geq 0$, $\mu = 0,\dots,\tilde{m}$, $\nu = 0,\dots,\tilde{n}$, $\varepsilon = \pm 1$

bzw

$\bar{\varepsilon}(-1)^{\mu+\nu}s(\xi_{\mu},\eta_{\mu\nu}) \geq 0,\ \mu = 0,\dots,\tilde{m},\ \nu = 0,\dots,\tilde{n},\ \bar{\varepsilon} = \pm 1$,

so verschwindet s auf R.

Aus diesem Hilfssatz folgt (sh. [15]):

1.4 HILFSSATZ: Es sei W ein Punktegitter wie im vorigen Hilfssatz. Dann gibt es kein $s \in V_{k,l}^{m,n}$ mit

$\varepsilon(-1)^{\mu+\nu}s(\xi_{\mu\nu},\eta_{\nu}) > 0,\ \mu = 0,\dots,\tilde{m},\ \nu = 0,\dots,\tilde{n},\ \varepsilon = \pm 1$

bzw

$\bar{\varepsilon}(-1)^{\mu+\nu}s(\xi_{\mu},\eta_{\mu\nu}) > 0,\ \mu = 0,\dots,\tilde{m},\ \nu = 0,\dots,\tilde{n},\ \bar{\varepsilon} = \pm 1.$

Wir erhalten aus diesem Hilfssatz eine hinreichende Bedingung für eine Minimallösung aus $V_{k,l}^{m,n}$.

1.2 SATZ: Sei $s \in V_{k,l}^{m,n}$. Es gebe 2 Zahlen $u \in \{0,\dots,k\}$, $v \in \{0,\dots,l\}$ mit $x_u < x_k$, $y_v < y_l$, so daß für f - s in $G = [x_u,x_k] \times [y_v,y_l]$ ein Alternantengitter

(1.3) $$W = \bigcup_{\nu=0}^{\tilde{n}+n_o-n_v-v} \{\xi_{o\nu},\dots,\xi_{\tilde{m}+m_o-m_u-u,\nu}\} \times \{\eta_{\nu}\}$$

(m_u - 1 sei die Anzahl der Knoten x_i ($i \neq u$), die in $[x_u,x_k]$ mit x_u übereinstimmen; n_v analog)

bzgl R existiert. Dann ist s eine Minimallösung für f bzgl $V_{k,l}^{m,n}$ auf R.

BEWEIS: Annahme: Es existiert ein $\tilde{s} \in V_{k,l}^{m,n}$ mit

$$||f-\tilde{s}||_R < ||f-s||_R$$

Wegen $\varepsilon(-1)^{\mu+\nu}(f-s)(\xi_{\mu\nu},\eta_{\nu}) = ||f-s||_R$

$> \varepsilon(-1)^{\mu+\nu}(f-\tilde{s})(\xi_{\mu\nu},\eta_{\nu})$

für $\varepsilon = \pm 1,\ \mu = 0,\dots,\tilde{m}+m_o-m_u-u,\ \nu = 0,\dots,\tilde{n}+n_o-n_v-v$

gilt $\varepsilon(-1)^{\mu+\nu}(\tilde{s}-s)(\xi_{\mu\nu},\eta_{\nu}) > 0$. Die Funktion $\tilde{s}$ - s läßt sich auf G folgendermaßen darstellen:

$$\text{rest}_G(\tilde{s}-s)(x,y) = \sum_{\mu=u+m_u-m}^{k-m_k} \sum_{\lambda=v+n_v-n}^{l-n_l} \alpha_{\mu\lambda} B_{m,\mu}(x) B_{n,\lambda}(y), \quad \alpha_{k-m_k,l-n_l} = 0$$

Die Indizierung ergibt sich wegen $B_{m,\mu}(x) \neq 0$ für $x \notin (x_\mu, x_{\mu+m})$. Die Funktion $\text{rest}_G(\tilde{s}-s)$ besitzt auf G ein Punktegitter mit $(\tilde{m}+m_0-m_u-u+1)(\tilde{n}+n_0-n_v-v+1)$ Punkten, für das $\varepsilon(-1)^{\mu+\nu}(\tilde{s}-s)(\xi_{\mu\nu}, \eta_\nu) > 0$ gilt. Da $\text{rest}_G(\tilde{s}-s)$ eine Funktion aus

$$\tilde{V} := \{s \mid s(x,y) = \sum_{\mu=u+m_u-m}^{k-m_k} \sum_{\lambda=v+n_v-n}^{l-n_l} \alpha_{\mu\lambda} B_{m,\mu}(x) B_{n,\lambda}(y), \quad \alpha_{k-m_k,l-n_l} = 0\}$$

ist, erhält man nach Hilfssatz 1.4 einen Widerspruch. Deshalb ist s eine Minimallösung für f bzgl $V_{k,l}^{m,n}$ auf R.

Im weiteren Verlauf arbeiten wir nur mit Alternantengittern der Gestalt (1.1). Alle Sätze gelten auch für Alternantengitter der Gestalt (1.2).

Wir wenden uns nun der Frage zu, wann es zu f nur eine Minimallösung gibt. Dazu beweisen wir unter schärferen Voraussetzungen die Aussage von Hilfssatz 1.3. Bis zum Ende dieses Kapitels fordern wir $m \geq 3$, $n \geq 3$. Außerdem seien nun von den Knoten x_i jeweils höchstens $m-2$ und von den Knoten y_j jeweils höchstens $n-2$ gleich. Dann gilt:

1.5 <u>HILFSSATZ</u>: Es sei in R ein Punktegitter W der Gestalt (1.1) mit $(\tilde{m}+1)(\tilde{n}+1)$ Punkten gegeben. Außerdem gelte

(1.4) $\quad x_{m_0-m+\mu+1} < \xi_{\mu\nu} < x_{m_0+\mu-1}\,, \quad \mu = 1,\ldots,\tilde{m}-1,\ \nu = 0,\ldots,\tilde{n}$

und

(1.5) $y_{n_o-n+\nu} < \eta_\nu < y_{n_o+\nu-1}$, $\nu = 1,\ldots,\tilde{n}$.

Dann gibt es kein $s \in V_{k,l}^{m,n}$ mit $s \not\equiv 0$ in R und

$$\varepsilon(-1)^{\mu+\nu} s(\xi_{\mu\nu},\eta_\nu) \geq 0 \text{ , } \mu = 0,\ldots,\tilde{m}, \ \nu = 0,\ldots,\tilde{n}, \ \varepsilon = \pm 1.$$

BEWEIS: Gegeben sei ein $s \in V_{k,l}^{m,n}$ mit $\varepsilon(-1)^{\mu+\nu}s(\xi_{\mu\nu},\eta_\nu) \geq 0$, $\mu = 0,\ldots,\tilde{m}$, $\nu = 0,\ldots,\tilde{n}$, $\varepsilon = \pm 1$. Nun sei $\eta := \eta_{\nu_o}$. Dann gilt für die Punkte $\xi_\mu := \xi_{\mu\nu_o}$, $\mu = 0,\ldots,\tilde{m}$:

$$\varepsilon(-1)^{\mu+\nu_o} s(\xi_\mu,\eta) \geq 0$$

Deshalb existieren $\tilde{m}$ Nullstellen $z_{\mu-1} \in [\xi_{\mu-1},\xi_\mu]$ von $s(\ ,\eta)$, wobei die Nullstellen doppelt gezählt werden, an denen $s(\ ,\eta)$ das Vorzeichen nicht wechselt. Diese Nullstellen erfüllen folgende Bedingungen:

Wegen $x_{m_o-m+\mu+1} < \xi_\mu < x_{m_o+\mu-1}$ für $\mu = 1,\ldots,\tilde{m}-1$ gilt

$$x_{m_o-m+\mu} < \xi_{\mu-1} \leq z_{\mu-1} \leq \xi_\mu < x_{m_o+\mu-1} \quad \text{für} \quad \mu = 1,\ldots,\tilde{m}-1.$$

Dabei beachte man, daß $x_{m_o-m+1} < x_o \leq \xi_0 < \xi_1$ gilt.

Dann erhält man

(1.6) $x_{m_o-m+\mu+1} < z_\mu < x_{m_o+\mu}$ für $\mu = 0,\ldots,\tilde{m}-1$.

Wir werden nun zeigen, daß $s(\ ,\eta)$ keine von $z_o,\ldots,z_{\tilde{m}-1}$ verschiedenen Nullstellen in (x_{m_o-m},x_{k-m_k+m}) besitzt und daß alle z_μ, die nur einmal auftreten, einfache Nullstellen sind oder daß $s(\ ,\eta) \equiv 0$ ist. Dazu nehmen wir zu $z_o,\ldots,z_{\tilde{m}-1}$ einen Punkt $t \in (x_{m_o-m},x_{k-m_k+m})$ hinzu, der in diese Punktmenge folgendermaßen eingeordnet wird:

$$t_o := z_o \leq t_1 := z_1 \leq \ldots \leq t_{i_o} := z_{i_o} \leq t_{i_o+1} := t \leq \ldots \leq t_{\tilde{m}} := z_{\tilde{m}-1}$$

Wegen (1.6) gilt für diese Punkte

$$x_{m_o-m+\mu} < t_\mu < x_{m_o+\mu} \quad \text{für} \quad \mu = 0,\ldots,\tilde{m}.$$

Nach Korollar 1.1 gilt deshalb

$$B_m^{*}\begin{pmatrix} m_o-m & \dots & k-m_k \\ t_o & \dots & t_{\tilde{m}} \end{pmatrix} > 0 ,$$

sofern die t_μ höchstens zweimal auftreten. Da t beliebig in (x_{m_o-m}, x_{k-m_k+m}) gewählt werden kann, bedeutet dies, daß entweder $s(\ ,\eta)$ keine von $z_o,\dots,z_{\tilde{m}-1}$ verschiedenen Nullstellen in (x_{m_o-m}, x_{k-m_k+m}) besitzt und daß alle z_μ, die nur einmal auftreten, einfache Nullstellen sind oder daß $s(\ ,\eta) \equiv 0$ ist. (Hier wurde benötigt, daß für $z_\mu = z_{\mu+1}$ für ein μ stets z_μ als doppelte Nullstelle von $s(\ ,\eta)$ auftritt). Falls $s(\ ,\eta) \not\equiv 0$ ist, gilt also $s(x,\eta) \neq 0$ für alle $x \in (\xi_{\tilde{m}}, x_{k-m_k+m})$. Daraus erhält man nach dem eben über die Art der Nullstellen Gezeigten:

$$\varepsilon(-1)^{\tilde{m}+\nu_o} s(x,\eta) > 0 \quad \text{für alle } x \in (\xi_{\tilde{m}}, x_{k-m_k+m})$$

$s(\ ,\eta)$ ist eindeutig bestimmt durch die Nullstellen $z_o,\dots,z_{\tilde{m}-1}$ und durch einen Funktionswert $s(t,\eta)$ mit $t \notin \{z_o,\dots,z_{\tilde{m}-1}\}$. Deshalb gilt

$$\varepsilon(-1)^{\tilde{m}+\nu_o} s(x,\eta) = c_o B_m^{*}\begin{pmatrix} m_o-m & & k-m_k \\ z_o & \dots & z_{\tilde{m}-1} \quad x \end{pmatrix} \text{ mit } c_o > 0.$$

Dann erhält man den Koeffizienten von $B_{m,k-m_k}$ durch Entwicklung nach der letzten Spalte zu

$$c_o\, B_m^{*}\begin{pmatrix} m_o-m & \dots & k-m_k-1 \\ z_o & \dots & z_{\tilde{m}-1} \end{pmatrix}$$

Dieser ist wegen (1.6) positiv. Wegen

$$s(x,y) = \sum_{\mu=m_o-m}^{k-m_k} \sum_{\lambda=n_o-n}^{l-n_l} \alpha_{\mu\lambda} B_{n,\lambda}(y) B_{m,\mu}(x) \quad \text{ist also}$$

$\varepsilon(-1)^{\tilde{m}+\nu_0} \sum_{\lambda=n_0-n}^{l-n_1} \alpha_{k-m_k,\lambda} B_{n,\lambda}(\eta) > 0$. Für $s(\ ,\eta) \equiv 0$ erhält man wegen der linearen Unabhängigkeit von $\{B_{m,\mu}\}$

$\sum_{\lambda=n_0-n}^{l-n_1} \alpha_{k-m_k,\lambda} B_{n,\lambda}(\eta) = 0$ und damit stets

$\varepsilon(-1)^{\tilde{m}+\nu_0} \sum_{\lambda=n_0-n}^{l-n_1} \alpha_{k-m_k,\lambda} B_{n,\lambda}(\eta) \geq 0$. Also gilt für $\rho = 0,\ldots,\tilde{n}$:

$$\varepsilon(-1)^{\tilde{m}+\rho} \sum_{\lambda=n_0-n}^{l-n_1} \alpha_{k-m_k,\lambda} B_{n,\lambda}(\eta_\rho) \geq 0$$

Die Funktion $\tilde{s}(y) := \sum_{\lambda=n_0-n}^{l-n_1} \alpha_{k-m_k,\lambda} B_{n,\lambda}(y)$ besitzt deshalb mindestens $\tilde{n}$ Nullstellen $\tilde{z}_{\rho-1} \in [\eta_{\rho-1},\eta_\rho]$, wobei die Nullstellen doppelt gezählt werden, an denen $\tilde{s}$ das Vorzeichen nicht wechselt. Wegen $y_{n_0-n+\rho} < \eta_\rho < y_{n_0+\rho-1}$ für $\rho = 1,\ldots,\tilde{n}$ ist $y_{n_0-n+\rho} < \tilde{z}_\rho < y_{n_0+\rho}$ für $\rho = 0,\ldots,\tilde{n}-1$. Wegen $\alpha_{k-m_k,l-n_1} = 0$ ist $\tilde{s}$ eine Linearkombination von $B_{n,n_0-n},\ldots,B_{n,l-n_1-1}$. Nach Korollar 1.1 ist deshalb $\tilde{s} \equiv 0$. Folglich ist

$$s(x,y) = \sum_{\mu=m_0-m}^{k-m_k-1} \sum_{\lambda=n_0-n}^{l-n_1} \alpha_{\mu\lambda} B_{n,\lambda}(y) B_{m,\mu}(x).$$

$s(\ ,\eta_\rho)$ wird mit den Funktionen $B_{m,m_0-m},\ldots,B_{m,k-m_k-1}$ gebildet und besitzt für $\rho = 0,\ldots,\tilde{n}$ $\tilde{m}$ Nullstellen $z_{\mu,\rho}$, wobei die Nullstellen, an denen $s(\ ,\eta_\rho)$ das Vorzeichen nicht wechselt, doppelt gezählt werden, mit

$x_{m_0-m+\mu+1} < z_{\mu,\rho} < x_{m_0+\mu}$ für $\mu = 0,\ldots,\tilde{m}-1$. Deshalb ist $s(\ ,\eta_\rho) \equiv 0$ für $\rho = 0,\ldots,\tilde{n}$. Daraus folgt:

$\sum_{\lambda=n_0-n}^{l-n_1} \alpha_{\mu\lambda} B_{n,\lambda}(\eta_\rho) = 0$ für $\mu = m_0-m,\ldots,k-m_k-1$, $\rho = 0,\ldots,\tilde{n}$

Wegen $y_{n_0-n+\rho} < \eta_\rho < y_{n_0+\rho-1}$ für $\rho = 0,\ldots,\tilde{n}$ erhält man $\alpha_{\mu\lambda} = 0$ für $\mu = m_0-m,\ldots,k-m_k-1$ und $\lambda = n_0-n,\ldots,l-n_1$.

Also ist $s \equiv 0$ in R.

Aus Hilfssatz 1.5 folgt sofort

1.3 SATZ: Sei $s_o \in V_{k,l}^{m,n}$. Zu $f - s_o$ existiere in R ein Alternantengitter der Gestalt (1.1), dessen Punkte die Bedingungen (1.4) und (1.5) erfüllen. Dann ist s_o die einzige Minimallösung für f bzgl $V_{k,l}^{m,n}$ auf R.

Schließlich wollen wir eine Abschätzung für die Minimalabweichung $A_{k,l}^{m,n} := \min\limits_{s \in V_{k,l}^{m,n}} ||f-s||_R$ angeben.

1.4 SATZ: Sei $s_o \in V_{k,l}^{m,n}$. Zu $f - s_o$ existiere in $G = [x_u, x_k] \times [y_v, y_l]$ ein Punktegitter der Gestalt (1.3) und es gebe $(\hat{m}+1)(\hat{n}+1)$ positive reelle Zahlen $m_{\mu\nu}$ mit

$$(f-s_o)(\xi_{\mu\nu}, \eta_\nu) = \varepsilon(-1)^{\mu+\nu} m_{\mu\nu}, \quad \mu = 0,\dots,\hat{m}, \nu = 0,\dots,\hat{n}, \varepsilon = \pm 1.$$

Dabei sei $\hat{m} = m+m_o-m_u-u$, $\hat{n} = n+n_o-n_v-v$.

Dann gilt: $\min\limits_{\mu,\nu} m_{\mu\nu} \leq A_{k,l}^{m,n} \leq ||f-s_o||_R$

BEWEIS: Annahme: Es existiert ein $s \in V_{k,l}^{m,n}$ mit

$$||f-s||_R < \min_{\mu,\nu} m_{\mu\nu}$$

Dann ist für $\mu = 0,\dots,\hat{m}$, $\nu = 0,\dots,\hat{n}$ und $\varepsilon = \pm 1$:

$$\varepsilon(-1)^{\mu+\nu}(s-s_o)(\xi_{\mu\nu}, \eta_\nu) > 0$$

Wie im Beweis von Satz 1.2 gezeigt, ergibt sich daraus ein Widerspruch.

2. Notwendige Bedingungen bei der Approximation auf halbdiskreten Teilmengen von R

Wir wählen $\tilde{n} + 1$ Zahlen η_ν mit $\gamma \leq \eta_o < \dots < \eta_{\tilde{n}} \leq$

Es sei $T := \bigcup\limits_{\nu=0}^{n} [\alpha, \beta] \times \{\eta_\nu\}$.

Bei der Approximation von stetigen reellen Funktionen auf T durch Elemente aus $V_{k,l;t}^{m,n}$ können nicht nur hinreichende, sondern auch notwendige Bedingungen für die Existenz von Minimallösungen gefunden werden. Nach EHLICH, HAUSSMANN [5] gilt folgender Satz:

2.1 SATZ: Es ist $s \in V_{k,l;t}^{m,n}$ genau dann Minimallösung für $f \in C(T)$ bzgl $V_{k,l;t}^{m,n}$ auf T, wenn eine der beiden folgenden Bedingungen erfüllt ist:

(i) Es existiert ein $\eta \in \{\eta_o, \ldots, \eta_1\}$, so daß $(f-s)(\,,\eta)$ eine Alternante mit mindestens $\tilde{m} + 2$ Punkten in $[\alpha, \beta]$ besitzt

(ii) Es existiert für $f - s$ ein Alternantengitter der Gestalt (1.1) mit $(\tilde{m}+1)(\tilde{n}+1)$ Punkten in T

Beim Übergang von $V_{k,l;t}^{m,n}$ zu $V_{k,l}^{m,n}$ geht die "genau dann"- Bedingung von Satz 2.1 verloren. Es läßt sich nur die Existenz einer Minimallösung zeigen, für die eine der beiden Bedingungen (i), (ii) gilt.

2.2 SATZ: Sei $f \in C(T)$. Dann existiert eine Minimallösung $s_0 \in V_{k,l}^{m,n}$ für f auf T, so daß eine der beiden folgenden Bedingungen erfüllt ist:

(i) Es existiert ein $\eta \in \{\eta_o, \ldots, \eta_1\}$, so daß $(f-s_o)(\,,\eta)$ eine Alternante mit mindestens $\tilde{m} + 2$ Punkten in $[\alpha, \beta]$ besitzt

(ii) Es existiert für $f - s_0$ ein Alternantengitter der Gestalt (1.1) mit $(\tilde{m}+1)(\tilde{n}+1)$ Punkten in T

BEWEIS: Nach Satz 2.1 gibt es eine monoton fallende Null-folge $(t_i)_{i\in\mathbb{N}}$ mit der Eigenschaft: Entweder besitzt $f - s_{t_i}$ (s_{t_i} ist Minimallösung für f bzgl $V^{m,n}_{k,l;t_i}$ auf T) für jedes $i \in \mathbb{N}$ mindestens $\tilde{m} + 2$ alternierende Extremalpunkte in einer Teilmenge $[\alpha,\beta]\times\{\eta_{t_i}\}$ von T oder $f - s_{t_i}$ besitzt für jedes $i \in \mathbb{N}$ in T ein Alternantengitter W_{t_i} mit $(\tilde{m}+1)(\tilde{n}+1)$ Punkten der Gestalt (1.1). Wir nehmen zunächst den 2. Fall an. Die Konstanten $\varepsilon_{t_i} =: \varepsilon$ mögen stets gleiches Vorzeichen besitzen. Ansonsten gehe man zu einer Teilfolge über. Nun gilt:

Wegen $||f-s_{t_i}||_R \leq ||f-0||_R = ||f||_R$ ist $||s_{t_i}||_R \leq 2||f||_R$. Deshalb sind die Koeffizienten $(\alpha_{\mu\nu}(t_i))_{i\in\mathbb{N}}$ von

$$s_{t_i}(x,y) = \sum_{\mu=m_o-m}^{k-m_k} \sum_{\nu=n_o-n}^{l-n_l} \alpha_{\mu\nu}(t_i) u_{m,\mu}(x,t_i) u_{n,\nu}(y,t_i)$$

beschränkt. Dann existiert eine konvergente Teilfolge - die wir ebenfalls mit $(t_i)_{i\in\mathbb{N}}$ bezeichnen - mit $\lim_{i\to\infty} \alpha_{\mu\nu}(t_i) =: \alpha_{\mu\nu}$. Deshalb gilt:

$$s_o(x,y) :=$$
$$= \lim_{i\to\infty} s_{t_i}(x,y) = \sum_{\mu=m_o-m}^{k-m_k} \sum_{\nu=n_o-n}^{l-n_l} \alpha_{\mu\nu} \lim_{i\to\infty} u_{m,\mu}(x,t_i) \lim_{i\to\infty} u_{n,\nu}(y,t_i)$$
$$= \sum_{\mu=m_o-m}^{k-m_k} \sum_{\nu=n_o-n}^{l-n_l} \alpha_{\mu\nu} B_{m,\mu}(x) B_{n,\nu}(y) \in V^{m,n}_{k,l}$$

Die Konvergenz von s_{t_i} gegen s_o ist gleichmäßig. Offenbar ist s_o Minimallösung für f bzgl $V^{m,n}_{k,l}$ auf T.

Wir wollen nun zeigen, daß für $f - s_o$ Bedingung (ii) gilt. W_{t_i} besitzt auf jedem η_ν $\tilde{m} + 1$ Punkte $\xi_{\mu\nu}(t_i), \mu = 0,\ldots,\tilde{m}$. Da die Punkte $(\xi_{\mu\nu}(t_i),\eta_\nu)$ in einer kompakten Menge liegen, gibt es eine Teilfolge - die wir wiederum mit $(t_i)_{i\in\mathbb{N}}$ bezeichnen - mit $\lim_{i\to\infty} (\xi_{\mu\nu}(t_i),\eta_\nu)) =:(\xi_{\mu\nu},\eta_\nu)$. Außerdem gilt:

$$\alpha \leq \xi_{0\nu} \leq \ldots \leq \xi_{\tilde{m}\nu} \leq \beta\ ,\ \nu = 0,\ldots,\tilde{n}$$

Wir zeigen nun, daß sogar

$$\alpha \leq \xi_{0\nu} < \ldots < \xi_{\tilde{m}\nu} \leq \beta \text{ gilt: Es ist}$$

$$||f-s_0||_R = \lim_{i\to\infty} ||f-s_{t_i}||_R =$$
$$= \lim_{i\to\infty} \varepsilon\{f(\xi_{\mu\nu}(t_i),\eta_\nu) - s_{t_i}(\xi_{\mu\nu}(t_i),\eta_\nu)\}(-1)^{\mu+\nu}$$
$$= \lim_{i\to\infty} \varepsilon\{f(\xi_{\mu\nu}(t_i),\eta_\nu) - s_0(\xi_{\mu\nu}(t_i),\eta_\nu)\}(-1)^{\mu+\nu}$$
$$+ \lim_{i\to\infty} \varepsilon\{s_0(\xi_{\mu\nu}(t_i),\eta_\nu) - s_{t_i}(\xi_{\mu\nu}(t_i),\eta_\nu)\}(-1)^{\mu+\nu}$$
$$= \varepsilon(f(\xi_{\mu\nu},\eta_\nu) - s_0(\xi_{\mu\nu},\eta_\nu))(-1)^{\mu+\nu}$$

Deshalb gilt: $\varepsilon(-1)^{\mu+\nu}(f-s_0)(\xi_{\mu\nu},\eta_\nu) = ||f-s_0||_R$,

$$\mu = 0,\ldots,\tilde{m},\ \nu = 0,\ldots,\tilde{n},\ \varepsilon = \pm 1$$

Also besitzt $f - s_0$ ein Alternantengitter der gewünschten Art in T.

Falls $f - s_{t_i}$ für jedes $i \in \mathbb{N}$ mindestens $\tilde{m} + 2$ alternierende Extremalpunkte in einer Teilmenge $[\alpha,\beta]\times\{\eta_{t_i}\}$ von T besitzt, führen analoge Untersuchungen zu folgendem Ergebnis: Es existiert eine Minimallösung $s_0 \in V_{k,l}^{m,n}$ für f auf T, so daß $f - s_0$ in einer Teilmenge $[\alpha,\beta]\times\{\eta_\nu\}$ von T mindestens $\tilde{m} + 2$ alternierende Extremalpunkte besitzt.

Im folgenden seien die η_ν durch

$y_{n_0-n+\nu} < \eta_\nu < y_{n_0+\nu-1}$, $\nu = 1,\ldots,\tilde{n}$ gegeben.

Nun sei $s_0 \in V_{k,l}^{m,n}$ eine Minimallösung für $f \in C(T)$ bzgl $V_{k,l}^{m,n}$ auf T. Wir werden zunächst zeigen. daß $f - s_0$ stets ein Alternantengitter mit $m(\tilde{n}+1)$ Punkten oder $m + 1$ alternierende Extremalpunkte in T besitzt. Anschließend behandeln wir das Approximationsproblem mit Splines aus $V_{k,l}^{m,n}$ mit einfachen Knoten. Wir werden zeigen, daß aus der Eindeutigkeit einer Minimallösung s_o die Existenz spezieller Alternanten für $f - s_o$ in T folgt. STRAUSS [16] hat solche notwendigen Bedingungen für eindimensionale Spline-Approximation gefunden.

Zunächst benötigen wir zwei Hilfssätze, die sich mit der Interpolation durch Elemente aus $V_{k,l}^{m,n}$ beschäftigen.

2.1 HILFSSATZ: Sei $\nu_o \in \{0,\dots,\tilde{n}\}$ gegeben. Für jedes $\rho \neq \nu_o$ seien $\tilde{m} + 1$ verschiedene Zahlen $\{\xi_{\mu\rho}\}_{\mu=0,\dots,\tilde{m}}$ durch $x_{\mu+m_o-m} < \xi_{\mu\rho} < x_{\mu+m_o}$ für $\mu = 0,\dots,\tilde{m}$ gegeben. Außerdem seien $\tilde{m}$ verschiedene Zahlen $\{\xi_\mu\}_{\mu=0,\dots,\tilde{m}-1}$ durch $x_{\mu+1+m_o-m} < \xi_\mu < x_{\mu+m_o}$ für $\mu = 0,\dots,\tilde{m}-1$ gegeben. Dann existiert zu jedem $c \in \mathbb{R}$, $c \neq 0$, genau ein $s \in V_{k,l}^{m,n}$ mit

$$s(\xi_{\mu\rho},\eta_\rho) = 0,\ \rho \neq \nu_o,\ \mu = 0,\dots,\tilde{m}$$

$$s(\xi_\mu,\eta_{\nu_o}) = 0,\ \mu = 0,\dots,\tilde{m}-2$$

$$s(\xi_{\tilde{m}-1},\eta_{\nu_o}) = c,$$

für das folgende Bedingungen gelten:

$$s(x,\eta_\rho) \equiv 0 \quad \text{für} \quad \rho \neq \nu_o$$

$$s(x,\eta_{\nu_o}) = 0 \quad \text{für} \quad x \geq x_{\tilde{m}+m_o-1}$$

$s(\ ,\eta_{\nu_o})$ besitzt in $(x_{m_o-m}, x_{\tilde{m}+m_o-1})$ nur die Nullstellen $\xi_o,\dots,\xi_{\tilde{m}-2}$. Diese sind einfach.

BEWEIS: Wir zeigen, daß das homogene Interpolationsproblem nur die triviale Lösung $s \equiv 0$ besitzt.

Wegen $s(\xi_{\mu\rho},\eta_\rho) = 0$ für $\rho \neq \nu_o$ und $x_{\mu+m_o-m} < \xi_{\mu\rho} < x_{\mu+m_o}$ für $\mu = 0,\dots,\tilde{m}$ ist nach Korollar 1.1

$s(x,\eta_\rho) = \sum_{\mu=m_o-m}^{k-m_k} \sum_{\lambda=n_o-n}^{l-n_l} \alpha_{\mu\lambda} B_{n,\lambda}(\eta_\rho) B_{m,\mu}(x) \equiv 0$ und damit

$\sum_{\lambda=n_o-n}^{l-n_l} \alpha_{\mu\lambda} B_{n,\lambda}(\eta_\rho) = 0$ für $\mu = m_o-m,\ldots,k-m_k$. Insbesondere

ist $\sum_{\lambda=n_o-n}^{l-n_l} \alpha_{k-m_k,\lambda} B_{n,\lambda}(\eta_\rho) = 0$. Wegen $\alpha_{k-m_k,l-n_l} = 0$ und

$y_{n_o-n+\rho} < \eta_\rho < y_{n_o+\rho-1}$ für $\rho = 1,\ldots,\tilde{n}$ ist nach

Korollar 1.1 $\sum_{\lambda=n_o-n}^{l-n_l-1} \alpha_{k-m_k,\lambda} B_{n,\lambda}(y) \equiv 0$ und deshalb

$\alpha_{k-m_k,n_o-n} = \ldots = \alpha_{k-m_k,l-n_l-1} = 0$. Damit wird

$s(x,\eta_{\nu_o}) = \sum_{\mu=m_o-m}^{k-m_k-1} \sum_{\lambda=n_o-n}^{l-n_l} \alpha_{\mu\lambda} B_{n,\lambda}(\eta_{\nu_o}) B_{m,\mu}(x)$. Da wir

$s(\xi_\mu,\eta_{\nu_o}) = 0$ für $\mu = 0,\ldots,\tilde{m}-1$ wählen, ist wegen

$x_{\mu+1+m_o-m} < \xi_\mu < x_{\mu+m_o}$ für $\mu = 0,\ldots,\tilde{m}-1$ und Korollar 1.1

$s(\ ,\eta_{\nu_o}) \equiv 0$. Deshalb ist $\sum_{\lambda=n_o-n}^{l-n_l} \alpha_{\mu\lambda} B_{n,\lambda}(\eta_\rho) = 0$ für

$\rho = 0,\ldots,\tilde{n}$ und wiederum nach Korollar 1.1

$\sum_{\lambda=n_o-n}^{l-n_l} \alpha_{\mu\lambda} B_{n,\lambda}(y) \equiv 0$ für $\mu = m_o-m,\ldots,k-m_k-1$. Daraus ergibt sich $\alpha_{\mu\lambda} = 0$ für $\mu = m_o-m,\ldots,k-m_k-1$,

$$\lambda = n_o-n,\ldots,l-n_l.$$

Also ist die Interpolationsaufgabe eindeutig lösbar.

Nun sei s das Element aus $V_{k,l}^{m,n}$, für das $s(\xi_{\mu\rho},\eta_\rho) = 0$, $\rho \neq \nu_o$, $\mu = 0,\ldots,\tilde{m}$, $s(\xi_\mu,\eta_{\nu_o}) = 0$, $\mu = 0,\ldots,\tilde{m}-2$ und $s(\xi_{\tilde{m}-1},\eta_{\nu_o}) = c \neq 0$ gilt. Wie eben gezeigt, ist $s(\ ,\eta_\rho) \equiv 0$ für $\rho \neq \nu_o$. Außerdem wurde gezeigt, daß

$s(x,\eta_{\nu_o}) = \sum_{\mu=m_o-m}^{k-m_k-1} \sum_{\lambda=n_o-n}^{l-n_l} \alpha_{\mu\lambda} B_{n,\lambda}(\eta_{\nu_o}) B_{m,\mu}(x)$ ist. Deshalb

ist $s(x,\eta_{\nu_o}) = 0$ für $x \geq x_{\tilde{m}+m_o-1}$.

Nun sei ξ eine weitere Nullstelle von $s(\ ,\eta_{\nu_o})$ in $(x_{m_o-m}, x_{\tilde{m}+m_o-1})$ mit $\xi_{i_o} \leq \xi \leq \xi_{i_o+1}$. Wir bilden

$$\tilde{\xi}_\mu := \begin{cases} \xi_\mu & \mu = 0,\dots,i_o \\ \xi & \mu = i_o+1 \\ \xi_{\mu-1} & \mu = i_o+2,\dots,m-1 \end{cases}$$

Dann gilt: $x_{\mu+m_o-m} < \tilde{\xi}_\mu < x_{\mu+m_o}$ für $\mu = 0,\dots,\tilde{m}-1$

Nach Korollar 1.1 folgt deshalb $s(\ ,\eta_{\nu_o}) \equiv 0$. Dies ist ein Widerspruch zu $s(\xi_{\tilde{m}-1},\eta_{\nu_o}) = c \neq 0$. Also besitzt $s(\ ,\eta_{\nu_o})$ nur die einfachen Nullstellen $\xi_o,\dots,\xi_{\tilde{m}-2}$.

2.2 <u>HILFSSATZ</u>: Es seien $\nu,\nu+1 \in \{0,\dots,\tilde{n}\}$. Für jedes $\rho \neq \nu,\nu+1$ seien $\tilde{m}+1$ verschiedene Zahlen $\{\xi_{\mu\rho}\}_{\mu=0,\dots,\tilde{m}}$ durch $x_{\mu+m_o-m} < \xi_{\mu\rho} < x_{\mu+m_o}$ für $\mu = 0,\dots,\tilde{m}$ gegeben. Außerdem seien $\tilde{m}+1$ verschiedene Zahlen $\{\xi_{\mu\nu}\}_{\mu=0,\dots,\tilde{m}}$ durch $x_{\mu+1+m_o-m} < \xi_{\mu\nu} < x_{\mu+m_o}$ für $\mu = 0,\dots,\tilde{m}$ und $\tilde{m}$ verschiedene Zahlen $\{\xi_{\mu,\nu+1}\}_{\mu=0,\dots,\tilde{m}-1}$ durch $x_{\mu+1+m_o-m} < \xi_{\mu,\nu+1} < x_{\mu+m_o}$ für $\mu = 0,\dots,\tilde{m}-1$ gegeben.
Dann existiert zu jedem $c \in \mathbb{R}$, $c \neq 0$, genau ein $s \in V_{k,l}^{m,n}$ mit

$$s(\xi_{\mu\rho},\eta_\rho) = 0,\ \mu = 0,\dots,\tilde{m},\rho \neq \nu,\nu+1$$
$$s(\xi_{\mu\nu},\eta_\nu) = 0,\ \mu = 0,\dots,\tilde{m}-1$$
$$s(\xi_{\tilde{m},\nu},\eta_\nu) = c$$
$$s(\xi_{\mu,\nu+1},\eta_{\nu+1}) = 0,\ \mu = 0,\dots,\tilde{m}-1,$$

für das folgende Bedingungen gelten:

$$s(x,\eta_\rho) \equiv 0 \quad \text{für } \rho \neq \nu,\nu+1$$

$s(\ ,\eta_\nu)$ und $s(\ ,\eta_{\nu+1})$ besitzen in $(x_{m_o-m}, x_{\tilde{m}+m_o})$ nur die Nullstellen $\xi_{o\nu},\dots,\xi_{\tilde{m}-1,\nu}$ bzw $\xi_{o,\nu+1},\dots,\xi_{\tilde{m}-1,\nu+1}$. Diese sind einfach.

Für $z \in (\xi_{\tilde{m}-1,\nu}, x_{\tilde{m}+m_o})$, $z' \in (\xi_{\tilde{m}-1,\nu+1}, x_{\tilde{m}+m_o})$ ist
$\operatorname{sgn} s(z,\eta_\nu) = \operatorname{sgn} s(z',\eta_{\nu+1})$.

Den Beweis dieses Hilfssatzes findet man in [15]. Er wird mit ähnlichen Methoden, wie sie in den Beweisen zu den Hilfssätzen 1.5 und 2.1 verwendet werden, durchgeführt.
Wir werden nun drei weitere Hilfssätze angeben, die sich durch Anwendung der Hilfssätze 2.1 und 2.2 beweisen lassen. Da die Beweise stets von ähnlicher Art sind - es wird jeweils ein $s \in V_{k,l}^{m,n}$ konstruiert, welches die Approximation von f bzgl $V_{k,l}^{m,n}$ verbessert - seien sie hier weggelassen. Man findet sie in [15]. Die Sätze dienen dazu, für eine Fehlerfunktion $f - s_o$, $s_o \in V_{k,l}^{m,n}$, die Existenz eines minimalen Alternantengitters bzw einer minimalen Alternanten in T nachzuweisen.

Dazu führen wir folgende Bezeichnung ein:
Für $\nu \in \{0,\ldots,\tilde{n}\}$ sei $||f||_\nu := \max_{x \in [x_o,x_k]} |f(x,\eta_\nu)|$

2.3 <u>HILFSSATZ</u>: Sei $f \in C(T)$ und $s_o \in V_{k,l}^{m,n}$. Es gebe Zahlen $\nu_o, \nu_1 \in \{0,\ldots,\tilde{n}\}$ mit

a) $||f-s_o||_{\nu_o} < ||f-s_o||_T$

b) $(f-s_o)(\ ,\eta_{\nu_1})$ besitze in $[x_o,x_k]$ eine Alternante mit den Punkten $\xi_1,\ldots,\xi_m$, jedoch keine mit $m + 1$ Punkten.

Dann gibt es ein $s \in V_{k,l}^{m,n}$ mit

$$||f-s_o-s||_{\nu_o} < ||f-s_o||_T$$
$$||f-s_o-s||_{\nu_1} < ||f-s_o||_T$$
$$||f-s_o-s||_T \leq ||f-s_o||_T$$

2.4 <u>HILFSSATZ</u>: Sei $f \in C(T)$ und $s_0 \in V_{k,l}^{m,n}$. Es gebe ein $\nu_0 \in \{0,\ldots,\tilde{n}\}$, so daß $(f-s_o)(\ ,\eta_{\nu_o})$ in $[x_0,x_k]$ höchstens $m-1$ alternierende Extremalpunkte besitzt. Dann gibt es ein $s \in V_{k,l}^{m,n}$ mit

$$||f-s_0-s||_{\nu_0} < ||f-s_0||_T$$

$$||f-s_0-s||_T \leq ||f-s_0||_T.$$

2.5 <u>HILFSSATZ</u>: Sei $f \in C(T)$ und $s_0 \in V_{k,l}^{m,n}$. Gibt es zwei Zahlen $\nu,\nu+1 \in \{0,\ldots,\tilde{n}\}$ und $2m$ Punkte $\{\xi_{\mu\rho}\}_{\substack{\mu=0,\ldots,m-1\\ \rho=\nu,\nu+1}}$ in $[x_0,x_k]$ mit

$\varepsilon(-1)^\mu (f-s_o)(\xi_{\mu\rho},\eta_\rho) = ||f-s_0||_T$ für $\mu = 0,\ldots,m-1$, $\rho = \nu,\nu+1$, $\varepsilon = \pm 1$, so existiert ein $s \in V_{k,l}^{m,n}$ mit

$$||f-s_0-s||_\rho < ||f-s_0||_T \quad \text{für } \rho = \nu,\nu+1 \text{ und}$$

$$||f-s_0-s||_T \leq ||f-s_0||_T.$$

Für den nächsten Satz verwenden wir folgende von EHLICH, HAUSSMANN eingeführte Bezeichnungen:

Für $f \in C(T)$ und $s_0 \in V_{k,l}^{m,n}$ sei $\Delta := f - s_0$.
Sei $\Gamma := \{0,\ldots,\tilde{n}\}$.

$\Gamma'_\Delta := \{\nu \in \Gamma \mid (f-s_o)(\ ,\eta_\nu)$ besitzt genau m alternierende Extremalpunkte in $[x_0,x_k]\}$

$\Gamma''_\Delta := \{\nu \in \Gamma \mid\ ||f-s_0||_\nu < ||f-s_0||_T\}$

$\Gamma'''_\Delta := \{\nu \in \Gamma \mid (f-s_o)(\ ,\eta_\nu)$ besitzt höchstens $m-1$ alternierende Extremalpunkte in $[x_0,x_k]\}$

2.3 <u>SATZ</u>: Sei $f \in C(T)$ und $s_0 \in V_{k,l}^{m,n}$ sei Minimallösung für f bzgl $V_{k,l}^{m,n}$ auf T. Dann gilt mindestens eine der folgenden Bedingungen:

(i) Es existiert ein $\nu_o \in \Gamma$, so daß $(f-s_o)(\ ,\eta_{\nu_o})$ mindestens $m+1$ alternierende Extremalpunkte in $[x_o, x_k]$ besitzt

(ii) $f - s_o$ besitzt in T ein Alternantengitter mit $m(\tilde{n}+1)$ Punkten

BEWEIS: Wenn (i) erfüllt ist, ist der Satz bewiesen. Wenn (i) nicht gilt, ist $\Gamma = \Gamma'_\Delta \cup \Gamma''_\Delta \cup \Gamma'''_\Delta$.

Sei $\Gamma''_\Delta \neq 0$. Dann findet man nach wiederholter Anwendung von Hilfssatz 2.3 ein $s_1 \in V^{m,n}_{k,1}$ mit $||f-s_o-s_1||_\nu < ||f-s_o||_T$ für $\nu \in \Gamma'_\Delta \cup \Gamma''_\Delta$ und $||s_1||_\nu = 0$ für $\nu \in \Gamma'''_\Delta$. Falls außerdem $\Gamma'''_\Delta \neq 0$ ist, finden wir nach wiederholter Anwendung von Hilfssatz 2.4 ein $s_2 \in V^{m,n}_{k,1}$ mit $||f-s_o-s_2||_\nu < ||f-s_o||_T$ für $\nu \in \Gamma'''_\Delta$ und $||s_2||_\nu = 0$ für $\nu \in \Gamma'_\Delta \cup \Gamma''_\Delta$. Dann ist $s_o + s_1 + s_2$ eine bessere Approximation für f bzgl $V^{m,n}_{k,1}$ auf T als s_o.

Für $\Gamma''_\Delta = 0$ und $\Gamma'''_\Delta \neq 0$ findet man ein $s_3 \in V^{m,n}_{k,1}$ mit $||f-s_o-s_3||_\nu < ||f-s_o||_T$ für $\nu \in \Gamma'''_\Delta$ und $||f-s_o-s_3||_T = ||f-s_o||_T$. Wir bilden für die Funktion $\bar{\Delta} := f - s_o - s_3$ die Punktmengen $\Gamma'_{\bar{\Delta}}$, $\Gamma''_{\bar{\Delta}}$, $\Gamma'''_{\bar{\Delta}}$. Dann ist $\Gamma''_{\bar{\Delta}} \neq 0$ und das obige Verfahren für $f - s_o$ wird auf $f - s_o - s_3$ angewendet.

Der letzte Fall ist $\Gamma'_\Delta = \Gamma$. Dann besitzt $f - s_o$ entweder ein Alternantengitter mit $m(\tilde{n}+1)$ Punkten in T oder es gibt zwei Zahlen $\nu, \nu+1 \in \Gamma$ und $2m$ Punkte $\{\xi_{\mu\rho}\}_{\substack{\mu=0,\dots,m-1\\ \rho=\nu,\nu+1}}$ mit

$\varepsilon(-1)^\mu (f-s_o)(\xi_{\mu\rho}, \eta_\rho) = ||f-s_o||_T$. Nach Hilfssatz 2.5 gibt es ein $s \in V^{m,n}_{k,l}$ mit $||f-s_o-s||_\rho < ||f-s_o||_T$ für $\rho = \nu, \nu+1$ und $||f-s_o-s||_T = ||f-s_o||_T$. Dann ist für $\overline{\Delta} := f - s_o - s$ $\Gamma^{\perp}_{\overline{\Delta}}{}' \neq 0$ und man kann wie im ersten Teil dieses Beweises ein $\tilde{s} \in V^{m,n}_{k,l}$ mit $||f-s_o-s-\tilde{s}||_T < ||f-s_o-s||_T = ||f-s_o||_T$ finden.

Im folgenden seien alle Knoten x_i bzw y_j einfach.

Wir werden zeigen, daß die Eindeutigkeit einer Minimallösung $s_o \in V^{m,n}_{k,l}$ für $f \in C(T)$ bzgl $V^{m,n}_{k,l}$ auf T die Existenz von Alternanten für $f - s_o$ auf Teilmengen von T zur Folge hat. Damit wird ein Ergebnis übertragen, das STRAUSS [16] für eindimensionale Splines gefunden hat.

Für eine Punktmenge $H = \{h_\mu\}_{\mu=1,\dots,r}$ aus $[x_o, x_k]$ sei $H_{\mu,\nu} := \{h \in H | h \in (x_\mu, x_\nu)\}$.

Unter Verwendung eines Hilfssatzes von STRAUSS [17] erhält man

2.6 <u>HILFSSATZ</u>: Gegeben sei eine Punktmenge $H = \{h_\mu\}_{\mu=1,\dots,r}$ in (x_o, x_k) mit $r \leq k - m$. Dann existiert entweder

(i) ein Intervall $[x_u, x_{u+m}] \subset [x_o, x_k]$ mit $H_{u,u+m} = \emptyset$

oder

(ii) ein Intervall $[x_u, x_{u+m+h}] \subset [x_o, x_k]$ mit

$H_{u,u+m+h} = \{\tau_\mu\}_{\mu=1,\dots,h}$ und $\tau_\mu \in (x_{u+\mu}, x_{u+\mu+m-1})$ für $\mu = 1,\dots,h$.

Nun gilt:

2.4 <u>SATZ</u>: Sei s_o die einzige Minimallösung für f bzgl $V^{m,n}_{k,l}$ auf T. Dann besitzt $f - s_o$ in jeder Teilmenge $[x_o, x_j] \times \{\eta_\nu\}$, $j = 1,\dots,k$, von T mindestens $j + 1$ und in jeder Teilmenge $[x_j, x_k] \times \{\eta_\nu\}$, $j = 1,\dots,k-2$, von T mindestens $k - j$ alternierende Extremalpunkte.

Falls $k > m$ ist, besitzt $f - s_0$ in jeder Teilmenge $[x_i, x_{i+j+m-1}] \times \{\eta_\nu\}$, $j \geq 1$, von T mindestens $j + 1$ alternierende Extremalpunkte.

BEWEIS: Zunächst sei $k > m$ und es existiere eine Teilmenge $[x_i, x_{i+j+m-1}] \times \{\eta_{\nu_o}\}$ von T, in der $f - s_0$ nur $l \leq j$ alternierende Extremalpunkte $\xi_1,\dots,\xi_l$ besitzt. Dann gibt es $l - 1$ verschiedene Punkte $t_1,\dots,t_{l-1}$ in $(x_i, x_{i+j+m-1})$ und ein $\varepsilon > 0$ mit

$\eta(-1)^\mu (f-s_0)(x,\eta_{\nu_o}) \leq \|f-s_0\|_T - \varepsilon$ für $x \in [t_{\mu-1}, t_\mu]$,

$\mu = 1,\dots,l$, $\eta = \pm 1$, wobei $t_0 := x_i$ und $t_l := x_{i+j+m-1}$ ist.

Nach Hilfssatz 2.6 existiert nun entweder ein Intervall $[x_u, x_{u+m}] \subset [x_i, x_{i+j+m-1}]$ mit $H_{u,u+m} = \emptyset$ oder es gibt ein Intervall $[x_u, x_{u+m+h}] \subset [x_i, x_{i+j+m-1}]$ mit

$H_{u,u+m+h} = \{t_\mu \in (x_u, x_{u+m+h}),\ \mu = 1,\dots,l-1\} =: \{\tau_\mu\}_{\mu=1,\dots,h}$

und $\tau_\mu \in (x_{u+\mu}, x_{u+\mu+m-1})$, $\mu = 1,\dots,h$.

Wir konstruieren nun einen Spline aus dem Raum

$$\hat{V} := \{s \mid s(x,y) = \sum_{\mu=u}^{u+h+1} \sum_{\lambda=1-n}^{l-1} \alpha_{\mu\lambda} B_{n,\lambda}(y) B_{m,\mu}(x),\ \alpha_{u+h+1,l-1} = 0\}.$$

Dabei ist stets der Fall $h = 0$ zugelassen.

Für $\rho \neq \nu_o$ wählen wir jeweils $h + 2$ verschiedene Punkte $\{t_{\mu\rho}\}_{\mu=0,\dots,h+1}$ mit $x_{u+\mu} < t_{\mu\rho} < x_{u+\mu+m}$ für $\mu = 0,\dots,h+1$. Nach Hilfssatz 2.1 existiert genau ein $s \in \hat{V}$ mit

$$s(t_{\mu\rho}, \eta_\rho) = 0,\ \rho \neq \nu_o,\ \mu = 0,\dots,h+1$$
$$s(\tau_\mu, \eta_{\nu_o}) = 0,\ \mu = 1,\dots,h$$

$$s(z,\eta_{\nu_o}) = \operatorname{sgn}\,(f-s_o)(\overline{\xi},\eta_{\nu_o}) ,$$

wobei für $h \neq 0$ $z \in (x_{u+m+h-1}, x_{u+m+h})$ und $\overline{\xi} := \min\{\xi_\mu \mid \tau_h < \xi_\mu , \mu = 1,\ldots,l\}$ sei. Für $h = 0$ sei $z \in (x_u, x_{u+m})$. Falls von den Punkten $\xi_1,\ldots,\xi_l$ einer in $[x_u, x_{u+m}]$ liegt, sei dieser $\overline{\xi}$. Andernfalls setzen wir $s(z,\eta_{\nu_o}) = 1$.

Für dieses s gilt nach Hilfssatz 2.1:

a) $s(\;,\eta_\rho) \equiv 0$ für $\rho \neq \nu_o$

b) $s(\;,\eta_{\nu_o})$ besitzt nur die Nullstellen $\tau_1,\ldots,\tau_h$ in (x_u, x_{u+m+h}). Diese sind einfach

c) $s(x,\eta_{\nu_o}) = 0$ für $x \geq x_{u+m+h}$

Wegen $\tilde{\eta}(-1)^\mu (f-s_o)(x,\eta_{\nu_o}) \leq ||f-s_o||_T - \varepsilon$ für $x \in [\tau_{\mu-1}, \tau_\mu]$, $\mu = 1,\ldots,h+1$, $\tilde{\eta} = \pm 1$, wobei $\tau_o := x_u$ und $\tau_{h+1} := x_{u+m+h}$ ist, und wegen der Wahl des Vorzeichens von s gilt:

$\tilde{\eta}(-1)^{\mu+1} s(x,\eta_{\nu_o}) \geq 0$ für $x \in [\tau_{\mu-1}, \tau_\mu]$, $\mu = 1,\ldots,h+1$.

Deshalb existiert ein $c \in \mathbb{R}$, $c \neq 0$ mit $||f-s_o-cs||_T \leq ||f-s_o||_T$.

Da $x_{u+h+1} < x_{u+h+m} \leq x_{i+j+m-1} \leq x_k$ ist, ist $\hat{V} \subset V^{m,n}_{k,l}$ und deshalb auch $s_o + cs \in V^{m,n}_{k,l}$. Also ist s_o nicht die einzige Minimallösung für f bzgl $V^{m,n}_{k,l}$ auf T.

Nun sei eine Teilmenge $[x_o, x_j] \times \{\eta_{\nu_o}\}$, $j = 1,\ldots,k$ von T gegeben, so daß $(f-s_o)(\;,\eta_{\nu_o})$ nur $l \leq j$ alternierende Extremalpunkte in $[x_o, x_j]$ besitzt. Wir betrachten das Intervall $[x_{1-m}, x_j]$. Wie im ersten Teil des Beweises gezeigt wurde, existiert ein Teilintervall $[x_u, x_{u+m+h}]$ von $[x_{1-m}, x_j]$ mit $h \geq 0$, so daß sich ein Spline $s \in \hat{V}$ konstruieren läßt, der in (x_u, x_{u+m+h}) nur h einfache Nullstellen besitzt und wegen $x_0 < x_{u+m+h}$ deshalb in T nicht identisch verschwindet. Dann ist aber für geeignetes $c \in \mathbb{R}$, $c \neq 0$, $s_0 + cs \in V_{k,l}^{m,n}$ eine Funktion, die auf T die Funktion f mindestens genauso gut approximiert wie s_0. Dies ist ein Widerspruch zur Voraussetzung.

Schließlich sei eine Teilmenge $[x_j, x_k] \times \{\eta_{\nu_o}\}$, $j = 1,\ldots,k-2$, von T gegeben, so daß $(f-s_o)(\ ,\eta_{\nu_o})$ nur $1 \leq k - j - 1$ alternierende Extremalpunkte in $[x_j, x_k]$ besitzt. Wir betrachten das Intervall $[x_j, x_{k+m-2}]$. Analog zu den schon behandelten Fällen existiert ein Teilintervall $[x_u, x_{u+m+h}]$, $h \geq 0$, von $[x_j, x_{k+m-2}]$ und ein $s \in \hat{V}$, so daß s in (x_u, x_{u+m+h}) nur h einfache Nullstellen besitzt und somit wegen $x_u < x_k$ in T nicht identisch verschwindet.
Wegen

$$s(x,y) = \sum_{\mu=u}^{u+h+1} \sum_{\lambda=1-n}^{l-1} \alpha_{\mu\lambda} B_{n,\lambda}(y) B_{m,\mu}(x), \quad \alpha_{u+h+1,l-1} = 0 \quad \text{und}$$

$x_{u+m+h+1} \leq x_{k+m-1}$ ist $s \in V_{k,l}^{m,n}$ und wie in den vorherigen Fällen erhält man einen Widerspruch zur Voraussetzung.

<u>Diskussionsbeitrag</u>

Herr Professor de Boor fragte nach einem Beispiel für eine reelle Funktion in zwei Variablen, die sich nicht als Tensorprodukt von reellen Funktionen in einer Variablen darstellen läßt und deren Fehlerfunktion bzgl $V_{k,l}^{m,n}$ ein Alternantengitter besitzt. Möglicherweise wird dies durch folgendes Beispiel beantwortet.

<u>BEISPIEL</u>: Seien m,n,k,l natürliche Zahlen, $m \geq 2$, $n \geq 2$. Sei $r := m + k - 1$, $s := n + l - 1$. Wir wählen

$R := [\sqrt{\pi}, r\sqrt{\pi}] \times [\sqrt{\pi}, s\sqrt{\pi}]$. R sei durch das Knotengitter

$\sqrt{\pi} = x_0 < x_1 < \ldots < x_{k-1} < x_k = r\sqrt{\pi}$

$\sqrt{\pi} = y_0 < y_1 < \ldots < y_{l-1} < y_l = s\sqrt{\pi}$

unterteilt. Die Funktion $f(x,y) := \cos 2xy \cos [(x+y)^2]$ besitzt in R ein Alternantengitter mit $r \cdot s$ Punkten $(\mu\sqrt{\pi}, \nu\sqrt{\pi})_{\substack{\mu=1,\ldots,r \\ \nu=1,\ldots,s}}$. Deshalb ist nach Satz 1.2 0 Minimallösung für f bzgl $V_{k,l}^{m,n}$ auf R.

Wir zeigen nun, daß es keine achsenparallele Strecke $[\sqrt{\pi}, r\sqrt{\pi}] \times \{\bar{y}\}$ in R gibt, auf der f mindestens $r + 1$ Extremalpunkte besitzt.

Sei $\bar{y} := c\sqrt{\pi}$, $c \in [1,s]$.

Für $c = i$, $i = 1,\ldots,s$ ist die Behauptung offenbar richtig.

Sei nun $c \in (i,i+1)$, $i \in \{1,\ldots,s-1\}$.

Damit f einen Extremalpunkt in $(x,\bar{y})$ besitzt, müssen folgende beiden Bedingungen gelten:

$2x\bar{y} = n \cdot \pi$, $n \in \mathbb{N}$ und $(x+\bar{y})^2 = m \cdot \pi$, $m \in \mathbb{N}$

Seien nun $x_0,\ldots,x_r$ $r + 1$ alternierende Extremalpunkte von $f(\ ,\bar{y})$. Dann gilt für diese Punkte:

(1) $x_i = \dfrac{(n_0+\ldots+n_i)\pi}{2\bar{y}}$, $n_i \in \mathbb{N}$, $i = 0,\ldots,r$

(2) $(x_i+\bar{y})^2 = (m_0+\ldots+m_i)\pi$, $m_i \in \mathbb{N}$, $i = 0,\ldots,r$

Wir schätzen nun $m_0 + m_1 + \ldots + m_r$ nach unten ab und zeigen, daß $x_r \geq (r+1)\sqrt{\pi}$ ist, also x_r nicht im angegebenen Intervall liegt.

Wegen $\bar{y} = c\sqrt{\pi}$ und der Gleichungen (1),(2) gilt:

$\frac{n_0}{2c} + c = \sqrt{m_0}$, $\frac{n_1}{2c} = \sqrt{m_0+m_1} - \sqrt{m_0}$, $\ldots$,

$$\frac{n_r}{2c} = \sqrt{m_o+\ldots+m_r} - \sqrt{m_o+\ldots+m_{r-1}}$$

Durch Division je zwei aufeinanderfolgender Gleichungen erhält man:

$$\sqrt{m_o(m_o+m_1)} \in \mathbb{N}\ ,\ \sqrt{(m_o+m_1)(m_o+m_1+m_2)} \in \mathbb{N}\ ,\ \ldots\ ,$$
$$\sqrt{(m_o+\ldots+m_{r-1})(m_o+\ldots+m_r)} \in \mathbb{N}$$

Wegen $c \in (i,i+1)$ und $x_o \geq \sqrt{\pi}$ ist $m_o > (i+1)^2$.

1. Fall: $m_o = h^2$ mit $h \in \mathbb{N}$

Dann ist $m_o \geq (i+2)^2$ und alle Zahlen $m_o+m_1,\ldots,m_o+\ldots+m_r$ sind Quadratzahlen. Deshalb gilt:

$$m_o+m_1 \geq (i+3)^2\ ,\ \ldots\ ,\ m_o+\ldots+m_r \geq (i+r+2)^2$$

Dann folgt $(x_r+\bar{y})^2 = (m_o+\ldots+m_r)\pi \geq (i+r+2)^2\pi$ und daraus

$$x_r \geq (i+r+2)\sqrt{\pi} - \bar{y} > ((i+r+2) - (i+1))\sqrt{\pi} = (r+1)\sqrt{\pi}\ .$$

2. Fall: m_o ist keine Quadratzahl

Da $\sqrt{m_o(m_o+m_1)}\ ,\ \ldots\ ,\sqrt{(m_o+m_1+\ldots+m_{r-1})(m_o+m_1+\ldots+m_r)}$ natürliche Zahlen sind, erhält man

$$m_o(m_o+m_1) \geq m_o^2\cdot 2^2 \geq m_o(4(i+1)^2+4)$$

$$m_o+m_1+m_2 \geq m_o\cdot 3^2 \geq 9(i+1)^2+9$$

$$\vdots$$

$$m_o+m_1+\ldots+m_r \geq m_o\cdot(r+1)^2 \geq (r+1)^2(i+1)^2+(r+1)^2$$

Dann folgt $(x_r+\bar{y})^2 = (m_o+\ldots+m_r)\pi \geq ((r+1)^2(i+1)^2+(r+1)^2)\pi$ und deshalb

$$x_r \geq (r+1)\sqrt{(i+1)^2+1}\,\sqrt{\pi} - \bar{y} > ((r+1)\sqrt{(i+1)^2+1} - (i+1))\sqrt{\pi}$$
$$> (r+1)\sqrt{\pi}\ .$$

In beiden Fällen erhält man sogar $x_{r-1} > r\sqrt{\pi}$ und somit gibt es für $c \neq i$, $i = 1,\ldots,s$ nicht einmal r Extremalpunkte von f auf einer zur x-Achse parallelen Strecke in R.

Diese Arbeit ist ein Teil der Dissertation des Autors, die dieser bei Herrn Professor Dr. G. Meinardus am Institut für Angewandte Mathematik der Universität Erlangen-Nürnberg angefertigt hat.

Literatur

[1] Barrar, R.B., Loeb, H.L.: Spline Functions with Free Knots as the Limits of Varisolvent Families. J. Approximation Theory 12(1974), 70-77.

[2] Buck, R.C.: Alternation Theorems for Functions of Several Variables. J. Approximation Theory 1(1968), 325-334.

[3] Cheney, E.W.: Introduction to Approximation Theory. Mc Graw-Hill New York 1966.

[4] Curry, H.B., Schoenberg, I.J.: On Pólya Frequency Functions and Their Limits. J. d'Anal. Math. 18, 71-107.

[5] Ehlich, H., Haußmann, W.: Tschebyscheff-Approximation stetiger Funktionen in zwei Veränderlichen. Math. Z. 117(1970), 21-34.

[6] Ehlich, H., Zeller, K.: Cebysev-Polynome in mehreren Veränderlichen. Math. Z. 93(1966), 142-143.

[7] Greville, T.N.E.: Introduction to Spline Functions in: Theory and Applications of Spline Functions. Academic Press New York 1969.

[8] Haußmann, W.: Alternanten bei mehrdimensionaler Tschebyscheff-Approximation. ZAMM 52(1972), T 206 - T 208.

[9] Karlin, S.: Total Positivity, Volume I. Stanford University Press, Stanford California 1968.

[10] Karlin, S., Studden, W.J.: Tschebyscheff Systems: With Applications in Analysis and Statistics. Interscience Publishers New York 1966.

[11] Meinardus, G.: Approximation of Functions: Theory and Numerical Methods. Springer-Verlag Berlin Heidelberg New York 1967.

[12] Rice, J.R.: The Approximation of Functions, Volume II. Addison-Wesley Publishing Company, Reading Massachusetts 1969.

[13] Schumaker, L.: Uniform Approximation by Tschebyscheffian Spline Functions. Journal of Mathematics and Mechanics 18(1968), 369-377.

[14] Schumaker, L.: Approximation by Splines in: Greville, T.N.E.: Theory and Applications of Spline Functions. Academic Press New York 1969.

[15] Sommer, M.: Gleichmäßige Approximation mit zweidimensionalen Splinefunktionen, Dissertation, Universität Erlangen-Nürnberg, 1975.

[16] Strauß, H.: Eindeutigkeit bei der gleichmäßigen Approximation mit Tschebyscheffschen Splinefunktionen. Erscheint in J. Approximation Theory.

[17] Strauß, H.: L_1-Approximation mit Splinefunktionen. ISNM 26 Birkhäuser Verlag, Basel und Stuttgart, 1975.

Dr. M. Sommer

Institut für Angewandte Mathematik
der Universität Erlangen-Nürnberg
852 Erlangen, Martensstraße 1

APPROXIMATION MIT SPLINEFUNKTIONEN UND QUADRATURFORMELN

HANS STRAUSS

This paper is concerned with the problem of approximating functions in the L_1-norm by spline functions with fixed and free knots and its applications to the approximation of linear functionals. For this best L_1-approximation characterizations are given which involve perfect splines. In addition, one-sided approximation is studied in more detail. The results are used to give another proof of the existence of a monospline with maximal number of zeros.

0. Einleitung

Die Approximation von Funktionen durch Splinefunktionen wurde in den letzten Jahren in der L_∞-Norm bzw. L_2-Norm untersucht (siehe Rice und Schumaker).

In dieser Arbeit wenden wir uns der Approximation in der L_1-Norm zu und interessieren uns vor allem für die Anwendungen auf die Approximation linearer Funktionale. Zunächst wird die L_1-Approximation durch Splinefunktionen mit festen Knoten an stetige Funktionen untersucht. Ein notwendiges und hinreichendes Kriterium für Minimallösungen kann man mit Hilfe von Perfektsplines angeben. Dabei zeigt sich, daß der Fehlerfunktion einer Minimallösung ein Perfektspline zugeordnet werden kann, der bestimmte Nullstelleneigenschaften erfüllt. Als Beispiel wird der Bernoullische Monospline betrachtet, der eine Minimaleigenschaft bezüglich der L_1-Norm besitzt (siehe Micchelli). Der zugeordnete Perfektspline ist ein Eulerspline. Bei diesen Untersuchungen lassen sich auch Eindeutigkeitsaussagen für Interpolationsprobleme bei Perfektsplines gewinnen.

Zur Bestimmung optimaler Quadraturformeln werden Monosplines mit minimaler L_1-Norm gesucht, wobei auch die Knoten als Variable zugelassen werden. Aus der nichtlinearen Approximationstheorie läßt sich ein notwendiges Kriterium für Minimallösungen verwenden. Es führt auf eine Charakterisierung von Lösungen des Problems mit freien Knoten, wobei die Überlegungen für das lineare Problem verwendet werden. Die Minimallösungen werden wieder mit Perfektsplines charakterisiert. Daraus kann man ein Iterationsverfahren ableiten, das einem Spline s_1 einen Spline s_2 zuordnet, der x^n besser approximiert als s_1, wenn s_1 nicht auf sich selbst abgebildet wird. Jeder Fixpunkt erfüllt die Bedingungen des Kriteriums.

Schließlich untersuchen wir noch die einseitige Approximation von Funktionen durch Splinefunktionen. Ein Zusammenhang zwischen einseitiger L_1-Approximation durch Tschebyscheffsysteme und Quadraturformeln wurde schon von DeVore gezeigt. Die Behandlung der einseitigen Approximation mit Splinefunktionen führt ebenfalls auf Quadraturformeln. Dies findet folgende Anwendung. Jeder Quadraturformel kann ein Monospline zugeordnet werden, der gewisse Nullstelleneigenschaften erfüllt. Damit ergibt sich auch ein anderer Existenzbeweis für ein Interpolationsproblem bei Monosplines als er von Karlin und Schumaker angegeben wird.

1. Quadraturformeln und Monosplines.

Es sei eine Funktion f Element der Differenzierbarkeitsklasse $C^{(n)}[a,b]$. Dann besteht folgende Relation, die man durch partielle Integration des Ausdrucks

$$\int_a^b M(x)f^{(n)}(x)dx$$

beweisen kann (siehe Karlin [18]):

$$\int_a^b f(x)dx = \sum_{i=0}^{n-1} \beta_i f^{(i)}(b) + \sum_{i=0}^{n-1} \alpha_i f^{(i)}(a)$$

$$(1.1) \qquad + \sum_{i=1}^{k} A_i f(x_i) + (-1)^n \int_a^b M(x)f^{(n)}(x)dx \,,$$

wobei $M \in M_{n,k}$ und

$$M_{n,k}(x_1,\ldots,x_k) =$$

$$\{M \mid M(x) = \frac{x^n}{n!} - \sum_{i=0}^{n-1} a_i x^i - \sum_{i=1}^{k} b_i (x-x_i)_+^{n-1}\}$$

ist. Die Konstanten besitzen folgende Werte

$$\alpha_i = (-1)^{i+1} M^{(n-1-i)}(a) \qquad i=0,\ldots,n-1$$

$$(1.2) \qquad \beta_i = (-1)^i M^{(n-1-i)}(b)$$

$$A_i = M^{(n-1)}(x_{i-o}) - M^{(n-1)}(x_{i+o}) \qquad i=1,\ldots,k \,.$$

Es sei Q eine Quadraturformel nach (1.1)

$$(1.3) \qquad Qf = \int_a^b f(x)dx - Rf$$

und $Rf = (-1)^n \int_a^b M(x)f^{(n)}dx$.

Nun legen wir folgenden Splineraum zu Grunde

$$S_{n-1,k}[a,b]=\{s\,|\,s(x)=\sum_{i=0}^{n-1} c_i x^i + \tag{1.4}$$
$$\sum_{i=1}^{r}\sum_{j=1}^{m_i} d_{ij}(x-y_i)_+^{n-j}\},$$

wobei $1\leq m_i<n$ und $\sum_{i=1}^{r} m_i=k$ ist. c_i und d_{ij} sind reelle Zahlen. Außerdem gilt für die Knoten

$$a<y_1<\ldots<y_r<b.$$

Durch partielle Integration von $\int_a^b M(x)s^{(n)}(x)dx=0$, wobei $s\in S_{n-1,k}$ und M wie oben definiert sind, ergibt sich

$$\int_a^b s(x)dx = \sum_{i=0}^{n-1} \beta_i s^{(i)}(b) + \sum_{i=0}^{n-1} \alpha_i s^{(i)}(a)+$$
$$\sum_{i=1}^{k} A_i s(x_i) + \tag{1.5}$$
$$\sum_{i=1}^{r}\sum_{j=0}^{m_i-1} (-1)^j s^{(n-j-1)}(x)M^{(j)}(x)\Big|_{y_i+0}^{y_i-0}$$

Wenn $M^{(j)}(y_i)=0$ für $i=1,\ldots,m$ und $j=0,\ldots,m_i-1$ gilt, verschwindet die Doppelsumme in (1.5) und die Elemente von $S_{n-1,k}$ werden durch die Quadraturformel (1.3) exakt integriert. Eine Umkehrung dieser Aussage werden wir später noch benötigen. Es sei

$$\bar{S}_{n-1,k}[a,b]=\{s\in S_{n-1,k}\,|\,s^{(i)}(a)=0 \quad i=0,\ldots,p-1 \tag{1.6}$$
$$s^{(i)}(b)=0 \quad i=0,\ldots,q-1\}$$

gegeben.

<u>Satz 1.1.</u> Die Quadraturformel

$$Qf = \sum_{i=1}^{m} A_i f(t_i) \qquad a<t_1<\ldots<t_m<b$$

integriere alle $s\in\bar{S}_{n-1,k}$ exakt. Dann existiert ein Monospline M, der die Stützstellen der Quadraturformel als Knoten hat und die Nullstellen

$$M^{(i)}(a) = 0 \qquad i=0,\dots,n-p-1$$

$$M^{(i)}(b) = 0 \qquad i=0,\dots,n-q-1$$

$$M^{(j)}(y_i) = 0 \qquad i=1,\dots,r$$

$$j=0,\dots,m_i-1$$

besitzt.

Beweis: Man gibt sich einen Monospline vor

$$\bar{M}(x) = \frac{x^n}{n!} - \sum_{i=0}^{n-1} a_i x^i - \sum_{i=1}^{m} \frac{A_i}{(n-1)!}(x-t_i)_+^{n-1}$$

und kann die Konstanten $(a_i)_{i=0}^{n-1}$ so festlegen, daß die genannten Aussagen gültig sind. Dies wird mit Hilfe von (1.5) gezeigt.

2. Ein Momentenproblem und Perfektsplines.

Bei der Berechnung von Interpolationsfunktionen, deren n-te Ableitung gewisse Minimaleigenschaften besitzt, wurden von Glaeser [11] und Karlin [19] die sogenannten Perfektsplines eingeführt. In unserer Arbeit werden diese Splines eine große Bedeutung erlangen.

Definition 2.1. Ein Perfektspline vom Grad n mit r-1 Knoten in [a,b] ist ein Polynomspline der speziellen Form

$$(2.1) \qquad P(x) = c(x^n+2\sum_{i=1}^{r-1} (-1)^i (x-u_i)_+^n) + \sum_{i=0}^{n-1} a_i x^i ,$$

wobei $c,a_o,\dots,a_{n-1}$ reelle Zahlen sind und die Knoten $(u_i)_{i=1}^{r-1}$ den Bedingungen $a<u_1<\dots<u_{r-1}<b$ unterliegen.

Es gilt der Interpolationssatz (siehe Karlin [19]).

Satz 2.2. Es sei die Punktmenge $(x_i)_{i=1}^{n+k}$ mit $a \leq x_1 \leq \ldots x_{n+k} \leq b$ gegeben, wobei $x_i < x_{i+n}$ für $i=1,\ldots,k$ ist. Es sei $(y_i)_{i=1}^{n+k}$ eine Menge von reellen Zahlen. Dann existiert ein Perfektspline P von der Form (2.1) mit höchstens k-1 Knoten in $[a,b]$, so daß

$$P(x_i)=y_i \quad i=1,\ldots,n+k$$

gilt.

Nun betrachten wir ein Momentenproblem, das für die L_1-Approximation eine große Bedeutung hat.

Sei $\alpha = (a_i)_{i=0}^{n+1}$ ein (n+2)-Tupel reeller Zahlen mit der Eigenschaft

$$a=a_0 \leq a_1 \leq \ldots a_n \leq a_{n+1}=b.$$

Dann hat die Signumsfunktion sgn (α) die Gestalt

$$\operatorname{sgn}(\alpha,x) = \begin{cases} (-1)^i & x\in(a_i,a_{i+1}) \quad i=0,\ldots,n \\ 0 & x\in a_i \quad i=1,\ldots,n \end{cases} .$$

Man sucht jetzt einen Vektor α , so daß

$$(2.2) \qquad \int_a^b s(x)\operatorname{sgn}(\alpha,x)\,dx = 0 \qquad s\in B_{n-1,k}$$

gilt. Es sei

$$(2.3) \qquad B_{n-1,k}[a,b] = \langle M_1,\ldots,M_k\rangle \qquad n\geq 2 ,$$

wobei M_i ein polynomialer B-Spline vom Grad n-1 mit den Knoten $x_i,\ldots,x_{i+n}$ ist und

$$a \leq x_1 \leq x_2 \leq \ldots x_{n+k-1} \leq x_{n+k} \leq b \qquad x_i < x_{i+n} .$$

Es zeigt sich, daß ein Zusammenhang zwischen der Interpolation mit Perfektsplines und der Lösung des Problems (2.2) besteht.

<u>Satz 2.3</u> (a) Besitzt der Perfektspline P in x_i für $i=1,\dots,n+k$ Nullstellen, dann hat $P^{(n)}$ die Eigenschaft

$$(2.4) \qquad \int_a^b s(x)\, P^{(n)}(x)\, dx = 0 \qquad s \in B_{n-1,k}.$$

(b) Gegeben sei eine Signumsfunktion sgn (α), die Problem (2.3) löst. Dann existiert ein Perfektspline P, so daß

$$(2.5) \qquad P^{(n)}(x) = \operatorname{sgn}(\alpha,x) \qquad x \in [a,b]$$

$$(2.6) \qquad P(x_i) = 0 \qquad i=1,\dots,n+k$$

gilt.

Beweis: (a) Sei $[x_i,\dots,x_{i+n}]P$ die n-te dividierte Differenz von P in den Punkten $x_i,\dots,x_{i+n}$. Nach Curry and Schoenberg gilt für jeden B-Spline $M_i \in B_{n-1,k}$

$$[x_i,\dots,x_{i+n}]P = \int_a^b M_i(x)\, P^{(n)}(x)\, dx.$$

Nachdem $P(x_i) = 0$ für $i=1,\dots,n+k$ ist, ergibt sich

$$\int_a^b M_i(x)\, P^{(n)}(x) dx = 0$$

und damit die Behauptung.

(b) sgn (α) habe die Unstetigkeitsstellen $u_1,\dots,u_r$.
Wegen (2.5) kann der Perfektspline P nur die Gestalt

$$P(x) = \sum_{i=o}^{n-1} a_i x^i + \frac{1}{n!} (x^n + 2 \sum_{i=1}^{r} (-1)^i (x-u_i)_+^n)$$

haben.

Man legt nun die Konstanten (a_i) fest und erhält die Aussagen unserer Behauptung, wobei man eine (1.5) entsprechende Relation für Perfektsplines aufstellen muß

<u>Satz 2.4</u> Es seien zwei Perfektsplines P_1 und P_2 vom Grad n gegeben, die das Interpolationsproblem

$$P_j(x_i) = y_i \qquad i=1,\ldots,n+k$$
$$j=1,2$$

lösen, wobei $a \leq x_1 \leq x_2 \leq \ldots \leq x_{n+k} \leq b$ mit $x_i < x_{i+n}$ für $i=1,\ldots,k$ ist und $(y_i)_{i=1}^{n+k}$ reelle Zahlen sind. Dann gilt

(2.7) $$\int_a^b s(x) \, (P_1^{(n)} - P_2^{(n)}) (x) \, dx = 0 \qquad s \in B_{n-1,k}$$

<u>Beweis:</u> Ebenso wie in Satz 2.3 (a).

Nun soll das Momentenproblem (2.2) näher untersucht werden. Hobby und Rice haben folgende Ergebnisse gezeigt:

Gegeben sei ein n-dimensionaler Unterraum V von C[a,b] mit der Basis $(g_1,\ldots,g_n)$. Es existiert stets ein (n+2)-Tupel $\alpha = (a_o \leq a_1 \leq \ldots \leq a_n \leq a_{n+1})$, so daß gilt

(2.8) $$\int_a^b g_i(x) \operatorname{sgn} (\alpha,x) \, dx = 0 \qquad i=1,\ldots,n.$$

sgn (α) kann dabei weniger als n Vorzeichenwechsel besitzen. Dieser Fall tritt auf, wenn a_i zusammenfallen.

Ist V jedoch ein Tschebyscheffsystem, so kann man nachweisen, daß die Lösung von (2.8) eindeutig bestimmt ist und sgn (α) genau n Vorzeichenwechsel hat.

Ein entsprechendes Ergebnis erhalten wir, wenn als Teilraum V der Raum von Splinefunktionen $B_{n-1,k}$ genommen wird. $B_{n-1,k}$ ist nur ein schwaches Tschebyscheffsystem.

Wir benötigen folgende Hilfssätze über Splinefunktionen.

Lemma 2.5. Ein Perfektspline P vom Grad n mit k Knoten sei gegeben.

(a) P besitzt höchstens n+k Nullstellen.

(b) P besitze $(u_i)_{i=1}^k$ als Knoten und die n+k Nullstellen $(z_i)_{i=1}^{n+k}$ mit $z_1 \le z_2 \le \ldots \le z_{n+k}$, wobei $z_i < z_{i+n}$ für i=1,...,k gilt. Dann ist folgende Relation erfüllt

$$z_i < u_i < z_{i+n} \qquad i=1,\ldots,k.$$

Lemma 2.6. Es seien die Punkte $(u_i)_{i=1}^k$ vorgegeben, die

$$u_i \in (x_i, x_{i+n}) \qquad i=1,\ldots,k$$

erfüllen. Ein Spline $s \in B_{n-1,k}$, der

$$s(u_i) = 0 \qquad i=1,\ldots,k \quad i \neq i_0$$

$$(2.9) \qquad s(u_{i_0}) = 1$$

interpoliert, besitzt ein Kernintervall $(x_\nu, x_\mu) \subset (x_1, x_{n+k})$, auf dem er als Nullstellen genau die Nullstellen von (2.9) besitzt, die in diesem Intervall (x_ν, x_μ) liegen. Außerhalb von (x_ν, x_μ) ist s gleich Null.

Satz 2.7. Es sei $B_{n-1,k}$ gegeben. Dann existiert genau ein $\alpha^o = (a<a_1\leq a_2\leq\ldots\leq a_k<b)$, so daß

$$(2.10)\qquad \int_a^b s(x)\,\mathrm{sgn}\,(\alpha^o,x)\,dx = 0 \qquad s\in B_{n-1,k}$$

gilt und außerdem

$$a_i < a_{i+1} \qquad i=1,\ldots,k-1$$

ist.

Beweis: (a) Nach (2.8) existiert eine Signumsfunktion sgn (α) mit höchstens k Vorzeichenwechseln. Nun wird gezeigt, daß sgn (α) wenigstens k Vorzeichenwechsel besitzen muß.

Nach Satz 2.3 ist sgn (α) ein Perfektspline vom Grad n zugeordnet, der in $(x_i)_{i=1}^{n+k}$ Nullstellen besitzt. Nach Lemma 2.5 muß P wenigstens k Knoten haben. Also besitzt P genau k Knoten und $P^{(n)}$ = sgn (α) k Vorzeichenwechsel.

(b) Für den Nachweis der Eindeutigkeit ist folgende Relation wichtig. Seien $u = (u_i)_{i=1}^{k}$ die Knoten von P. Dann folgt nach Lemma 2.5 (b)

$$(2.11)\qquad x_i < u_i < x_{i+n} \qquad i=1,\ldots,k.$$

Angenommen, es gebe zwei Signumsfunktionen p_1 = sgn (u) und p_2 = sgn (v), die (2.10) lösen. $u = (u_i)_{i=1}^{k}$, bzw.

$v=(v_i)_{i=1}^k$ seien die Punkte, an denen p_1 bzw. p_2 das Vorzeichen wechseln. Sei weiterhin $u_0=v_0=a$ und $u_{k+1}=v_{k+1}=b$. Es gibt ein kleinstes ν, so daß $u_\nu \neq v_\nu$ gilt. Sei $u_\nu < v_\nu$ und für ein hinreichend kleines $\varepsilon>0$ gelte nun $(p_1-p_2)(x)>0$ für $x\in(u_\nu,u_\nu+\varepsilon)$. Also gilt $(-1)^\mu(p_1-p_2)(x)\geq 0$ für $x\in[u_{\nu+\mu},u_{\nu+\mu+1}]$ mit $\mu=0,\ldots,k-\nu$. Nun konstruieren wir einen Spline $s_0\in B_{n-1,k}$ mit den Eigenschaften

$$s_0(u_i)=0 \qquad i=1,\ldots,k \quad i\neq\nu \qquad s_0(u_\nu)=1.$$

Aus Lemma 2.6 folgt $(p_1-p_2)(x)s_0(x)\geq 0$ für $x\in[a,b]$ und damit

$$\int_a^b (p_1-p_2)(x)s_0(x)dx \geq \int_{u_\nu}^{u_\nu+\varepsilon} (p_1-p_2)(x)s_0(x)dx > 0$$

Dies widerspricht der Voraussetzung

$$\int_a^b (p_1-p_2)(x)s_0(x)dx = 0.$$

Daraus folgt die eindeutige Lösbarkeit unseres Problems. Aufgrund der Beziehung,die zwischen dem Momentenproblem und der Interpolation mit Perfektsplines besteht, kann man Ergebnisse über das Momentenproblem dazu verwenden, um Aussagen über Perfektsplines zu machen.

<u>Satz 2.8.</u> Es seien die Punkte

$$(2.12) \qquad a\leq x_1\leq \ldots \leq x_{n+k}\leq b \qquad x_i<x_{i+n}$$

vorgegeben. Dann existiert (bis auf die Konstate ± 1) genau ein Perfektspline P mit höchstens k Knoten, der den Bedingungen

$$(2.13)\qquad \begin{aligned} P(x_i) &= 0 \qquad i=1,\dots,n+k \\ ||P^{(n)}|| &= 1 \end{aligned}$$

genügt. Dieser Perfektspline besitzt genau k Knoten.

Beweis: Der Punktmenge (2.12) ist nach (2.3) der Raum $B_{n-1,k}$ zugeordnet. Es existiert nun eine Signumsfunktion, die das Momentenproblem (2.10) löst. Aus Satz 2.3 (b) folgt die Existenz eines Perfektsplines, der die Bedingungen (2.13) erfüllt. Damit ist die Existenz unter Verwendung der Sätze von Hobby und Rice gezeigt. Sie folgt natürlich auch aus Satz 2.2. Hier soll vor allem festgestellt werden, daß mit der Lösung von (2.10) die Knotenmenge unseres Perfektsplines gefunden ist.

Man nimmt nun an, daß es zwei Funktionen p_1 und p_2 gibt, die (2.13) lösen.

Besitzen beide die gleichen Knoten $(u_i)_{i=1}^{k}$, dann ist $(p_1-p_2)^{(n)} \equiv 0$ und p_1-p_2 ein Polynom vom Höchstgrad n. Aufgrund der Nullstellenbedingungen gilt $p_1 = p_2$.

Wenn p_1 und p_2 nicht die gleichen Knoten besitzen, dann sind $p_1^{(n)}$ und $p_2^{(n)}$ verschiedene Signumsfunktionen, die (2.10) lösen. Damit ist ein Widerspruch zu Satz 2.8 gezeigt und die Eindeutigkeit bewiesen.

Satz 2.9. Sei $P(x) = \sum_{i=o}^{n-1} a_i x^i + c[\, x^n + 2 \sum_{i=1}^{k-1} (-1)^i (x-u_i)_+^n \,]$

ein Perfektspline, der die Interpolationsbedingungen

(2.14) $$P(x_i) = y_i \qquad i=1,\ldots,n+k$$

erfüllt und dessen Knoten den Bedingungen

(2.15) $$x_{i+1} < u_i < x_{i+n} \qquad i=1,\ldots,k-1$$

genügen. Dann ist P der einzige Perfektspline mit höchstens k-1 Knoten, der (2.14) genügt.

Beweis: Es gebe einen weiteren Perfektspline mit den genannten Eigenschaften

$$P_1(x) = \sum_{i=0}^{n-1} \bar{a}_i x^i + \bar{c}\,[x^n + 2 \sum_{i=1}^{r} (-1)^i (x-\bar{u}_i)_+^n] \quad r \leq k-1.$$

Nach Satz 2.4 hat $P-P_1$ die Eigenschaft

(2.16) $$\int_a^b s(x) \quad (P^{(n)} - P_1{}^{(n)})(x)\, dx = 0 \quad s \in B_{n-1,k}.$$

Es zeigt sich, daß für $P^{(n)} - P_1{}^{(n)}$ die Aussage gilt

$$\eta(-1)^i (P^{(n)} - P_1{}^{(n)})(x) \geq 0 \qquad x \in (u_i, u_{i+1})$$

$$u_o = x_1,\ u_k = x_{n+k},\ \eta = \pm 1,\ i=0,\ldots,k-1.$$

Wegen (2.15) existiert ein Spline $s_1 \in B_{n-1,k}$, der in (x_1, x_{n+k}) kein identisch verschwindendes Intervall besitzt und in $(u_i)_{i=1}^{k-1}$ einfache Nullstellen hat. Dies folgt aus einem Schluß, wie er in Lemma 2.4 in Strauß [38] durchgeführt wird. Damit erhält man

$$\int_a^b s_1(x) \ (P^{(n)} - P_1{}^{(n)})(x)dx \neq 0,$$

falls $P^{(n)} \neq P_1{}^{(n)}$ ist, also ein Widerspruch zu (2.16). Man bekommt $P^{(n)} - P_1{}^{(n)} \equiv 0$. Aufgrund der Nullstelleneigenschaften von $P-P_1$ bedeutet es, daß $P=P_1$ gilt.

3. Approximation mit festen Knoten.

Perfektsplines können dazu verwendet werden, um Minimallösungen bei der L_1-Approximation zu charakterisieren.

Satz 3.1. Es sei $s_o \in B_{n-1,k}$ gegeben, wobei $B_{n-1,k}$ wie in (2.3) definiert ist und in $C[a,b]$ enthalten sein soll $(n>2)$. Für alle $s \in B_{n-1,k}$ gelte $\int_{Z(f-s_o)} |s(x)| dx = 0$, wenn $Z(f-s_o)$ die Menge der Nullstellen von $f-s_o$ ist.

Die Funktion s_o ist genau dann Minimallösung an f bezüglich der Approximation in der L_1-Norm, falls

$$(3.1) \qquad \int_a^b s(x) \operatorname{sgn} (f(x) - s_o(x)) dx = 0 \quad s \in B_{n-1,k}$$

gilt.

Existenz und Eindeutigkeit der Lösung sind gesichert (siehe Strauß [41]). Satz 3.1 kann man aufgrund der Beziehung des Momentenproblems zu einer Interpolationsaufgabe mit Perfektsplines auch folgendermaßen formulieren.

Satz 3.2. Die Voraussetzungen seien wie in Satz 3.1. Außerdem besitze $f-s_o$ nur endlich viele Nullstellen und wechsle in den Punkten $(u_i)_{i=1}^r$ das Vorzeichen.
Der Spline $s_o \in B_{n-1,k}$ ist genau dann eine Minimallösung an f, wenn ein Vektor $\alpha = (c_o, a_o, \ldots, a_{n-1})$ existiert, so daß der Perfektspline

$$(3.2) \qquad P(x) = \sum_{i=o}^{n-1} a_i x^i + c_o \Big(x^n + \sum_{i=1}^{r} 2(-1)^i (x-u_i)_+^n\Big)$$

die Nullstellen

(3.3) $P(x_i) = 0 \qquad i=1,\dots,n+k$

besitzt.

Beweis: Wenn s_0 Minimallösung ist, gilt

$$\int_a^b s(x)\ \mathrm{sgn}\ (f(x) - s_0(x)dx = 0.$$

Nach Satz 2.3 (b) existiert ein Perfektspline P mit den genannten Eigenschaften.

Existiert umgekehrt ein Perfektspline P mit den Eigenschaften (3.2) und (3.3), dann gilt nach Satz 2.3 (a)

$$\int_a^b s(x)\ P^{(n)}(x)dx = 0.$$

Außerdem folgt aus den Voraussetzungen

$$P^{(n)}(x) = \mathrm{sgn}(f(x) - s_0(x)) \qquad x \in [a,b]$$

bis auf höchstens endlich viele Punkte, in denen $f-s_0$ Nullstellen besitzen muß. Dies bedeutet aber nach Satz 3.1, daß s_0 eine Minimallösung ist.

Dieses Kriterium soll noch in einem sehr einfachen Fall angegeben werden.

Es sei

$$S_{n-1,k}[a,b] = \{s \mid s(x) = \sum_{i=0}^{n-1} a_i x^i + \sum_{i=1}^{k} b_i (x-x_i)_+^{n-1}\}$$

gegeben.

Korollar: Der Spline $s_0 \in S_{n-1,k}$ ist genau dann eine Minimallösung an die Funktion f bezüglich $S_{n-1,k}$ auf dem Intervall [a,b], wenn der Perfektspline

$$P(x) = x^n + \sum_{i=1}^{r} 2(-1)^i (x-u_i)_+^n$$

die Nullstellen

$$P^{(i)}(a) = P^{(i)}(b) = 0 \qquad i=0,\ldots,n-1$$

$$P(x_i) = 0 \qquad i=1,\ldots,k$$

besitzt, wobei $(u_i)_{i=1}^r$ die Punkte sind, an denen $f-s_0$ das Vorzeichen wechselt und falls $\int_{Z(f-s_0)} |s(x)|dx=0$ für $s \in S_{n-1,k}$.
Die Aussage läßt sich aus Satz 3.2 folgern.

Um nachzuprüfen, ob eine vorgegebene Funktion $s \in S_{n-1,k}$ Minimallösung an f ist, muß man zunächst die Vorzeichenwechsel von $f-s_0$ bestimmen und beim zugeordneten Perfektspline nachrechnen, ob er die geforderten Nullstellen am Rand, bzw. in den Knoten von $S_{n-1,k}$ besitzt.

Es ist häufig auch nötig, Randbedingungen zuzulassen.
Ein sehr bekanntes Beispiel liefert der Bernouillische Monospline, dessen Minimaleigenschaften schon Micchelli untersucht hat.

Dieser Monospline ist die Fehlerfunktion eines L_1-Approximationsproblems und ein etwas modifizierter Eulerspline ist der zugeordnete Perfektspline im Sinne von Satz 3.2.
Der n-te Bernouillische Monospline $\overline{B}_n$ wird definiert als

$$\overline{B}_n(x) = B_n(x) \qquad 0 \leq x \leq 1$$

$$\overline{B}_n(x+1) = \overline{B}_n(x),$$

wobei B_n das n-te Bernouillische Polynom sein soll.

Folgende Eigenschaften werden für uns Bedeutung haben.

Lemma 3.3. (a) Für den Monospline $\overline{B}_{2t+1}$, wobei t eine natürliche Zahl sei, gilt auf dem Intervall $[0,m+1]$

(α) $\overline{B}_{2t+1}$ hat die Nullstellen $(\frac{1}{2}\, i)_{i=0}^{2m+2}$

(β) $\overline{B}_{2t+1}^{(j)}(0) = \overline{B}_{2t+1}^{(j)}(m+1) = 0 \qquad j=0,\ldots,2t-2$

(b) Für den Monospline $\overline{B}_{2t}$, wobei t eine natürliche Zahl sein soll, gilt auf dem Intervall $[0,m+1]$

(α) $\overline{B}_{2t}$ hat die Nullstellen $(\frac{1}{2}i + \frac{1}{4})_{i=0}^{2m+1}$

(α) $\overline{B}_{2t}^{(j)}(0) = \overline{B}_{2t}^{(j)}(m+1) = 0 \qquad j=1,\ldots,2t-3$

Beweis: (α) folgt jeweils aus der Definition.

(β) ergibt sich aus $B_n'(x) = nB_{n-1}(x)$ für $n=1,2,3..$ und der Definition der Bernouillischen Polynome.

Für die weiteren Überlegungen benötigen wir eine transformierte Form des Eulersplines.

Es sei $E_n(x)$ das Eulerpolynom vom Grad n auf dem Intervall $[0,1]$. $\overline{E}_n$ sei die Fortsetzung

$$\overline{E}_n(x) = E_n(x) \qquad 0 \le x \le 1$$

$$\overline{E}_n(x+1) = -\overline{E}_n(x).$$

$\overline{E}_n$ wird auch Eulerspline genannt. Wir definierten folgende Transformationen:

$$\widetilde{E}_{2t+1}(x) = (-1)^t \overline{E}_{2t+1}(2x) + (-1)^{t+1} E_{2t+1}(0)$$

$$\tilde{E}_{2t}(x) = (-1)^t \bar{E}_{2t}(2x-\tfrac{1}{2}) + (-1)^t E_{2t}(\tfrac{1}{2})$$

Lemma 3.4. $\tilde{E}_n$ hat folgende Eigenschaften auf dem Intervall $[0,m+1]$:

(a) Es sei $n = 2t+1$ und t eine natürliche Zahl.

(α) $\tilde{E}_{2t+1}$ hat Knoten in $(\frac{1}{2}i)_{i=1}^{2m+1}$

(β) doppelte Nullstellen in $(i)_{i=1}^{m}$

(γ) $\tilde{E}_{2t+1}^{(j)}(0) = \tilde{E}_{2t+1}^{(j)}(m+1) = 0 \quad j=0,1,3,\ldots,2t-1$

(b) Es sei $n = 2t$ und t eine natürliche Zahl.

(α) $\tilde{E}_{2t}$ hat Knoten in $(\frac{1}{2}i + \frac{1}{4})_{i=0}^{2m+1}$

(β) doppelte Nullstellen in $(i)_{i=1}^{m}$

(γ) $\tilde{E}_{2t}^{(j)}(0) = \tilde{E}_{2t}^{(j)}(m+1) = 0 \qquad j=0,1,3,\ldots,2t-1.$

Beweis: (α) und (β) ergeben sich aus der Definition.
(γ) folgt aus $E_n'(x) = nE_{n-1}(x)$ und den Nullstelleneigenschaften der Eulerpolynome.

Nun haben wir alle Voraussetzungen zusammengestellt, um unser Approximationsproblem betrachten zu können. Die Funktion x^n wird durch Splinefunktionen approximiert. Dabei ist der gerade und der ungerade Fall von n zu unterscheiden.

(I) Es sei $n = 2t+1$ und sei folgender Splineraum definiert

$$S_{2t,m}[0,m+1] = \{s \mid s(x) = \sum_{i=0}^{2t} a_i x^i + \sum_{i=1}^{m} b_i (x-i)_+^{2t} + \sum_{i=1}^{m} c_i (x-i)_+^{2t-1}\}.$$

Satz 3.5. Auf dem Intervall $[0,m+1]$ wird die Funktion

x^{2t+1} durch Elemente des Raumes $S_{2t,m}$ in der L_1-Norm approximiert, wobei noch die Randbedingungen

$$M^{(i)}(0) = M^{(i)}(m+1) = 0 \quad i=0,2,\ldots,2t-2$$

erfüllt sein sollen mit $M \in \bar{M}$

$$\bar{M} = \{M(x)=x^{2t+1}-s(x) \mid s \in S_{2t,m}\}.$$

Eine Minimallösung wird durch

$$s_0(x) = x^{2t+1} - \bar{B}_{2t+1}(x)$$

geliefert, wobei $\bar{B}_{2t+1}$ ein Bernouillischer Monospline ist.

<u>Beweis:</u> $\bar{B}_{2t+1}$ werde durch Elemente aus

$$\bar{S}_{2t,m} = \{s \in S_{2t,m} \mid s^{(i)}(0) = s^{(i)}(m+1) = 0$$

$$i=0,2,\ldots,2t-2 \}$$

approximiert. Es zeigt sich, daß die Nullfunktion Minimallösung aus $\bar{S}_{2t,m}$ an B_{2t+1} ist. Daraus folgt die Aussage des Satzes. Es muß also

$$\int_0^{m+1} s(x) \operatorname{sgn} (x^{2t+1} - s_0(x))dx = 0 \quad s \in \bar{S}_{2t,m}$$

gezeigt werden. Aufgrund der Eigenschaften des Eulersplines $\tilde{E}_{2t+1}$ zeigt sich nach Lemma 3.4(a), daß

$$\tilde{E}^{(2t+1)}_{2t+1}(x) = \operatorname{sgn} (\bar{B}_{2t+1}(x)) \quad x \in [0,m+1]$$

gilt, denn die Knoten von $\tilde{E}_{2t+1}$ liegen genau in Punkten, in denen $\bar{B}_{2t+1}$ das Vorzeichen wechselt. Nun gilt für alle $s \in \bar{S}_{2t,m}$

$$\int_0^{m+1} s(x)\tilde{E}_{2t+1}^{(2t+1)}(x)dx = \sum_{i=0}^{2t} (-1)^i s^{(i)}(x)\tilde{E}_{2t+1}^{(2t-i)}(x)\Big|_0^{m+1}$$

$$+ \sum_{i=1}^{m} \sum_{j=0}^{1} (-1)^{(2t-j)} s^{(2t-j)}(x)\, \tilde{E}_{2t+1}^{(j)}(x)\Big|_{i+0}^{i-0} .$$

Nach den Aussagen von Lemma 3.4(a) verschwinden alle Ausdrücke der rechten Seite. Damit verschwindet das Integral für alle $s \in \bar{S}_{2t,m}$ und der Satz ist gezeigt.

(II) Sei $n=2t$ und

$$S_{2t-1,m}[0,m+1] = \{s \mid s(x) = \sum_{i=0}^{2t-1} a_i x^i + \sum_{i=1}^{m} b_i (x-i)_+^{2t-1}$$
$$+ \sum_{i=1}^{m} c_i (x-i)_+^{2t-2} \}.$$

Satz 3.6. Auf dem Intervall $[0,m+1]$ wird die Funktion x^{2t} durch Elemente des Raumes $S_{2t-1,m}$ in der L_1-Norm approximiert, wobei noch die Randbedingungen

$$M^{(i)}(0) = M^{(i)}(m+1) = 0 \quad i=1,3,\dots,2t-3$$

erfüllt sein sollen mit $M \in \bar{M}$ und

$$\bar{M} = \{M(x) = x^{2t} - s(x) \mid s \in S_{2t-1,m}\}.$$

Eine Minimallösung wird durch

$$s_o(x) = x^{2t} - \bar{B}_{2t}(x)$$

geliefert, wobei $\bar{B}_{2t}$ ein Bernouillischer Monospline ist.

Beweis: Der Beweis verläuft entsprechend Satz 3.5, nur werden hier statt Lemma 3.4(a) die Aussagen von Lemma 3.4(b) verwendet.

4. Approximation mit freien Knoten.

Jetzt wollen wir uns dem Approximationsproblem mit freien Knoten zuwenden. Es sei

$$(4.1)\quad S^{o}_{n-1,k}[a,b] = \{s \mid s(x) = \sum_{i=0}^{n-1} a_i x^i + \sum_{i=1}^{k} b_i (x-x_i)_+^{n-1}\}$$

gegeben, wobei $a<x_1<x_2<\ldots<x_k<b$ ist und a_i, b_i reelle Zahlen sind. Die Variablen sind x_i, a_i und b_i. Gesucht wird die beste Approximation an eine Funktion $f \in C[a,b]$. Die Parametermenge ist offen. Daher läßt sich ein notwendiges Kriterium für Minimallösungen aus der nichtlinearen Approximationstheorie anwenden (siehe Brosowski).

Satz 4.1. Es sei eine Funktion $f \in C[a,b]$ gegeben und

$$s_o(x) = \sum_{i=0}^{n-1} \alpha_i x^i + \sum_{i=1}^{k} \beta_i (x-x_i^o)_+^{n-1}$$ eine Minimallösung an f

in der L_1-Norm für das Approximationsproblem mit freien Knoten. Dann ist s_o auch Minimallösung bezüglich folgender linearer Approximationsaufgabe:

Der Spline s_o ist Minimallösung an f, wenn f durch den Raum

$$(4.2)\quad S_{n-1,2k}[a,b] = \{s \mid s(x) = \sum_{i=0}^{n-1} a_i x^i + \sum_{i=1}^{k} b_i (x-x_i^0)_+^{n-1} + \sum_{i=1}^{k} c_i (x-x_i^0)_+^{n-2}\}$$

in der L_1-Norm approximiert wird (mit festen Knoten).

Aus dieser Eigenschaft läßt sich folgende Charakterisierung ableiten.

<u>Satz 4.2.</u> Die Funktion $s_0 \in S^0_{n-1,k}$ löse das Approximationsproblem mit freien Knoten für die Funktion $f(x)=x^n$. Dann gilt:

(a) Die Fehlerfunktion $x^n - s_0(x)$ besitzt genau $n+2k$ Nullstellen $(u_i)_{i=1}^{n+2k}$ mit

$$a<u_1< \ldots <u_{n+2k}<b .$$

(b) Der Perfektspline P

$$P(x) = x^n + 2 \sum_{i=1}^{n+2k} (-1)^i (x-u_i)_+^n$$

besitzt die Nullstellen

$$P^{(i)}(a) = P^{(i)}(b) = 0 \qquad i=0,\ldots,n-1$$

$$P(x_i) = P'(x_i) = 0 \qquad i=1,\ldots,k.$$

<u>Beweis:</u> Da s_0 Minimallösung ist, gilt nach Satz 4.1 und Satz 3.1

$$\int_a^b s(x) \operatorname{sgn} (x^n - s_0(x))dx = 0 \qquad s \in S_{n-1,2k},$$

wobei $S_{n-1,2k}$ wie in (4.2) definiert sei. Nach Satz 2.3 (b) folgt die Existenz eines Perfektsplines mit den Eigenschaften von (b).

Das Approximationsproblem ist auf einfache Knoten beschränkt. Dies ist keine wirkliche Einschränkung, denn in einem Satz in §6 werden wir zeigen, daß zu jeder Splinefunktion s_o, die mehrfache Knoten besitzt, stets ein Spline s_1 mit einfachen Knoten angegeben werden kann, der x^n besser approximiert als s_o. Aus beweistechnischen Gründen wird es erst später gezeigt. Splinefunktionen, die die Eigenschaften von Satz 4.2 erfüllen, gewinnt man durch ein Iterationsverfahren.
In der Charakterisierung tauchte ein Perfektspline auf, mit dem man lineare Funktionale definieren kann, die in den folgenden Paragraphen wichtig sein werden.
Es sei P ein Perfektspline vom Grad n der Gestalt

$$P(x) = \frac{(-1)^n}{n!} \left(x^n + 2 \sum_{i=1}^{n+2k} (-1)^i (x-u_i)_+^n \right)$$

mit

$$a<u_1<u_2< \ldots <u_{n+2k}<b,$$

der

$$P^{(i)}(a) = P^{(i)}(b) = 0 \qquad i=0,\ldots,n-1$$

$$P(x_i) = P'(x_i) = 0 \qquad i=1,\ldots,k$$

erfüllt.
Dieser Spline existiert und ist eindeutig bestimmt nach Satz 2.2 und Satz 2.8.

Lemma 4.3. Es gilt $(-1)^n P(x) \geq 0$ für $x \in [a,b]$ und $P(x)=0$ für $x \in (a,b)$ genau dann, wenn $x \in (x_1,\ldots,x_k)$.

Beweis: P hat in (a,b) doppelte Nullstellen in x_i und darf nach einem Nullstellensatz keine weiteren Nullstellen besitzen. Also hat P keine Vorzeichenwechsel. Aufgrund des Faktors $(-1)^n$ folgt, daß $(-1)^n P(x)$ in einer Umgebung von a nicht-negativ ist, also auf $[a,b]$ nichtnegativ.

Nun definieren wir ein lineares Funktional. Es sei

$$Lf = \int_a^b f(x) P^{(n)}(x)dx.$$

Dieses lineare Funktional besitzt folgende Eigenschaften

Lemma 4.4. (a) $L(x^i) = 0 \quad i=0,\ldots,n-1$

(b) $L(x-t)_+^{n-1} = (-1)^n (n-1)! P_n(t) \geq 0$ und

$L(x-t)_+^{n-1} = 0$ genau dann für $t \in (a,b)$, wenn $t \in (x_i)_{i=1}^k$.

Beweis: Durch partielle Integration erhält man

$$L((x-t)_+^{n-1}) = \int_a^b (x-t)_+^{n-1} P^{(n)}(x)dx =$$

$$= \sum_{i=0}^{n-2} (-1)^i \frac{(n-1)!}{(n-i-1)!} (x-t)_+^{n-1-i} P^{(n-i-1)}(x)\Big|_a^b$$

$$+ (-1)^n (n-1)! P(t)$$

$$= (-1)^n (n-1)! P(t).$$

Daraus folgen die Aussagen von (a) und (b) unter Benutzung von Lemma 4.3.

5. Ein Iterationsverfahren.

In diesem Paragraphen wird ein Iterationsverfahren definiert. Es gibt eine Vorschrift an, wie aus einem gegebenen Spline

$s \in S^{o}_{n-1,k}$ (siehe(4.1)) ein Spline s_1 gewonnen werden kann, der x^n besser approximiert als der ursprüngliche Spline s. Es sollen dabei auch noch Randbedingungen zugelassen werden.

Iterationsverfahren:

(a) Gegeben sei ein Spline $s_\nu \in S^{o}_{n-1,k}$, so daß für $M_\nu(x) = x^n - s_\nu(x)$ folgendes gilt:

$$M_\nu^{(i)}(a) = 0 \qquad i=0,\dots,p-1$$

(5.1)

$$M_\nu^{(i)}(b) = 0 \qquad i=0,\dots,q-1,$$

wobei p,q natürliche Zahlen mit $0 \leq p,q \leq n$ sind und außerdem

$$s_\nu(x) = \sum_{i=0}^{n-1} a_{i\nu} x^i + \sum_{i=1}^{k} b_{i\nu}(x-x_{i\nu})_+^{n-1}$$

gilt.

(b) Ein Iterationsschritt liefert ausgehend von einem Spline $s_\nu \in S^{o}_{n-1,k}$, für den $M_\nu(x) = x^n - s_\nu(x)$ die Randbedingungen (5.1) erfüllt, einen neuen Spline $s_{\nu+1}$, wobei $M_{\nu+1}(x) = x^n - s_{\nu+1}(x)$ ebenfalls (5.1) erfüllt.

Es sei also s_ν vorgegeben:

(α) Dann wird der nach Satz 2.8 eindeutig bestimmte Perfektspline vom Grad n mit N = 2k+n-p-q Knoten gebildet

(5.2) $$P_\nu(x) = \sum_{i=0}^{n-1} c_{i\nu} x^i + c_\nu\Big(x^n + 2\sum_{i=1}^{N} (-1)^i (x-u_{i\nu})_+^n\Big),$$

wobei

$$P_\nu^{(i)}(a) = 0 \qquad i=0,\ldots,n-p-1$$

$$P_\nu^{(i)}(b) = 0 \qquad i=0,\ldots,n-q-1$$

(5.3) $$P_\nu(x_{i\nu}) = P_\nu'(x_{i\nu}) = 0 \qquad i=1,\ldots,k$$

$$||P_\nu^{(n)}|| = 1$$

$$(-1)^n P_\nu(x) \geq 0 \qquad x\in[a,b]$$

sein soll.

(β) Man bestimmt nun einen Spline $s_{\nu+1}\in S^o_{n-1,k}$ durch $s_{\nu+1}(x) = x^n - M_{\nu+1}(x)$, wobei $M_{\nu+1}$ ein Monospline

$$M_{\nu+1}(x) = x^n - \sum_{i=0}^{n-1} a_{i,\nu+1}x^i - \sum_{i=1}^{k} b_{i,\nu+1}(x-x_{i,\nu+1})_+^{n-1}$$

ist, der

$$M_{\nu+1}^{(i)}(a) = 0 \qquad i=0,\ldots,p-1$$

(5.4) $$M_{\nu+1}^{(i)}(b) = 0 \qquad i=0,\ldots,q-1$$

$$M_{\nu+1}(u_{i\nu}) = 0 \qquad i=1,\ldots,N$$

erfüllt.

Im folgenden soll gezeigt werden, daß $\int_a^b |M_\nu(x)|dx \geq \int_a^b |M_{\nu+1}(x)|dx$ gilt. Das Gleichheitszeichen ist genau dann erfüllt, wenn $s_\nu = s_{\nu+1}$ ist. Dazu benötigen wir einige Hilfssätze.

<u>Lemma 5.1.</u> Es gilt

(5.5) $$L_\nu(x^n - s_\nu(x)) = \int_a^b M_\nu(x)P_\nu^{(n)}(x)dx =$$

$$= \sum_{i=1}^{k} (-1)^{n-1} b_{i\nu}(n-1)!P_\nu(x_{i\nu}) + (-1)^n n! \int_a^b P_\nu(x)dx.$$

Beweis: Durch partielle Integration beweist man die Beziehung

$$\int_a^b M_\nu(x)P_\nu^{(n)}(x)dx = \sum_{i=0}^{n-1} (-1)^i M_\nu^{(i)}(x)P_\nu^{(n-i-1)}(x)\Big|_a^b$$

$$+ \sum_{i=1}^{k} (-1)^{n-1} M_\nu^{(n-1)}(x)P_\nu(x)\Big|_{x_{i\nu}+0}^{x_{i\nu}-0}$$

$$+ (-1)^n n! \int_a^b P_\nu(x)dx \; .$$

Aufgrund der Randbedingungen von M_ν und P_ν, die in (5.3) und (5.4) beschrieben sind, fallen die Ausdrücke der ersten Summe der rechten Seite weg. Außerdem gilt

$$M_\nu^{(n-1)}(x)\Big|_{x_{i\nu}+0}^{x_{i\nu}-0} = b_{i\nu}(n-1)! .$$

Daraus folgt die Behauptung (5.5).

In den nächsten beiden Sätzen werden Aussagen über das Vorzeichenverhalten von Perfekt- und Monosplines gemacht.

Lemma 5.2. Es sei P ein Perfektspline über $[a,b]$ vom Grad n mit $N=n+2k-p-q$ Knoten. Er besitze genau $n+N$ Nullstellen. Außerdem gelte $(-1)^n P(x) \geq 0$ für $x \in [a,b]$ und $P^{(i)}(a)=0$ für $i=0,\dots,n-p-1$ und $P^{(n-p+i)}(a) \neq 0$ für $i=0,..,p$. Dann gilt

$$(-1)^{n-p} P^{(n)}(a) > 0 \; .$$

Beweis: Da $(-1)^n P(x) \geq 0$ für $x \in [a,b]$ ist, gilt

$(-1)^n P^{(n-p)}(a)>0$. Für die weiteren Ableitungen hat man

$$P^{(n-p+i)}(a)\, P^{(n-p+i+1)}(a) < 0 \qquad i=0,\dots,p-1.$$

Dies ergibt sich folgendermaßen. Angenommen es sei $P^{(n-p+i)}(a)>0$ für ein $0\le i<p$. Nach Lemma 2.1 in [21] hat der Spline $P^{(n-p+i)}$ höchstens $N+p-i$ Nullstellen. Nach dem Satz von Rolle muß er aber wenigstens $N+p-i$ Nullstellen haben, denn P hatte $N+n$ Nullstellen. Also hat P genau $N+p-i$ Nullstellen. Sei u_1 die erste Nullstelle von $P^{(n-p+i)}$. Dann gilt $P^{(n-p+i+1)}(x)<0$ für $x\in[a,u_1)$. Um dies zu beweisen, nimmt man an, daß $P^{(n-p+i+1)}$ eine Nullstelle in $[a,u_1)$ besitzt. Nach dem Satz von Rolle muß $P^{(n-p+i+1)}$ in $[u_1,b]$ weitere $N+p-i-1$ Nullstellen besitzen, also insgesamt $N+p-i$ Nullstellen. Dies ist ein Widerspruch zum Nullstellensatz für Polynomsplines. Daher gilt

$$P^{(n-p+i)}(a)\, P^{(n-p+i+1)}(a) < 0.$$

Ebenso kann man die Aussage zeigen, wenn $P^{(n-p+i)}(a)<0$ angenommen wird.

Unter Berücksichtigung der Tatsache, daß $(-1)^n P^{(n-p)}(a)>0$ ist, folgt die Behauptung des Satzes.

<u>Lemma 5.3.</u> Es sei ein Monospline M vom Grad n mit k Knoten und $n+2k$ Nullstellen gegeben. Weiterhin sei $M^{(i)}(a)=0$ für $i=0,\dots,p-1$ und $M^{(p)}(a)\neq 0$. Dann folgt, daß ein $\varepsilon>0$ existiert mit

$$(-1)^{n-p}M(x) > 0 \quad \text{für } x\in(a,\varepsilon).$$

Beweis: Nach Voraussetzung gilt $M^{(i)}(a)=0$ für $i=0,..,p-1$. Mit den gleichen Argumenten wie im vorangegangenen Satz läßt sich zeigen, daß

$$M^{(p+i)}(a)\,M^{(p+i+1)}(a) < 0 \quad i=1,\dots,n-p-1$$

ist. Da M die maximale Anzahl an zulässigen Nullstellen besitzt, gilt nach einem Schluß, wie er in Lemma 2.4 in [21] verwendet wird

$$-M^{(n-1)}(a) > 0.$$

Daraus folgt $(-1)^{n-p}M^{(p)}(a)>0$.

Lemma 5.4. Es sei ein Monospline

$$M(x) = x^n - \sum_{i=0}^{n-1} a_i x^i - \sum_{i=1}^{k} b_i (x-x_i)_+^{n-1}$$

gegeben, der die Nullstellen

$$M^{(i)}(a) = 0 \quad i=0,\dots,p-1$$

$$M^{(i)}(b) = 0 \quad i=0,\dots,q-1$$

$$M(u_i) = 0 \quad i=1,\dots,N$$

besitzt, wobei $a<u_1<\dots<u_N<b$ und $N=n+2k-p-q$ ist. Dann folgt

$$b_i>0 \quad i=1,\dots,k.$$

Beweis: Man kann den Beweis von [21] in Lemma 2.4 verwenden. Der einzige Unterschied zu unserem Satz besteht darin, daß hier am Rand des Intervalls mehrfache Nullstellen zugelassen sind.

Mit diesen Hilfssätzen können wir jetzt folgendes Ergebnis zeigen.

Satz 5.5. Es sei (s_ν) eine Folge des Iterationsverfahrens, wobei $s_\nu \in S^o_{n-1,k}$ ist. Dann gilt

$$\int_a^b |x^n - s_{\nu+1}(x)|dx \leq \int_a^b |x^n - s_\nu(x)|dx.$$

Das Gleichheitszeichen gilt genau dann, wenn $s_\nu = s_{\nu+1}$ ist.

Beweis: Es gelten die Beziehungen

$$\begin{aligned}(5.6)\quad &\int_a^b |x^n - s_\nu(x)|dx = \int_a^b (x^n - s_\nu(x)) \operatorname{sgn}(x^n - s_\nu(x)) \geq \\ &\int_a^b (x^n - s_\nu(x))\, P_\nu^{(n)}(x)dx = (-1)^n n! \int_a^b P_\nu(x)dx.\end{aligned}$$

Die letzte Aussage folgt aus Lemma 5.1. Weiterhin ist

$$(5.7)\quad \int_a^b |x^n - s_{\nu+1}(x)|dx = \int_a^b (x^n - s_{\nu+1}(x))\, P_\nu^{(n)}(x)dx$$

nach Lemma 5.2 und Lemma 5.3. Lemma 5.1 ergibt

$$\begin{aligned}&\int_a^b (x^n - s_{\nu+1}(x))P_\nu^{(n)}(x)dx = \\ (5.8)\quad &\sum_{i=1}^k (-1)^{n-1} b_{i,\nu+1}\, (n-1)!\, P_\nu(x_{i,\nu+1}) + \\ &(-1)^n n! \int_a^b P_\nu(x)dx.\end{aligned}$$

Nun gilt $b_{i\nu} > 0$ für $i=1,\ldots,k$ nach Lemma 5.4 und

$$(-1)^n P_\nu(x) \geq 0 \qquad x\in[a,b].$$

Also ist $(-1)^{n-1}P_\nu(x_{i,\nu+1})\leq 0$. Dies bedeutet, daß die erste Summe auf der rechten Seite von (5.8) nichtpositiv ist. Durch Zusammenfassung von (5.6), (5.7) und (5.8) folgt

$$\int_a^b |x^n - s_\nu(x)|dx \geq \int_a^b |x^n - s_{\nu+1}(x)|dx - \sum_{i=1}^{k} (-1)^{n-1} b_{i,\nu+1}(n-1)!P_\nu(x_{i,\nu+1}). \tag{5.9}$$

Wenn $x_{i\nu}=x_{i,\nu+1}$ für alle $i=1,\ldots,k$ gilt, dann ist $P_\nu=P_{\nu+1}$, d.h. auch $s_\nu=s_{\nu+1}$ und damit

$$\int_a^b |x^n - s_\nu(x)|dx = \int_a^b |x^n - s_{\nu+1}(x)|dx.$$

Gilt wenigstens für ein μ, daß $x_{\mu,\nu} \neq x_{\mu,\nu+1}$ ist, dann hat man

$$(-1)^{n-1}P_\nu(x_{\mu,\nu+1}) < 0.$$

Daraus folgt nun

$$\int_a^b |x^n - s_\nu(x)|dx > \int_a^b |x^n - s_{\nu+1}(x)|dx ,$$

denn der letzte Ausdruck von (5.9) wird positiv.

<u>Bemerkung:</u> Man kann mit Hilfe des Iterationsverfahrens die Splinefunktionen gewinnen, die die Eigenschaften besitzen, wie sie die in Satz 4.2 beschriebene Minimallösung hat.

6. Eine Reduktion des Approximationsproblems.

Eigentlich müßte man mehrfache Knoten beim Approximationsproblem mit freien Knoten zulassen. In diesem Paragraphen wird gezeigt, daß man darauf verzichten kann.

Es wird der Splineraum $S_{n-1,k}$ von (1.4) zugrunde gelegt.

Satz 6.1. Es sei $s_1 \in S_{n-1,k}$ gegeben, für das die Funktion $M_1(x) = x^n - s_1(x)$ die Bedingungen

$$(6.1)\quad \begin{aligned} M_1^{(i)}(a) &= 0 \qquad i=0,\ldots,p-1 \quad 0 \le p,q \le n \\ M_1^{(i)}(b) &= 0 \qquad i=0,\ldots,q-1 \end{aligned}$$

erfüllt. Dabei besitze der Spline höchstens k (nicht notwendigerweise einfache) Knoten. Dann kann man einen Spline s_2 angeben, so daß $M_2(x) = x^n - s_2(x)$ die Randbedingungen (6.1) wie M_1 erfüllt, s_2 höchstens k einfache Knoten besitzt und

$$(6.2)\quad \int_a^b |x^n - s_1(x)|\,dx \ge \int_a^b |x^n - s_2(x)|\,dx$$

ist.

Beweis: Der Spline s_1 sei durch

$$s_1(x) = \sum_{i=0}^{n-1} c_i x^i + \sum_{i=1}^{r} \sum_{j=1}^{m_i} d_{ij} (x-y_i)_+^{n-j}$$

dargestellt und $\sum_{i=1}^{r} m_i \le k$. Man konstruiert einen Perfektspline P vom Grad n, der die folgenden Eigenschaften hat

$$P^{(i)}(a) = 0 \quad i=0,\dots,n-q-1$$

(6.3) $$P^{(i)}(b) = 0 \quad i=0,\dots,n-q-1$$

$$P^{(j)}(y_i) = 0 \quad 0 \le j \le h_i - 1 ,$$

wobei $h_i = m_i$, falls m_i gerade

$h_i = m_i + 1$, falls m_i ungerade.

Es sind insgesamt $2n+h-p-q$ Nullstellen vorgegeben, wenn

$h = \sum_{i=1}^{r} h_i$ ist. Nach Satz 2.8 gibt es genau einen Perfekt-spline P vom Grad n mit $t=n+h-p-q$ Knoten, der noch folgenden Bedingungen genügt

$$(-1)^n P(x) \ge 0 \qquad x \in [a,b]$$

$$||P^{(n)}|| = 1 .$$

Nach den gleichen Überlegungen wie in Lemma 4.3 besitzt P keine Vorzeichenwechsel. P ist gegeben durch

$$P(x) = \sum_{i=0}^{n-1} a_i x^i + c(x^n + 2 \sum_{i=1}^{t} (-1)^i (x-u_i)_+^n).$$

Wie in Satz 5.5 erhält man

(6.4) $$\int_a^b |x^n - s_1(x)| dx \ge \int_a^b (x^n - s_1(x)) P^{(n)}(x) dx = = (-1)^n n! \int_a^b P(x) dx.$$

Der Monospline

$$M_2(x) = x^n - \sum_{i=0}^{n-1} a_i x^i - \sum_{i=1}^{h'} b_i (x-x_i)_+^{n-1}$$

erfülle die Bedingungen

$$M_2(u_i) = 0 \qquad i=1,\dots,t$$

$$M_2^{(i)}(a) = 0 \qquad i=0,\dots,p-1$$

$$M_2^{(i)}(b) = 0 \qquad i=0,\dots,q-1 \ ,$$

wobei $h' = \frac{h}{2}$ ist. Es wird nun gezeigt, daß

$$s_2(x) = \sum_{i=0}^{n-1} a_i x^i - \sum_{i=1}^{h'} b_i (x-x_i)_+^{n-1}$$

die gesuchte Funktion ist. Es gilt

$$\int_a^b |M_2(x)|dx = \int_a^b M_2(x)\ P^{(n)}(x)dx =$$

(6.5)

$$\sum_{i=1}^{h'} (-1)^{n-1} b_i (n-1)!\ P(x_i) + (-1)^n n! \int_a^b P(x)dx.$$

Nach Lemma 5.4 ist $b_i > 0$ für $i=1,\dots,h'$ und damit

$$\sum_{i=1}^{h'} (-1)^{n-1} b_i (n-1)! P(x_i) \leq 0.$$

Ein Vergleich von (6.4) und (6.5) ergibt

$$\int_a^b |M_1(x)|dx \geq \int_a^b |M_2(x)|dx$$

gilt.

7. Einseitige Approximation.

Das einseitige Approximationsproblem, das dem letzten Paragraphen zu Grunde liegt, lautet:
Es sei eine Funktion f aus C[a,b], dem Raum der stetigen, reellwertigen Funktionen auf dem Intervall [a,b] , gegeben. Außerdem sei der Splineraum $B_{n-1,k}$ vorgegeben ($n \geq 2$), der wie in (2.3) definiert sei.
Gesucht ist ein Element $s_o \in B_{n-1,k}$, für das

$$(7.1) \qquad \int_a^b (f(x)-s_o(x))dx \leq \int_a^b (f(x)-s(x))dx \qquad s \in B_{n-1,k}(f)$$

gilt, wobei s, s_o Elemente von

$$(7.2) \qquad B_{n-1,k}(f) = \{ s \in B_{n-1,k} \mid (f(x)-s(x)) \geq 0 \ , \ x \in [a,b] \ \}$$

sind.

Satz 7.1. Das genannte Approximationsproblem besitzt stets eine Lösung.

Für unsere weiteren Betrachtungen wird die Nullstellenmenge der Fehlerfunktion von Bedeutung sein.
Hier sollen nur solche Approximationsprobleme betrachtet werden, die für die Approximation linearer Funktionale von Bedeutung sind.

Definition 7.2. Es sei $f-s_o$ eine Fehlerfunktion für das Problem (7.1). Jeder Punkt $\bar{x} \in [a,b]$, für den $(f-s_o)(\bar{x})=0$ gilt, heiße Kontaktpunkt.

Für unsere Betrachtungen werden Kontaktpunktmengen mit einer bestimmten Eigenschaft wichtig sein.

Definition 7.3. Es sei die Knotenmenge $a \leq x_1 \leq \ldots \leq x_{n+k} \leq b$ gegeben. Dann besitze die Menge $U = (u_i)_{i=1}^r$ folgende Eigenschaft:

Es existiert eine Menge von Punkten $(v_i)_{i=1}^k$, so daß

$$(7.3) \qquad \begin{array}{ll} (u_i)_{i=1}^r \subset (v_i)_{i=1}^k & v_i < v_{i+1} \\ v_i \in (x_i, x_{i+n}) & i=1,\ldots,k \end{array}$$

gilt. Falls in a bzw. b n-fache Knoten vorliegen, wird

$$v_1 \in [x_1, x_{n+1}) \quad \text{bzw.} \quad v_k \in (x_k, x_{n+k}]$$

zugelassen.

Falls eine Minimallösung s_0 an f, die das Approximationsproblem (7.1) löst, eine Fehlerfunktion besitzt, deren Kontaktpunktmenge die Eigenschaft von Def. 7.3 hat, kann man daraus eine Quadraturformel ableiten. Ein Beweis, der bei DeVore angegeben ist, läßt sich auf unser Problem übertragen.

Satz 7.4. Es sei $s_0 \in B_{n-1,k}$ eine Minimallösung an $f \in C[a,b]$. Die Fehlerfunktion $f-s_0$ habe die Kontaktpunkte $U = (u_i)_{i=1}^r$, die die Eigenschaften von Def. 7.3 besitzen. Dann existiert eine Quadraturformel

$$Qs = \sum_{i=1}^{r} A_i s(u_i) \quad A_i > 0 \quad i=1,\dots,r,$$

die alle Elemente $s \in B_{n-1,k}$ exakt integriert.

Beweis: Die Menge der Kontaktpunkte U von $f-s_o$ ist in einer Menge $(v_i)_{i=1}^{k}$ enthalten, für die (7.3) gilt. Dann kann man eine Quadraturformel bestimmen

$$\int_a^b g(x)dx \sim Qg = \sum_{i=1}^{k} A_i g(v_i),$$

die alle $s \in B_{n-1,k}$ exakt integriert. Die Gewichte ergeben sich folgendermaßen:

Es sei $s_j \in B_{n-1,k}$

$$s_j(v_i) = 0 \quad i=1,\dots,k, \quad i \neq j$$

$$s_j(v_j) = 1.$$

Wegen (7.3) ist s_j eindeutig festgelegt. Dann gilt

$$A_j = \int_a^b s_j(x)dx.$$

Mit den gleichen Argumenten wie in Satz 4.1 bei DeVore folgt, daß $A_k = 0$ für $v_k \notin U$ und $A_k > 0$ für $v_k \in U$.

Bemerkung: Der Satz 7.4 zeigt, daß es sinnvoll ist, Approximationsprobleme zu betrachten, wo die Kontaktpunktmenge der Minimallösung die Eigenschaften von Def. 7.3 besitzt. Die Kontaktpunktmenge liefert die Stützstellen einer Quadraturformel. Betrachtet man nun einen Teilraum von $B_{n-1,k}$, nämlich die Menge aller Funktionen, die außerhalb von (x_i, x_{i+j+n}) gleich Null sind, dann hat man einen $(j+1)$-dimensionalen Raum vorliegen. Die Elemente dieses Raumes kann man durch eine Quadraturformel mit höchstens $j+1$ Stützstellen exakt integrieren. Die Bedingung von Def. 7.3 ist daher eine sehr natürliche Bedingung, denn sie stellt sicher, daß die erhaltenen Quadraturformeln nicht mehr als $j+1$ Stützstellen in (x_i, x_{i+j+n}) enthalten.

Im folgenden werden eine Reihe von Sätzen nur noch zitiert. Einzelheiten sind in Strauß [41] zu finden.
Falls f und s differenzierbare Funktionen sind, wollen wir folgende Nullstellenzählung verwenden: In (a,b) wird eine Nullstelle von f-s doppelt, in a und b einfach gezählt. Es gilt folgender Satz.

Satz 7.5. Die Funktion $f \in C^1[a,b]$ besitze eine Minimallösung s_o bei einseitiger Approximation durch $B_{n-1,k} \subset C^1[a,b]$, so daß die Fehlerfunktion $f-s_o$ eine Kontaktpunktmenge nach Def. 7.3 besitzt. Dann enthält jedes Intervall $(x_i, x_{i+j+n}) \subset [x_1, x_{n+k}]$ wenigstens $j+1$ Nullstellen. Falls n-fache Knoten in a bzw. b liegen, ist (x_1, x_{1+j+n}) durch $[x_1, x_{1+j+n})$ bzw. (x_{k-j}, x_{n+k}) durch $(x_{k-j}, x_{n+k}]$ zu ersetzen.

Diese Aussage gilt nicht, wenn die Fehlerfunktion $f-s_o$ keine Kontaktpunktmenge nach Def. 7.3 besitzt.

Aus Satz 7.5 kann man leicht folgern:

Folgerung 7.6. Die Funktion $f \in C^1[a,b]$ werde durch Elemente aus $B_{n-1,k} \subset C^1[a,b]$ einseitig approximiert. Jede Fehlerfunktion f-s mit $s \in B_{n-1,k}$ besitze eine Kontaktpunktmenge mit den Eigenschaften von Def. 7.3. Dann existiert genau eine Minimallösung.

Diese Aussagen sollen nun angewendet werden. Es werden zunächst Funktionen angegeben, deren Minimallösung bei einseitiger Approximation die im letzten Paragraphen geforderten Eigenschaften besitzt.

Definition 7.7. Gegeben sei eine Knotenverteilung

$$a=x_0<x_1<\ldots<x_k<x_{k+1}=b.$$

Es sei nun $g \in C^2[a,b]$ eine Funktion mit der Eigenschaft

(7.4) $\quad (-1)^{i+1} g''(x) > 0 \quad x \in (x_i, x_{i+1}).$

Satz 7.8. Für jeden Polygonzug $s(x)=a_0+a_1x+\sum_{i=1}^{k} b_i(x-x_i)_+$ hat g-s nicht mehr als k+2 Nullstellen, wobei g eine Funktion wie in Def. 7.7 ist.

Daraus folgt, wenn man folgende Funktionenmenge verwendet:

$$F = \{f \in C^{(n)}[a,b] \,|\, (-1)^{i+1} f^{(n)}(x)>0 \quad x \in (x_{i-1}, x_i) \quad i=1,\ldots,k+1\} , \qquad (7.5)$$

wobei $a=x_0<x_1<\ldots<x_k<x_{k+1}=b$ ist.

Satz 7.9. Es sei eine Funktion $f \in F$ vorgegeben und ein Spline

$$s(x) = \sum_{i=o}^{n-1} a_i x^i + \sum_{i=1}^{k} b_i (x-x_i)_+^{n-1},$$

wobei die Knoten (x_i) wie im Funktionenraum F sein sollen. Dann besitzt die Funktion f-s höchstens n+k Nullstellen.

Beweis: Anwendung des Satzes von Rolle und von Satz 7.8.

Es seien die Punkte $x_1 < x_2 < \ldots < x_{n+2k}$ vorgegeben. Gesucht wird der Monospline M vom Grad n mit k Knoten, für den

$$M(x_i) = 0 \quad i=1,\ldots,n+2k$$

gilt.

Sei $B_{n-1,2k}$ der 2k-dimensionale Raum, der von den B-Splines vom Grad n-1 zu den Knoten $(x_i)_{i=1}^{n+2k}$ aufgespannt wird.
Sei weiterhin

$$S_{n-1,n+2k} = \{s \mid s(x) = \sum_{i=o}^{n-1} a_i x^i + \sum_{i=1}^{n+2k} b_i (x-x_i)_+^{n-1}\}.$$

Die Knotenmenge des gesuchten Monosplines wird sich als die Kontaktpunktmenge einer Fehlerfunktion ergeben, die ein einseitiges Approximationsproblem löst. Dabei werden Funktionen approximiert, wie sie in der folgenden Definition festgelegt werden.

Definition 7.10. Es sei $x_o < x_1 < x_2 < \ldots < x_{n+2k} < x_{n+2k+1}$ und g eine Funktion mit den Eigenschaften von (7.5) zu den

Punkten $(x_i)_{i=o}^{n+2k+1}$. Außerdem sei $\bar{s}\in S_{n-1,n+2k}$ ein Spline mit

$$g^{(i)}(x_o) = \bar{s}^{(i)}(x_o)$$
$$g^{(i)}(x_{n+2k+1}) = \bar{s}^{(i)}(x_{n+2k+1}) \qquad i=0,\dots,n-1.$$

Dann definieren wir

(7.6) $$f := g-\bar{s}.$$

Diese Funktionen besitzen die folgenden wichtigen Eigenschaften.

Lemma 7.11. Für jede Funktion f, die wie in Def. 7.10 festgelegt ist, gilt:

Auf den Intervallen $[x_o,x_{i+1}]$, bzw. $[x_{n+2k-i},x_{n+2k+1}]$ für i=0,...,n+2k hat f-s höchstens n+i Nullstellen. (In x_o und x_{n+2k+1} liegen n-fache Nullstellen). Dabei ist $s\in B_{n-1,2k}$.

Dies kann man mit Hilfe des Satzes von Rolle und Satz 7.8 beweisen.

Lemma 7.12. Gegeben sei die Punktmenge $X = (x_i)_{i=1}^{n+m}$

$$x_1 \le x_2 \le \dots \le x_{n+m-1} \le x_{n+m} \quad x_i < x_{i+n-1}$$

und $V = (v_i)_{i=1}^{r}$ mit $r \le m$ und $v_1<v_2<\dots<v_m$. Es gelte weiterhin, daß $V \subset (x_1,x_{n+m})$ und jedes Intervall $[x_1,x_{1+i}]$ bzw. $[x_{n+m-i},x_{n+m}]$ für i=1,...,m höchstens i Punkte von V

enthält.

Dann bildet V eine Menge mit den Eigenschaften von Def. 7.3 auf X.

Dies kann man durch eine Induktion über m zeigen.

Nun läßt sich folgendes Ergebnis formulieren.

Satz 7.13. Auf dem Intervall $[x_1, x_{n+2k}]$ werde die Funktion f einseitig durch Elemente aus $B_{n-1,2k}$ in der L_1-Norm approximiert. Der Spline $s_0 \in B_{n-1,2k}$ sei eine Minimallösung. Man wählt dabei $f = \bar{f}|[x_1, x_{n+2k}]$ und $\bar{f}$ sei eine Funktion nach Def. 7.10.

(a) Der Spline s_0 ist eindeutig bestimmt.

(b) Die Fehlerfunktion $f - s_0$ besitzt genau k Kontaktpunkte $(u_i)_{i=1}^k$, die in (x_1, x_{n+2k}) liegen.

(c) Die Punkte $(u_i)_{i=1}^k$ sind die Knoten eines Monosplines

$$M(x) = x^n - \sum_{i=0}^{n-1} a_i x^i - \sum_{i=1}^{k} b_i (x - u_i)_+^{n-1} ,$$

der

$$M(x_i) = 0 \qquad i = 1, \ldots, n+2k$$

erfüllt.

Beweis: Es wird zunächst gezeigt, daß für alle $s \in B_{n-1,2k}$ mit $(f-s)(x) \geq 0$ für $x \in [x_1, x_{n+2k}]$ die Menge der Kontaktpunkte von f-s eine Menge mit den Eigenschaften von Def. 7.3 ist. Aus Lemma 7.11 folgt, daß in x_1 und x_{n+2k} keine Nullstellen von f-s liegen; denn f-s enthält in $[x_0, x_1]$ und $[x_{n+2k}, x_{n+2k+1}]$ höchstens jeweils n Nullstellen. Sie liegen in x_0 und x_{n+2k+1}. Unser Approximationsproblem ist auf $[x_1,x_{n+2k}]$ beschränkt. Daher liegen alle Kontaktpunkte im Innern von $[x_1,x_{n+2k}]$. Weiterhin erhält man aus Lemma 7.11, daß jedes Intervall $[x_1,x_{i+1}]$ und $[x_{n+2k-i},x_{n+2k}]$ höchstens i Nullstellen von f-s hat. Nach Lemma 7.12 besitzt deshalb die Kontaktpunktmenge für jede Funktion f-s die erforderlichen Eigenschaften.

Wegen Satz 7.1 existiert eine Minimallösung s_0; nach Folgerung 7.6 ist s_0 eindeutig bestimmt.

Satz 7.5 ergibt, daß $f-s_0$ wenigstens 2k Nullstellen besitzen muß. Nach Lemma 7.11 kann $f-s_0$ aber höchstens 2k Nullstellen haben. Damit hat $f-s_0$ genau 2k Nullstellen und genau k Kontaktpunkte. Nachdem die Kontaktpunktmenge die Eigenschaften von Def. 7.3 erfüllt, sind die Kontaktpunkte Stützstellen einer Quadraturformel, die die Elemente des Raumes $B_{n-1,2k}$ exakt integriert (siehe Satz 7.4).

Satz 1.1 ordnet dieser Quadraturformel einen Monospline zu, der die gewünschten Nullstellenbedingungen erfüllt.

<u>Bemerkung 7.14.</u> Es wurde eine Menge von Funktionen definiert, so daß die einseitige Approximation die Knoten eines Monosplines liefert. Man sieht also auch, daß die Lage der Kontaktpunkte für alle Funktionen dieser Klasse gleich ist.

Hier wird gleichzeitig ein Existenzbeweis für das Interpolationsproblem angegeben und eine Möglichkeit, die Knoten mit Hilfe eines Approximationsproblems zu bestimmen.

Das Problem wurde nur für einfache Nullstellen behandelt. Die Überlegungen lassen sich auch auf mehrfache Nullstellen übertragen.

Einseitige Approximation kann noch in folgendem Zusammenhang eine Anwendung finden.
Das Restglied bei Quadraturformeln hat die Gestalt

$$R(f) = (-1)^n \int_a^b M(x)\, f^{(n)}(x)dx.$$

Wenn M das Vorzeichen nicht wechselt, dann gibt es ein $\eta \in [a,b]$, so daß

$$R(f) = (-1)^n f^{(n)}(\eta) \int_a^b M(x)\, dx$$

gilt. Es liegt daher nahe, die Funktion x^n einseitig durch Splinefunktionen zu approximieren, um beste Quadraturformeln für sogenannte semidefinite Kerne zu erhalten (siehe Schmeißer).
Nähere Einzelheiten und Beweise, die hier nicht ausgeführt wurden,sind in Strauß [41] zu finden.
Nach Abschluß der Arbeit habe ich im August 1975 das Manuskript [42] von C.A. Micchelli und A. Pinkus erhalten. Dort werden auch Probleme wie in diesem Paragraphen behandelt.
Herrn Prof. DeVore möchte ich für seine Hilfe herzlich danken.

Literatur

[1] Barrodale I.: On computing best L_1 approximations. In: Approximation theory (edit. by A. Talbot), London - New York, Academic Press 1970.

[2] Bojanic R. and R. DeVore: On polynomials of best one-sided approximation, L'Enseignement Math. 12 (1966), 139 - 144.

[3] De Boor C.: A remark concerning perfect splines, Bull. Amer. Math. Soc. 80 (1974), 724 - 727.

[4] Braess D.: Chebyshev Approximation by Spline Functions with free knots, Num. Math. 17 (1971), 357 - 366.

[5] Brosowski B.: Nichtlineare Approximation in normierten Vektorräumen, In "Abstract Spaces and Approximation", ISNM 10, 140 - 159, Birkhäuser-Verlag, 1970.

[6] Cavaretta A. S., Jr.: On Cardinal Perfect Splines of Least Sup-Norm on the Real Axis, J. Approx. Theory 8 (1973), 285 - 303.

[7] Curry H. B. and I. J. Schoenberg: On spline distributions and their limits: The Polya distribution functions, Bull.Amer.Math.Soc. 53 (1947), 1114.

[8] DeVore R.: One-sided approximation of functions, J. Approx. Theory 1 (1968), 11 - 25.

[9] Elsner L.: Ein Optimierungsproblem der Analysis, Bericht 018 des Instituts für Angewandte Mathematik I, Erlangen.

[10] Favard J.: Sur l'interpolation, J. Math. Pures Appl. (9) 19 (1940), 281 - 306.

[11] Glaeser G.: Prolongement extrêmal de fonctions différentiables d'une variable, J. Approx. Theory 8 (1973), 249 - 261.

[12] Handscomb D. C.: Characterization of best spline approximations with free knots, In: Approximation Theory (edit. by A. Talbot), London-New York, Academic Press 1970.

[13] Hobby C.R. and J.R. Rice: A moment problem in L_1 approximation, Proceed. Amer. Math. Soc., 16 (1965), 665 - 670.

[14] Jetter K.: Splines und optimale Quadraturformeln, Dissertation Tübingen, 1973.

[15] Karlin S.: Total Positivity, Stanford University Press, Stanford, California, 1968

[16] __________: Best quadrature formulas and interpolation by splines satisfying boundary conditions. In: Approximations with special emphasis on spline functions, ed. by I.J. Schoenberg, 447 - 466. Academic Press, New York, 1969.

[17] __________: The fundamental theorem of algebra for monosplines satisfying certain boundary conditions and applications to optimal quadrature formulas. In: Approximations with special emphasis on spline functions, ed. by I.J. Schoenberg, 467 - 484. Academic Press, New York, 1969.

[18] __________: Best quadrature formulas and splines, J. Approx. Theory 4 (1971), 59 - 90.

[19] ________:Some variational problems on certain Sobolev spaces and perfect splines, Bull.Amer. Math.Soc., 79 (1973),124 - 128.

[20] Karlin S. and C. Micchelli: The fundamental theorem of algebra for monosplines satisfying boundary conditions, Israel J. Math. 11 (1972), 405 - 451.

[21] Karlin S. and L. Schumaker: The fundamental theorem of algebra for Tchebyscheffian monosplines, J. Analyse Math. 20 (1967), 233 - 270.

[22] Kripke B. R. and T. J. Rivlin: Approximation in the metric $L^1(X,\mu)$, Trans.Amer.Math.Soc. 119 (1965), 101 - 122.

[23] Meinardus G.: Approximation von Funktionen und ihre numerische Behandlung, Berlin - Heidelberg-New York, 1964.

[24] Micchelli C.A.: Some minimum problem for spline functions to quadrature formulas. In: Approximation Theory (edit. by G. Lorentz) London-New York, Academic Press, 1973.

[25] ________________: Best quadrature formulas at equally spaced nodes, J. Math. Anal. Applic., 47 (1974 232 - 249.

[26] Micchelli C.A. and T.J. Rivlin: Quadrature Formulae and Hermite-Birkhoff Interpolation, Advances in Math. 11 (1973), 93 - 112.

[27] Nörlund N. E.: "Vorlesungen über Differenzenrechnung", Springer-Verlag Berlin 1924.

[28] Rice J. R.: "The Approximation of Functions", Addison-Wesley, vol. I 1964, vol. II 1969.

[29] Schmeisser G.: Optimale Quadraturformeln mit semidefiniten Kernen, Num. Math. 20 (1973), 32 - 53.

[30] Schoenberg I.J.: On best approximations of linear operators, Nederl. Akad.Wetensch.Indag. Math. 26 (1964), 155 - 163.

[31] ______________: On monosplines of least deviation and best quadrature formulae, J.Siam Numer.Anal.Ser. B 2 (1965), 144 - 170.

[32] ______________: On monosplines of least square deviation and best quadrature formulae II, J. Siam Numer.Anal. 3 (1966), 321 - 328.

[33] ______________: Monosplines and quadrature formulae. In: Theory and applications of spline functions, ed. by T.N.E. Greville, New York, Academ.Press 1969, 157 - 207.

[34] ______________: A second look at approximate quadrature formulae and spline interpolation, Advances in Math. 4 (1970), 277 - 300.

[35] ______________: The perfect B-splines and a time optimal control problem, Israel J.Math. 8 (1971) 261 - 275.

[36] Schumaker L.: Uniform Approximation by Tchebycheffian spline functions, J.Math.Mech. 18 (1968) 369-378.

[37] ____________: Uniform approximation by chebyshev spline functions II: free knots, Siam J.Numer.Anal. 5 (1968), 647 - 656.

[38] Strauss H.: L_1-Approximation mit Splinefunktionen, ISNM 26, Birkhäuser-Verlag, 1975.

[39] __________: Eindeutigkeit bei der gleichmäßigen Approximation mit Tschebyscheffschen Splinefunktionen, erscheint in Journ.Approx.Theory.

[40] Strauss H.: Nichtlineare L_1-Approximation, Bericht 020 des Instituts für Angewandte Mathematik I, Erlangen, 1973.

[41] __________: Approximation mit Splinefunktionen und Anwendungen auf die Approximation linearer Funktionale, Bericht 023 des Instituts für Angewandte Mathematik I, Erlangen, 1975.

[42] Micchelli C.A. and A. Pinkus: Moment theory for weak chebyshev systems with applications to monosplines, quadrature formulae and best one-sided L^1-approximation by spline functions with fixed knots.

Dr. H. Strauß
Institut für Angewandte Mathematik
der Universität Erlangen-Nürnberg
852 Erlangen, Martensstr. 1

ANSCHRIFTEN DER AUTOREN

C. de BOOR
University of Wisconsin - Madison
Madison 53 706
Mathematical Research Center
610 Walnut Street
USA

F.J. DELVOS
Gesamthochschule Siegen
Lehrstuhl für Mathematik I
D-5900 Siegen 21
Hölderlinstr. 3

G. HÄMMERLIN
Mathematisches Institut der
Ludwig - Maximilians - Universität
D-8000 München 2
Theresienstr. 39

G. JENTZSCH
Institut für Geophysik der
Technischen Universität
D-3392 Clausthal - Zellerfeld
Postfach 230

H. JOHNEN
Fakultät für Mathematik
Universität Bielefeld
D-4800 Bielefeld
Kurt - Schumacher - Str. 6

H.-W. KÖSTERS
Rechenzentrum
Ruhr-Universität
D-4630 Bochum
Universitätsstraße 150 NA

G. LANGE
Rechenzentrum der
Technischen Universität
D-3392 Clausthal - Zellerfeld
Erzstraße 51

T. LYCHE
Department of Mathematics
University of Oslo
Oslo 3, Norway

G. MEINARDUS
Gesamthochschule Siegen
Lehrstuhl für Mathematik IV
D-5900 Siegen 21
Hölderlinstr. 3

G. MICULA
Faculty of Mathematics
University of Cluj
3400 Cluj - Napoca (Romania)

H. ter MORSCHE
Department of Mathematics
Technological University
Eindhoven
Eindhoven, The Netherlands

O. ROSENBACH
Institut für Geophysik
der Technischen Universität
D-3392 Clausthal - Zellerfeld
Postfach 230

A. SARD
Department of Mathematics
University of California
La Jolla, California 92037
USA

W. SCHÄFER
Gesamthochschule Siegen
Lehrstuhl für Mathematik I
D-5900 Siegen 21
Hölderlinstr. 3

W. SCHEMPP
Gesamthochschule Siegen
Lehrstuhl für Mathematik I
D-5900 Siegen 21
Hölderlinstr. 3

K. SCHERER
Fakultät für Mathematik
Universität Bielefeld
D-4800 Bielefeld
Kurt - Schumacher - Str. 6

K.-H. SCHLOSSER
Rechenzentrum
Ruhr-Universität
D-4630 Bochum
Universitätsstraße 150 NA

I.J. SCHOENBERG
Mathematical Research Center
University of Wisconsin - Madison
Madison 53 70 6
610 Walnut Street
USA

L.L. SCHUMAKER
Department of Mathematics
The University of Texas
Austin, Texas 78712
USA

F. SCHURER
Department of Mathematics
Technological University
Eindhoven,
Eindhoven, The Netherlands

M. SOMMER
Institut für angewandte
Mathematik der
Universität Erlangen - Nürnberg
D-8520 Erlangen
Martensstr. 1

F.W. STEUTEL
Department of Mathematics
Technological University
Eindhoven,
Eindhoven, The Netherlands

H. STRAUSS
Institut für angewandte
Mathematik der
Universität Erlangen - Nürnberg
D-8520 Erlangen
Martensstr. 1

ol. 399: Functional Analysis and its Applications. Proceedings 973. Edited by H. G. Garnir, K. R. Unni and J. H. Williamson. 584 pages. 1974.

ol. 400: A Crash Course on Kleinian Groups. Proceedings 1974. dited by L. Bers and I. Kra. VII, 130 pages. 1974.

ol. 401: M. F. Atiyah, Elliptic Operators and Compact Groups. , 93 pages. 1974.

ol. 402: M. Waldschmidt, Nombres Transcendants. VIII, 277 ages. 1974.

ol. 403: Combinatorial Mathematics. Proceedings 1972. Edited y D. A. Holton. VIII, 148 pages. 1974.

ol. 404: Théorie du Potentiel et Analyse Harmonique. Edité par Faraut. V, 245 pages. 1974.

ol. 405: K. J. Devlin and H. Johnsbråten, The Souslin Problem. III, 132 pages. 1974.

ol. 406: Graphs and Combinatorics. Proceedings 1973. Edited y R. A. Bari and F. Harary. VIII, 355 pages. 1974.

ol. 407: P. Berthelot, Cohomologie Cristalline des Schémas de aracteristique p > o. II, 604 pages. 1974.

ol. 408: J. Wermer, Potential Theory. VIII, 146 pages. 1974.

ol. 409: Fonctions de Plusieurs Variables Complexes, Séminaire ançois Norguet 1970–1973. XIII, 612 pages. 1974.

ol. 410: Séminaire Pierre Lelong (Analyse) Année 1972–1973. 181 pages. 1974.

l. 411: Hypergraph Seminar. Ohio State University, 1972. lited by C. Berge and D. Ray-Chaudhuri. IX, 287 pages. 1974.

ol. 412: Classification of Algebraic Varieties and Compact omplex Manifolds. Proceedings 1974. Edited by H. Popp. V, 33 pages. 1974.

ol. 413: M. Bruneau, Variation Totale d'une Fonction. XIV, 332 ges. 1974.

l. 414: T. Kambayashi, M. Miyanishi and M. Takeuchi, Unitent Algebraic Groups. VI, 165 pages. 1974.

l. 415: Ordinary and Partial Differential Equations. Proceedings 74. XVII, 447 pages. 1974.

l. 416: M. E. Taylor, Pseudo Differential Operators. IV, 155 ges. 1974.

l. 417: H. H. Keller, Differential Calculus in Locally Convex aces. XVI, 131 pages. 1974.

l. 418: Localization in Group Theory and Homotopy Theory d Related Topics. Battelle Seattle 1974 Seminar. Edited by P. J. lton. VI, 172 pages 1974.

l. 419: Topics in Analysis. Proceedings 1970. Edited by O. E. hto, I. S. Louhivaara, and R. H. Nevanlinna. XIII, 392 pages. 1974.

l. 420: Category Seminar. Proceedings 1972/73. Edited by G. M. lly. VI, 375 pages. 1974.

l. 421: V. Poénaru, Groupes Discrets. VI, 216 pages. 1974.

l. 422: J.-M. Lemaire, Algèbres Connexes et Homologie des paces de Lacets. XIV, 133 pages. 1974.

l. 423: S. S. Abhyankar and A. M. Sathaye, Geometric Theory Algebraic Space Curves. XIV, 302 pages. 1974.

. 424: L. Weiss and J. Wolfowitz, Maximum Probability timators and Related Topics. V, 106 pages. 1974.

. 425: P. R. Chernoff and J. E. Marsden, Properties of Infinite mensional Hamiltonian Systems. IV, 160 pages. 1974.

. 426: M. L. Silverstein, Symmetric Markov Processes. X, 287 ges. 1974.

. 427: H. Omori, Infinite Dimensional Lie Transformation oups. XII, 149 pages. 1974.

. 428: Algebraic and Geometrical Methods in Topology, Proedings 1973. Edited by L. F. McAuley. XI, 280 pages. 1974.

Vol. 429: L. Cohn, Analytic Theory of the Harish-Chandra C-Function. III, 154 pages. 1974.

Vol. 430: Constructive and Computational Methods for Differential and Integral Equations. Proceedings 1974. Edited by D. L. Colton and R. P. Gilbert. VII, 476 pages. 1974.

Vol. 431: Séminaire Bourbaki – vol. 1973/74. Exposés 436–452. IV, 347 pages. 1975.

Vol. 432: R. P. Pflug, Holomorphiegebiete, pseudokonvexe Gebiete und das Levi-Problem. VI, 210 Seiten. 1975.

Vol. 433: W. G. Faris, Self-Adjoint Operators. VII, 115 pages. 1975.

Vol. 434: P. Brenner, V. Thomée, and L. B. Wahlbin, Besov Spaces and Applications to Difference Methods for Initial Value Problems. II, 154 pages. 1975.

Vol. 435: C. F. Dunkl and D. E. Ramirez, Representations of Commutative Semitopological Semigroups. VI, 181 pages. 1975.

Vol. 436: L. Auslander and R. Tolimieri, Abelian Harmonic Analysis, Theta Functions and Function Algebras on a Nilmanifold. V, 99 pages. 1975.

Vol. 437: D. W. Masser, Elliptic Functions and Transcendence. XIV, 143 pages. 1975.

Vol. 438: Geometric Topology. Proceedings 1974. Edited by L. C. Glaser and T. B. Rushing. X, 459 pages. 1975.

Vol. 439: K. Ueno, Classification Theory of Algebraic Varieties and Compact Complex Spaces. XIX, 278 pages. 1975

Vol. 440: R. K. Getoor, Markov Processes: Ray Processes and Right Processes. V, 118 pages. 1975.

Vol. 441: N. Jacobson, PI-Algebras. An Introduction. V, 115 pages. 1975.

Vol. 442: C. H. Wilcox, Scattering Theory for the d'Alembert Equation in Exterior Domains. III, 184 pages. 1975.

Vol. 443: M. Lazard, Commutative Formal Groups. II, 236 pages. 1975.

Vol. 444: F. van Oystaeyen, Prime Spectra in Non-Commutative Algebra. V, 128 pages. 1975.

Vol. 445: Model Theory and Topoi. Edited by F. W. Lawvere, C. Maurer, and G. C. Wraith. III, 354 pages. 1975.

Vol. 446: Partial Differential Equations and Related Topics. Proceedings 1974. Edited by J. A. Goldstein. IV, 389 pages. 1975.

Vol. 447: S. Toledo, Tableau Systems for First Order Number Theory and Certain Higher Order Theories. III, 339 pages. 1975.

Vol. 448: Spectral Theory and Differential Equations. Proceedings 1974. Edited by W. N. Everitt. XII, 321 pages. 1975.

Vol. 449: Hyperfunctions and Theoretical Physics. Proceedings 1973. Edited by F. Pham. IV, 218 pages. 1975.

Vol. 450: Algebra and Logic. Proceedings 1974. Edited by J. N. Crossley. VIII, 307 pages. 1975.

Vol. 451: Probabilistic Methods in Differential Equations. Proceedings 1974. Edited by M. A. Pinsky. VII, 162 pages. 1975.

Vol. 452: Combinatorial Mathematics III. Proceedings 1974. Edited by Anne Penfold Street and W. D. Wallis. IX, 233 pages. 1975.

Vol. 453: Logic Colloquium. Symposium on Logic Held at Boston, 1972–73. Edited by R. Parikh. IV, 251 pages. 1975.

Vol. 454: J. Hirschfeld and W. H. Wheeler, Forcing, Arithmetic, Division Rings. VII, 266 pages. 1975.

Vol. 455: H. Kraft, Kommutative algebraische Gruppen und Ringe. III, 163 Seiten. 1975.

Vol. 456: R. M. Fossum, P. A. Griffith, and I. Reiten, Trivial Extensions of Abelian Categories. Homological Algebra of Trivial Extensions of Abelian Categories with Applications to Ring Theory. XI, 122 pages. 1975.

Vol. 342: Algebraic K-Theory II, "Classical" Algebraic K-Theory, and Connections with Arithmetic. Edited by H. Bass. XV, 527 pages. 1973.

Vol. 343: Algebraic K-Theory III, Hermitian K-Theory and Geometric Applications. Edited by H. Bass. XV, 572 pages. 1973.

Vol. 344: A. S. Troelstra (Editor), Metamathematical Investigation of Intuitionistic Arithmetic and Analysis. XVII, 485 pages. 1973.

Vol. 345: Proceedings of a Conference on Operator Theory. Edited by P. A. Fillmore. VI, 228 pages. 1973.

Vol. 346: Fučík et al., Spectral Analysis of Nonlinear Operators. II, 287 pages. 1973.

Vol. 347: J. M. Boardman and R. M. Vogt, Homotopy Invariant Algebraic Structures on Topological Spaces. X, 257 pages. 1973.

Vol. 348: A. M. Mathai and R. K. Saxena, Generalized Hypergeometric Functions with Applications in Statistics and Physical Sciences. VII, 314 pages. 1973.

Vol. 349: Modular Functions of One Variable II. Edited by W. Kuyk and P. Deligne. V, 598 pages. 1973.

Vol. 350: Modular Functions of One Variable III. Edited by W. Kuyk and J.-P. Serre. V, 350 pages. 1973.

Vol. 351: H. Tachikawa, Quasi-Frobenius Rings and Generalizations. XI, 172 pages. 1973.

Vol. 352: J. D. Fay, Theta Functions on Riemann Surfaces. V, 137 pages. 1973.

Vol. 353: Proceedings of the Conference. on Orders, Group Rings and Related Topics. Organized by J. S. Hsia, M. L. Madan and T. G. Ralley. X, 224 pages. 1973.

Vol. 354: K. J. Devlin, Aspects of Constructibility. XII, 240 pages. 1973.

Vol. 355: M. Sion, A Theory of Semigroup Valued Measures. V, 140 pages. 1973.

Vol. 356: W. L. J. van der Kallen, Infinitesimally Central Extensions of Chevalley Groups. VII, 147 pages. 1973.

Vol. 357: W. Borho, P. Gabriel und R. Rentschler, Primideale in Einhüllenden auflösbarer Lie-Algebren. V, 182 Seiten. 1973.

Vol. 358: F. L. Williams, Tensor Products of Principal Series Representations. VI, 132 pages. 1973.

Vol. 359: U. Stammbach, Homology in Group Theory. VIII, 183 pages. 1973.

Vol. 360: W. J. Padgett and R. L. Taylor, Laws of Large Numbers for Normed Linear Spaces and Certain Fréchet Spaces. VI, 111 pages. 1973.

Vol. 361: J. W. Schutz, Foundations of Special Relativity: Kinematic Axioms for Minkowski Space-Time. XX, 314 pages. 1973.

Vol. 362: Proceedings of the Conference on Numerical Solution of Ordinary Differential Equations. Edited by D.G. Bettis. VIII, 490 pages. 1974.

Vol. 363: Conference on the Numerical Solution of Differential Equations. Edited by G. A. Watson. IX, 221 pages. 1974.

Vol. 364: Proceedings on Infinite Dimensional Holomorphy. Edited by T. L. Hayden and T. J. Suffridge. VII, 212 pages. 1974.

Vol. 365: R. P. Gilbert, Constructive Methods for Elliptic Equations. VII, 397 pages. 1974.

Vol. 366: R. Steinberg, Conjugacy Classes in Algebraic Groups (Notes by V. V. Deodhar). VI, 159 pages. 1974.

Vol. 367: K. Langmann und W. Lütkebohmert, Cousinverteilungen und Fortsetzungssätze. VI, 151 Seiten. 1974.

Vol. 368: R. J. Milgram, Unstable Homotopy from the Stable Point of View. V, 109 pages. 1974.

Vol. 369: Victoria Symposium on Nonstandard Analysis. Edited by A. Hurd and P. Loeb. XVIII, 339 pages. 1974.

Vol. 370: B. Mazur and W. Messing, Universal Extensions and One Dimensional Crystalline Cohomology. VII, 134 pages. 1974.

Vol. 371: V. Poenaru, Analyse Différentielle. V, 228 pages. 197

Vol. 372: Proceedings of the Second International Conferenc on the Theory of Groups 1973. Edited by M. F. Newman. V 740 pages. 1974.

Vol. 373: A. E. R. Woodcock and T. Poston, A Geometrical Stud of the Elementary Catastrophes. V, 257 pages. 1974.

Vol. 374: S. Yamamuro, Differential Calculus in Topologic Linear Spaces. IV, 179 pages. 1974.

Vol. 375: Topology Conference. Edited by R. F. Dickman Jr. an P. Fletcher. X, 283 pages 1974.

Vol. 376: I. J. Good and D. B. Osteyee, Information, Weight c Evidence. The Singularity between Probability Measures and Sign Detection. XI, 156 pages. 1974.

Vol. 377: A. M. Fink, Almost Periodic Differential Equation VIII, 336 pages. 1974.

Vol. 378: TOPO 72 - General Topology and its Application Proceedings 1972. Edited by R. A. Alò, R. W. Heath and J. Nagat XIV, 651 pages. 1974.

Vol. 379: A. Badrikian et S. Chevet, Mesures Cylindrique Espaces de Wiener et Fonctions Aléatoires Gaussiennes. 383 pages. 1974.

Vol. 380: M. Petrich, Rings and Semigroups. VIII, 182 page 1974.

Vol. 381: Séminaire de Probabilités VIII. Edité par P. A. Mey IX, 354 pages. 1974.

Vol. 382: J. H. van Lint, Combinatorial Theory Seminar Ein hoven University of Technology. VI, 131 pages. 1974.

Vol. 383: Séminaire Bourbaki - vol. 1972/73. Exposés 418-43 IV, 334 pages. 1974.

Vol. 384: Functional Analysis and Applications, Proceedin 1972. Edited by L. Nachbin. V, 270 pages. 1974.

Vol. 385: J. Douglas Jr. and T. Dupont, Collocation Methods Parabolic Equations in a Single Space Variable (Based on (Piecewise-Polynomial Spaces). V, 147 pages. 1974.

Vol. 386: J. Tits, Buildings of Spherical Type and Finite B Pairs. X, 299 pages. 1974.

Vol. 387: C. P. Bruter, Eléments de la Théorie des Matroïd V, 138 pages. 1974.

Vol. 388: R. L. Lipsman, Group Representations. X, 166 pag 1974.

Vol. 389: M.-A. Knus et M. Ojanguren, Théorie de la Descer et Algèbres d' Azumaya. IV, 163 pages. 1974.

Vol. 390: P. A. Meyer, P. Priouret et F. Spitzer, Ecole d'Eté Probabilités de Saint-Flour III - 1973. Edité par A. Badrik et P.-L. Hennequin. VIII, 189 pages. 1974.

Vol. 391: J. W. Gray, Formal Category Theory: Adjointness for Categories. XII, 282 pages. 1974.

Vol. 392: Géométrie Différentielle, Colloque, Santiago Compostela, Espagne 1972. Edité par E. Vidal. VI, 225 pag 1974.

Vol. 393: G. Wassermann, Stability of Unfoldings. IX, 164 pag 1974.

Vol. 394: W. M. Patterson, 3rd, Iterative Methods for the Solut of a Linear Operator Equation in Hilbert Space - A Surv III, 183 pages. 1974.

Vol. 395: Numerische Behandlung nichtlinearer Integrodiffer tial- und Differentialgleichungen. Tagung 1973. Herausgeget von R. Ansorge und W. Törnig. VII, 313 Seiten. 1974.

Vol. 396: K. H. Hofmann, M. Mislove and A. Stralka, The Pon agin Duality of Compact O-Dimensional Semilattices and Applications. XVI, 122 pages. 1974.

Vol. 397: T. Yamada, The Schur Subgroup of the Brauer Grc V, 159 pages. 1974.

Vol. 398: Théories de l'Information, Actes des Rencontres Marseille-Luminy, 1973. Edité par J. Kampé de Fériet et C.-F. Pic XII, 201 pages. 1974.